U0223144

MATERIALS CHARACTERIZATION SERIES

SERIES EDITORS: **C. Richard Brundle** and **Charles A. Evans, Jr.**

材料表征原版系列丛书

聚合物的表征

CHARACTERIZATION OF
Polymers

Ned J. Chou

Stephen P. Kowalczyk

Ravi Saraf

Ho-Ming Tong

哈尔滨工业大学出版社
HARBIN INSTITUTE OF TECHNOLOGY PRESS

黑版贸审字08-2013-089号

Ned J.Chou, Stephen P.Kowalczyk, Ravi Saraf, Ho-Ming Tong

Characterization of Polymers

9781606500538

Copyright © 2010 by Momentum Press, LLC

All rights reserved.

Originally published by Momentum Press, LLC

English reprint rights arranged with Momentum Press, LLC through McGraw-Hill Education (Asia)

图书在版编目（CIP）数据

聚合物的表征：英文 /（美）布伦德尔（Brundle C. R.），（美）埃文斯（Evans C. A.），（美）周（Chou N. J.）主编 .—哈尔滨：哈尔滨工业大学出版社，2014.1

（材料表征原版系列丛书）

ISBN978-7-5603-4284-9

Ⅰ.①聚…Ⅱ.①布…②埃…③周…Ⅲ.①聚合物–研究–英文Ⅳ.①O63

中国版本图书馆CIP数据核字（2013）第273407号

责任编辑 许雅莹 张秀华 杨 桦

出版发行 哈尔滨工业大学出版社

社　　址 哈尔滨市南岗区复华四道街10号 邮编 150006

传　　真 0451-86414749

网　　址 http://hitpress.hit.edu.cn

印　　刷 哈尔滨市石桥印务有限公司

开　　本 660mm×980mm 1/16 印张 21.5

版　　次 2014年1月第1版 2014年1月第1次印刷

书　　号 ISBN 978-7-5603-4284-9

定　　价 118.00元

（如因印刷质量问题影响阅读，我社负责调换）

CHARACTERIZATION OF POLYMERS

EDITORS

Ho-Ming Tong, Stephen P. Kowalczyk,
Ravi Saraf, and Ned J. Chou

SERIES EDITORS

C. Richard Brundle and Charles A. Evans, Jr.

MOMENTUM PRESS, LLC, NEW YORK

MATERIALS CHARACTERIZATION SERIES
Surfaces, Interfaces, Thin Films

Series Editors: C. Richard Brundle and Charles A. Evans, Jr.

Series Titles

Encyclopedia of Materials Characterization, C. Richard Brundle, Charles A. Evans, Jr., and Shaun Wilson

Characterization of Metals and Alloys, Paul H. Holloway and P.N. Vaidyanathan

Characterization of Ceramics, Ronald E. Loehman

Characterization of Polymers, Ned J. Chou, Stephen P. Kowalczyk, Ravi Saraf, and Ho-Ming Tong

Characterization in Silicon Processing, Yale Strausser

Characterization in Compound Semiconductor Processing, Yale Strausser

Characterization of Integrated Circuit Packaging Materials, Thomas M. Moore and Robert G. McKenna

Characterization of Catalytic Materials, Israel E. Wachs

Characterization of Composite Materials, Hatsuo Ishida

Characterization of Optical Materials, Gregory J. Exarhos

Characterization of Tribological Materials, William A. Glaeser

Characterization of Organic Thin Films, Abraham Ulman

We dedicate this volume to the memory of our dear colleague, Ned J. Chou, who passed away during the course of this project. Ned's kindness and wisdom will be greatly missed by all who have known him.

Contents

POLYMER STRUCTURES AND SYNTHESIS METHODS

POLYMER FABRICATION TECHNIQUES

CHEMICAL COMPOSITION OF POLYMERS

CHARACTERIZATION OF THE MORPHOLOGY OF POLYMER SURFACES, INTERFACES, AND THIN FILMS BY MICROSCOPY TECHNIQUES

STRUCTURE AND MORPHOLOGY OF INTERFACES AND THIN FILMS BY SCATTERING TECHNIQUES

SURFACE THERMODYNAMICS

SURFACE MODIFICATION OF POLYMERS

ADHESION

CHEMISTRY, REACTIVITY, AND FRACTURE OF POLYMER INTERFACES

Preface to the Reissue of the Materials Characterization Series

The 11 volumes in the *Materials Characterization Series* were originally published between 1993 and 1996. They were intended to be complemented by the *Encyclopedia of Materials Characterization*, which provided a description of the analytical techniques most widely referred to in the individual volumes of the series. The individual materials characterization volumes are no longer in print, so we are reissuing them under this new imprint.

The idea of approaching materials characterization from the material user's perspective rather than the analytical expert's perspective still has great value, and though there have been advances in the materials discussed inl each volume, the basic issues involved in their characterization have remained largely the same. The intent with this reissue is, first, to make the original information available once more, and then to gradually update each volume, releasing the changes as they occur by on-line subscription.

C. R. Brundle and C. A. Evans, October 2009

Preface to Series

This Materials Characterization Series attempts to address the needs of the practical materials user, with an emphasis on the newer areas of surface, interface, and thin film microcharacterization. The Series is composed of the leading volume, *Encyclopedia of Materials Characterization*, and a set of about 10 subsequent volumes concentrating on characterization of individual materials classes.

In the *Encyclopedia*, 50 brief articles (each 10 to 18 pages in length) are presented in a standard format designed for case of reader access, with straightforward technique descriptions and examples of their practical use. In addition to the articles, there are one-page summaries for every technique, introductory summaries to groupings of related techniques, a complete glossary of acronyms, and a tabular comparison of the major features of all 50 techniques.

The 10 volumes in the Series on characterization of particular materials classes include volumes on silicon processing, metals and alloys, catalytic materials, integrated circuit packaging, etc. Characterization is approached from the materials user's point of view. Thus, in general, the format is based on properties, processing steps, materials classification, etc., rather than on a technique. The emphasis of all volumes is on surfaces, interfaces, and thin films, but the emphasis varies depending on the relative importance of these areas for the materials class concerned. Appendixes in each volume reproduce the relevant one-page summaries from the *Encyclopedia* and provide longer summaries for any techniques referred to that are not covered in the *Encyclopedia*.

The concept for the Series came from discussion with Marjan Bace of Manning Publications Company. A gap exists between the way materials characterization is often presented and the needs of a large segment of the audience—the materials user, process engineer, manager, or student. In our experience, when, at the end of talks or courses on analytical techniques, a question is asked on how a particular material (or processing) characterization problem can be addressed the answer often is that the speaker is "an expert on the technique, not the materials aspects, and does not have experience with that particular situation." This Series is an attempt to bridge this gap by approaching characterization problems from the side of the materials user rather than from that of the analytical techniques expert.

We would like to thank Marjan Bace for putting forward the original concept, Shaun Wilson of Charles Evans and Associates and Yale Strausser of Surface Science Laboratories for help in further defining the Series, and the Editors of all the individual volumes for their efforts to produce practical, materials user based volumes.

C. R. Brundle   C. A. Evans, Jr.

Preface to the Re-issue of *Characterization of Polymers*

There were 22 authors originally involved in writing this volume, which presents an integrated overview of the surface, interface, and thin film properties of polymers, coupled with the analysis methods used to characterize these properties. In addition to the introductory chapters on polymer structure, synthesis, and fabrication, the areas of chemical composition, morphology, surface modification, adhesion, interface chemistry and reactivity, and friction and wear, are all addressed, together with the appropriate techniques to characterize them. Of course, there have been advances in all these areas since the volume was originally published, but the underlying principles and the basics of the characterization techniques and methodology remain valid. Following the reissue of the volume in close to its original form, it is our intention that updates covering advances made will be released as they become available.

C. R. Brundle and C. A. Evans, December 2009

Preface

Although a polymer specialist typically has the background needed to appreciate the subtleties of the polymer which are important for its application, he or she may not have the training to extract all the information available from the analysis of the polymeric material. Often, a polymer analyst has the opposite strengths and weaknesses. As a result, it is important for both the specialist and analyst to understand some aspects of the work of the other. But, whereas there are a number of books on surface and interfacial analysis written for the polymer analyst, there are few books written for the polymer specialist with a focus on surface and interfacial properties rather than the analysis. This series of books by Manning Publication Co., copublished with Butterworth-Heinemann, is intended to rectify this situation for polymers as well as ceramics, metals, semiconductors, and other materials.

The present volume, *Characterization of Polymers*, is intended for general polymer scientists and engineers without much experience in analysis but who have to deal with polymer surface and interface problems on a day-to-day basis. Because of the aforementioned focus on properties, and because many of the analytical techniques used for different types of materials are the same, this volume does not emphasize the characteristics of different techniques; this is accomplished in the lead volume of this Series, *Encyclopedia of Materials Characterization*. Instead, the present volume uses a case study approach to illustrate the importance of polymer problems and how they have been solved using surface and interfacial analyses. Each case study is carefully selected so that the whole range of important properties of polymers is adequately covered. The objective of this volume is not to make the reader an expert in analysis, but to get him or her started to become one. More importantly, the volume provides the reader with the knowledge to select appropriate characterization techniques which will help solve the problems at hand, and to plant questions and new ideas in the mind of the expert analyst with whom he or she is working. The techniques chosen are widely accepted techniques; novel techniques which have a high chance of acquiring wide acceptance in the near future are also mentioned.

Scope and Organization

The first two chapters give an overview of polymer chain structure and synthesis and fabrication techniques encountered commercially. The next five chapters discuss surface properties and modification techniques in the order of chemical composition, microstructures and related properties, structures and morphology of interfaces and thin films, thermodynamics, and surface modification. These are followed by four chapters dealing with interfacial properties involving at least one polymer.

Topics covered are adhesion, polymer–metal, polymer–ceramic, and polymer–polymer interfaces, as well as friction and wear pertaining to polymer damages caused by two contacting materials in relative motion. Chapter 7, on surface modification, bridges these two parts of the book because such techniques affect both surface and interfacial properties. Finally, in Chapter 12, references (e.g., books and handbooks) are given to aid in future studies. A brief description of all chapters is provided below to assist the reader in planning his or her studies:

Chapter 1, on chain structures and polymer synthesis, describes the chain structures in natural and synthetic polymers and the various routes of polymer synthesis. Also included is a brief description of polymer surface structure. Emphasis is placed on commodity structural polymers, engineering thermoplastics, high-temperature polymers, typical copolymers, and polymer blends.

Chapter 2 gives an overview on commercial polymers, their applications, and commonly used fabrication techniques for their processing. Polymer fabrication techniques discussed include foam processing, film forming/casting, composite processing, extrusion, and molding. Examples on the application of various fabrication techniques are described in some detail. When applicable, analytical methods are mentioned as they pertain to property studies or tailoring. The effects of processing on surface properties and others (e.g., mechanical and physical properties) are also discussed.

Chapter 3, on chemical composition, discusses the techniques and methodologies used to characterize the chemical composition of polymers. The main techniques used to determine chemical composition are summarized, and the types of information obtained from each are described. Chemical composition aspects that are addressed include degree of stoichiometry, functionality determination, completeness of reactions, reactive moieties, contaminations, dopants, surface segregation, and multilayer integrity.

A critical review of various microscopy techniques to measure the morphology of polymer surface and thin films is given in Chapter 4. This chapter describes contrast mechanisms and optics for various microscopy techniques, with specific examples from mesomorphic polymer systems to give an appreciation of the techniques. Apart from visualization, the chapter explains microscopy as a tool to obtain quantitative structural information when complemented with other methods. Recent scanning probe techniques such as atomic force microscopy and scanning tunneling microscopy, and modern developments in scanning and high-resolution electron microscopy, are also covered.

Surface morphology study using scattering methods is described in Chapter 5. The chapter outlines the theoretical and experimental aspects of the methods to provide an appreciation of their limitations and advantages. Less common methods such as surface wave probes and surface second-harmonic-generation methods that are established techniques but are not commonly used on polymer interfaces are also covered to highlight recent developments. Moreover, other surface methods such as grazing incidence X-ray scattering, reflectivity, and optical wave guiding covering length scale form 10^3 μm to 1 nm are covered.

Chapter 6, on surface thermodynamics, discusses the physical chemistry of polymer surface in terms of specific and dispersive interactions. This chapter outlines the theory of wetting and (work of) adhesion in terms of acid–base and London attraction. It provides the details on contact angle measurement techniques, adsorption isotherm methods and calorimetry, which are important in obtaining quantitative information on surface energy of interfaces with polymers. The chapter also touches upon ellipsometry serum replacement, inverse gas chromatography and infrared spectroscopy for the characterization of polymer surfaces.

Surface modification of polymers is covered in Chapter 7. Surface modification is an indispensable processing tool for the fabrication of a wide variety of polymer and plastic products. This chapter covers surface modification by mechanical means, by wet chemical treatments, and by dry processing techniques. This chapter also describes how surface modifications are monitored, as well as some of the tools that are employed in dry processing.

The successful application of polymers is quite often dependent on their adhesion to other materials. Chapter 8, on adhesion, reviews the basic concepts of adhesion. Some of the techniques to measure adhesion are described, along with their pitfalls. This chapter also discusses the issues of locus of failure, surface modification, and environmental stability.

Chapter 9, on primarily polymer–metal and polymer–ceramic interfaces, deals with the chemistry involved at these interfaces. Besides addressing the reactivity of these interfaces, the chapter deals with thermal stresses and interfacial fracture. These properties are particularly important for multilevel structures such as would be found in multichip modules important in microelectronics.

Chapter 10 details various techniques to characterize polymer–polymer interfaces. This chapter discusses the relevance of concentration profile of the polymers across the interface—a crucial property for adhesion and other interfacial properties. It also describes techniques such as forward recoil spectroscopy, neutron reflectivity, and dynamic SIMS nuclear reaction for quantitatively measuring interfaces of compatible and incompatible polymer systems.

In the preceding three chapters, ensuring strong adhesive bonding between materials is the aim. Chapter 11, on friction and wear, addresses how to minimize polymer damages caused by two contacting materials in relative motion. This chapter discusses the mechanisms and measurement of tribological properties and correlations between these properties with chemical, physical, and mechanical properties.

Chapter 12 provides readers with references for their future studies in the various subject areas covered by this book. Also given are references dealing with the investigation of polymers in general, polymer characterization and test methods, the advances in the use of computers for polymer research, and environmental effects on polymers. These references are included to help the reader formulate a well-rounded approach even though they may not be specifically related to polymer surfaces and interfaces.

We take this opportunity to thank the authors of the present volume, as well as Dr. Marjan Bace, Ms. Lee Fitzpatrick, and the editorial staff at both Manning Publications Co. and Butterworth-Heinemann for making this volume a reality. Thanks are also due to Dr. Richard Brundle, one of the two series editors, for many stimulating and informative discussions.

Ho-Ming Tong, Steven P. Kowalczyk,
Ravi Saraf, and *Ned J. Chou*

Contributors

Steven G. H. Anderson
Motorola, Inc.
Austin, TX

Chemistry, Reactivity, and Fracture
of Polymer Interfaces

Chin-An Chang
IBM Research Division
Yorktown Heights, NY

Surface Modification of Polymers

Ned J. Chou
IBM Research Division
Yorktown Heights, NY

Surface Modification of Polymers

Russell J. Composto
University of Pennsylvania
Philadelphia, PA

The Polymer–Polymer Interface

David W. Dwight
Polar Materials, Inc.
Martins Creek, PA

Surface Thermodynamics

Norman S. Eiss, Jr.
Virginia Polytechnic Institute and
State University
Blacksburg, VA

Friction and Wear (Tribology)

F. Emmi
IBM Corporation
Endicott, NY

Polymer Fabrication Techniques

Paul S. Ho
University of Texas at Austin
Austin, TX

Chemistry, Reactivity, and Fracture of
Polymer Interfaces

Jung-Ihl Kim
Anam Industrial Company
Seoul

Adhesion

Steven P. Kowalczyk
IBM Research Division
Yorktown Heights, NY

Chemical Composition of Polymers;
Adhesion

Thomas B. Lloyd
Lehigh University
Bethlehem, PA

Surface Thermodynamics

Todd Mansfield
University of Massachusetts
Amherst, MA

The Polymer–Polymer Interface

L. J. Matienzo
IBM Corporation
Endicott, NY

Polymer Fabrication Techniques

Charles E. Rogers
Case Western Reserve University
Cleveland, OH

Polymer Structures and Synthesis
Methods

Ravi F. Saraf
IBM Research Division
Yorktown Heights, NY

Structure and Morphology
of Interfaces and Thin Films by
Scattering Techniques

Dwight W. Schwark
Eastman Kodak Company
Rochester, NY

Characterization of the Morphology
of Polymer Surfaces, Interfaces, and
Thin Films by Microscopy Techniques

Da-Yuan Shih
IBM Research Division
Yorktown Heights, NY

References for Future Study

Robert Simha
Case Western Reserve University
Cleveland, OH

Polymer Structures and Synthesis
Methods

Richard S. Stein
University of Massachusetts
Amherst, MA

The Polymer–Polymer Interface

Edwin L. Thomas
Massachusetts Institute of Technology
Cambridge, MA

Characterization of the Morphology
of Polymer Surfaces, Interfaces, and
Thin Films by Microscopy Techniques

Ho-Ming Tong
IBM Corporation
Hopewell Junction, NY

References for Future Study

D. W. Wang
IBM Corporation
Endicott, NY

Polymer Fabrication Techniques

1

Polymer Structures and Synthesis Methods

CHARLES E. ROGERS and ROBERT SIMHA

Contents

1.1 Introduction

The purpose of this chapter is twofold. First, to introduce briefly certain basic notions and ideas in polymer science and the characterization of molecular structure, important for physical properties. Second, to sketch the various modes of preparation of synthetic polymers. Several of the topics mentioned here are given more detailed consideration and application in subsequent chapters.

1.2 Chain Structures in Natural and Synthetic Polymers

The Nature of Polymeric Materials

The essential difference between high polymers and most organic compounds resides in the word "high." Polymeric materials are characterized by high molecular weights and by a distribution of molecular weights in any given sample. Chemically, they are identical with their low-molecular-weight analogues. It is the very high molecular weights of polymeric materials that give them their properties, especially their mechanical properties, and, conversely, lead to difficulties in their characterization.

Most polymers are composed of long sequences of identical repeat units, the few exceptions being some naturally occurring and structurally complex polymers such as proteins and nucleic acids. The overall chemical composition of a polymer and its distribution of molecular weights are major factors affecting its structure and properties.

The chain configuration—that is, the arrangement of the atomic groups along the chain, which cannot be changed except by breaking and reforming chemical bonds—is established by the synthesis process. It, and the environment of the polymer chain, determine the arrangement of the atomic chain in space, the chain conformation. That arrangement can be changed by rotations about single bonds. Intrachain and interchain interactions between groups on polymer chains significantly affect chain conformational structures.

Highly regular chain configurations allow solid polymers to form crystalline regions. The size and complexity of polymer chains usually limits the attainment of completely crystalline solids, in contrast to low-molecular-weight materials of the same general composition. Crystalline regions of polymers also contain many more defects. The topic of morphology considers the semicrystalline texture of polymeric materials.

A characteristic feature is the greater or lesser degree of chain flexibility, depending on structure. The chain conformations of polymers are almost always different in solution from what they are in the dry solid state. The changes in chain conformations induced by the presence of a solvent in various concentrations not only change its properties but also provide a valuable degree of freedom in characterization of the polymer. General and specific interactions in solution affect chain conformations in predictable and measureable ways.

The effects of temperature and added solvents on polymer conformational rearrangements, as related to the segmental mobilities of various groups and chain sequences of different size in the polymer, are important both for understanding structure-property relationships and for characterization. Transitions (melting, glass transition, secondary transitions) can be well-defined experimentally as functions of temperature, frequency, and solvent type and content.

Depending on their chemical compositions and consequent physical structures, polymers are familiar to us as elastomers, plastics, fibers, composites, coatings, adhesives, etc. Many polymeric materials are combinations of several different materials such as ceramics, metals, and other polymers. Low-molecular-weight additives serve as plasticizers, lubricants, antioxidants, flame retarders, etc. All of these components modify the polymeric material to some extent and must be accounted for in characterization of structure and properties.

A major categorization is into the classes of thermoplastic and thermosetting polymers. Thermoplastic polymers can melt or soften so that they flow and can be formed from one shape into another. In concept, this procedure could be repeated indefinitely. However, in practice, the polymer is subject to degradation at high temperatures, limiting the number of times it can be processed and still retain useful properties. Thermoplastics also can be dissolved, if a suitable solvent can be found. In both cases, after melting and cooling and after dissolution and precipitation, the thermoplastic material is nominally unchanged in its chemical composition.

Thermosets are polymers cross-linked into one giant molecule. After cross-linking into a three-dimensional network structure, they cannot be melted or dissolved, but only swollen by suitable solvents to form gels. They can be destroyed, either thermally, chemically, or mechanically, into compounds of composition and structure different from the starting material. Thermoplastics may be converted into thermosets, but we cannot go in the reverse direction. A vulcanized rubber tire is a good example: it can be burned or ground up, but it cannot be returned to a mobile melt state capable of flow to fill a mold.

The Composition of Polymeric Materials

The chemical composition of polymers is the single most important factor determining their structure and properties. Our ability to control the composition is the major factor which has led to the use of synthetic polymers in ever-expanding areas of application.

Polymeric materials may be based on the polymerization of a single starting material, a monomer, or on the simultaneous polymerization of two or more monomers in the same polymerization reaction. In the first case, the resultant polymer is termed a "homopolymer." The polymerization of polyethylene is an example:

$$n\text{CH}_2{=}\text{CH}_2 \rightarrow -(-\text{CH}_2{-}\text{CH}_2{-})_n{-}$$

Repeat structures for typical homopolymers are illustrated in Tables 1.1–1.3. Table 1.1 lists the most widely used thermoplastic materials, characterized by moderate price and moderate to good properties. The first five polymers listed (to include a wide range of polymers based on polystyrene) comprise about 85% of the total amount of all polymers used for applications. Table 1.2 lists so-called engineering polymers, which generally are more expensive but have good to excellent properties. Polymers with good to excellent properties at high temperatures over a useful time span are given in Table 1.3.

When more than one monomer is used, the resulting polymer is a "copolymer," which has a composition reflecting the proportion of the component comonomers in the mixture and the relative reactivities of the different components. This dependency leads to sequential chain distributions (i.e., the sequence of occurrence of the different components along the main chain) of the polymerized components, which have a great effect upon the properties of the copolymer material.

The four general types of copolymer that can be made are classified as follows:

random: ~ABBAABAAABBABBBA~

alternating: ~ABABABABABABAB~

block: ~AAAAAAAABBBBBBBB~

graft: ~AAAAAAAAAAAAA~
        └─ BBBBBBBBBBBBBB~

Polymer	Chain Repeat Unit	T_g/T_m, °C
Polyethylene, linear density = 0.96 g/ml	$-CH_2-CH_2-$	$-80\pm10/138$
Polyethylene, branched density = 0.92 g/ml	$-CH_2-CH_2-$	$-80\pm10/105$
Polypropylene, isotactic	$-CH_2-CH(CH_3)-$	$-10/165$
Poly(vinyl chloride)	$-CH_2-CH(Cl)-$	$82/$
Polystyrene, atactic	$-CH_2-CH(\varnothing)-$	$100/$
Polymethylacrylate	$-CH_2-CH(COOCH_3)-$	$6/$
Polymethylmethacrylate	$-CH_2-C(CH_3)-$ $\qquad\ \ COOCH_3$	$105/$
Polyacrylonitrile	$-CH_2-CH(CN)-$	$\sim100/\sim319d$
Poly(vinyl acetate)	$-CH_2-CH-$ $\qquad\ OCOCH_3$	$28/$
Poly(vinyl alcohol)	$-CH_2-CH(OH)-$	$85/$
Poly(vinyl butyral) (13.5 mole % of residual $-OH$)	$-CH_2-CH-CH_2-CH-CH_2-CH-$ $\qquad\quad O\qquad\qquad O\qquad\qquad OH$ $\qquad\qquad\ \diagdown\ CH\diagup$ $\qquad\qquad\qquad C_3H_7$	$49/$
Cellulose	CH_2OH $\ CH-O$ $HC\ OH\quad HC\diagup\ O$ $\qquad CH-CH$ $\qquad\quad OH$	
Cellulose acetate	$CH_2-OCOCH_3$ $\ CH-O$ $HC\ OH\quad HC\diagup\ O$ $\qquad CH-CH$ $\qquad\quad OCOCH_3$	

Table 1.1   Commodity structural polymers.

Polymer	Chain Repeat Unit	T_g/T_m, °C
Poly(hexamethylene adipamide) (Nylon 66)	$-NH-(CH_2)_6-NH-CO-(CH_2)_4-CO-$	~57/267
Poly(ε-caprolactam) (Nylon 6)	$-NH-(CH_2)_5-CO-$	55/221
Polycarbonate	$-O-C\langle\substack{CH-CH\\CH=CH}\rangle C-\overset{CH_3}{\underset{CH_3}{C}}-C\langle\substack{CH-CH\\CH=CH}\rangle C-O-\overset{}{\underset{O}{C}}-$	149/
Poly(2,6-dimethyl phenylene oxide) (PPO)	$-CH\langle\substack{CH-\overset{CH_3}{C}\\CH=\underset{CH_3}{C}}\rangle C-O-$	135/261
Poly(ether sulphone)	$-C\langle\substack{CH-CH\\CH=CH}\rangle C-\overset{CH_3}{\underset{CH_3}{C}}-C\langle\substack{CH-CH\\CH=CH}\rangle C-O-C\langle\substack{CH-CH\\CH=CH}\rangle C-\overset{O}{\underset{O}{S}}-C\langle\substack{CH-CH\\CH=CH}\rangle C-$	190/
Polyoxymethylene	$-CH_2-O-$	−85/180
Polytetrafluoroethylene	$-CF_2-CF_2-$	−150/327
Poly(ethylene terephthalate)	$-O-\overset{O}{\underset{}{C}}-C\langle\substack{CH-CH\\CH=CH}\rangle C-\overset{O}{\underset{}{C}}-O-CH_2-CH_2-$	70/265
Poly(butylene terephthalate)	$-O-\overset{O}{\underset{}{C}}-C\langle\substack{CH-CH\\CH=CH}\rangle C-\overset{O}{\underset{}{C}}-O-CH_2-CH_2-CH_2-CH_2-$	17/240
Polyarylate	$-O-C\langle\substack{CH-CH\\CH=CH}\rangle C-\overset{CH_3}{\underset{CH_3}{C}}-C\langle\substack{CH-CH\\CH=CH}\rangle C-O-\overset{O}{\underset{}{C}}-C\langle\substack{CH-CH\\CH=CH}\rangle C-\overset{O}{\underset{}{C}}-$	190/
Poly(phenylene sulfide)	$-C\langle\substack{CH-CH\\CH=CH}\rangle C-S-$	185/285

Table 1.2 Engineering thermoplastics.

Polymer	Chain Repeat Unit	T_g / T_m, °C
Poly(ether ether ketone) (PEEK)		144/335
Poly(amide imide)		>290/
Polyimide		$T_d \approx 400$
Polyetherimide (Ar = 4,4'-biphenyl)		247/ $T_d \approx 460$
Poly[2,2'-(m-phenylene)-5,5'-bibenzimidazole] (PBI)		~430 (500)/ $T_d \approx 420$ (480)
Poly(p-phenylene benzobisoxazole) (PBO)		$T_d > 400$
Poly(p-phenylene benzobisthiazole) (PBT)		$T_d > 400$

Table 1.3 Polymers for high-temperature applications. T_d is the temperature (in °C) of onset of thermal decomposition (~1% weight loss per day).

Polymer	Chain Repeat Unit	T_g / T_m, °C

Copolyester of 2,6-hydroxynaphthoic acid
   and 1,4-hydroxybenzoic acid

$$\sim\!-O-C\!\!\begin{array}{c} CH-CH \\ \\ CH=CH \end{array}\!\!C-\overset{O}{\overset{\|}{C}}-O-C\!\!\begin{array}{c} CH=CH \\ \\ CH-C \end{array}\!\!C-CH\!\!\begin{array}{c} \\ \\ CH=CH \end{array}\!\!\overset{O}{\overset{\|}{C}}-\sim$$

Polyphenyl

$$\sim\!-C\!\!\begin{array}{c} CH-CH \\ \\ CH=CH \end{array}\!\!C-\sim$$

$T_d > 530$

Table 1.3 *continued*

The chain sequence distribution in "random" copolymers is statistical, determined by the relative reactivities of the comonomer reactants (see the section on polymer chain configurations). The resultant structures can show a bias toward blockiness or alternation, depending on the comonomers involved. The final overall composition of a copolymer can be controlled effectively by control of the polymerization conditions.

Copolymers constitute an important amount of polymers in present use. A listing of some of the more important copolymers is given in Table 1.4.

Another major category of useful polymer systems, and one currently popular, is the mixture of two or more individual polymers (either homopolymers or copolymers or both) into a "polymer blend" or polyblend. Such systems may have exceptional combinations of properties that would be difficult to obtain from singular homopolymer or copolymer compositions. Since they may be made using existing polymers, they also may be more economical to produce compared with the synthesis and consequent production scale-up of a new homopolymer or copolymer with comparable properties. Some typical polymer blends are

- High-impact polystyrene (HIPS): polystyrene/polybutadiene/poly(butadiene-*graft*-styrene)

- ABS: acrylonitrile/butadiene//styrene

- Xenoy: polycarbonate/copolyester or poly(ethylene terephthalate) or poly(butylene terephthalate)

- Noryl: poly(2,6-dimethyl phenylene oxide)/polystyrene

- Noryl GTX, Vydyne: poly(2,6-dimethyl phenylene oxide)/nylon

- Modified nylon: nylon/rubber (e.g., EPDM)

Comonomers (Typical % Composition)	Common Name or Use
Random Copolymers	
Butadiene$_{75}$/styrene$_{25}$	SBR rubber
Butadiene$_{80-60}$/acrylonitrile$_{20-40}$	NBR rubber, nitrile rubber
Vinyl chloride$_{85}$/vinyl acetate$_{15}$	"PVC"
Vinylidene chloride$_{85}$/vinyl chloride$_{15}$	Saran
Ethylene/propylene/diene terpolymer	EPDM, ethylene-propylene-rubber
Ethylene/vinyl acetate	EVA
Fluorinated ethylene/propylene	FEP
Isobutene$_{98-99}$/isoprene$_{2-1}$	Butyl rubber
Block Copolymers	
Butadiene/styrene	Thermoplastic elastomers
Hydrogenated butadiene/styrene	Thermoplastic elastomers
Isoprene/styrene	Thermoplastic elastomers
Hydrogenated isoprene/styrene	Thermoplastic elastomers
Polyether/urethane	Thermoplastic elastomers
Polyester/urethane	Thermoplastic elastomers
Graft Copolymers	
Polybutadiene/styrene	Component in high-impact polystyrene (HIPS)
Alternating Copolymers	
Butene-1/sulfur dioxide	Positive resist
Styrene/maleic anhydride	

Table 1.4   Some typical copolymers.

- Plasticizers for PVC: nitrile rubber, EVA copolymers, polyester polyurethanes, chlorinated polyethylene, poly(butylene terephthalate)-polytetrahydrofuran block copolymers (Hytrel), ethylene-vinyl acetate-carbon dioxide terpolymers

- Other polyblends: chlorinated PVC/styrene maleic anhydride, ABS/polycarbonate, polymethylmethacrylate/poly(vinylidene fluoride), poly(ethylene terephthalate)/poly(butylene terephthalate), and polyelectrolyte complexes.

Polymer blends are usually composed of immiscible individual polymers leading to the formation of phase-separated bulk materials. There are a few (about 150) binary polymer pairs that exhibit miscibility over limited ranges of composition and temperature. There are even fewer that show apparently complete miscibility over their entire range of composition. One of these is the blend of poly(2,6-dimethyl phenylene oxide)(GE's PPO) and polystyrene, known commercially as a member of the Noryl™ (manufactured by GE) family of polymers. Even here there is controversy as to the true extent of miscibility down to the segmental and molecular level.

Miscible, nominally one-phase polyblends offer the possibility of obtaining certain synergistic properties where the behavior of the polyblend is better than would be expected from a rule of mixtures based on the properties of the component polymers. For example, a 50:50 mixture of PPO and polystyrene has nearly as good mechanical properties as does PPO itself, but it is more easily processed and is less expensive.

Immiscible or partially miscible polymer blends are more common. A distinction must be made here between the terms "miscible" and "compatible." Miscibility is denoted as a thermodynamic criterion in the terms of solution thermodynamics defining a true solution of a solute (polymer A) in a solvent (polymer B). The difficulty here is in the experimental confirmation that the solution truly forms a single phase down to the molecular level.

A "compatible" polyblend is defined much less rigorously. It is one in which the degree of mixing is on a fine scale, approaching the molecular scale (down to the molecular scale in colloquial practice), or it indicates a more or less useful degree of partial miscibility in which the phase boundaries between the phase-separated domains are more or less diffuse. Such systems often exhibit useful property characteristics.

Compatible polymer blends—also often denoted, unfortunately, polymer "alloys"—comprise a large number of useful polymer blend systems. In most cases, the degree of dispersion of one polymeric component in a continuous matrix of the other polymeric component is obtained under strict conditions of mixing and other variables in the product manufacture process. It is a matter of the control of kinetics to overcome the thermodynamic preferences of the total polymeric system.

Thermodynamics usually desires a complete phase separation of two immiscible polymeric components, tempered by some entropic intermixing at the phase boundary (thus we have an "interphase" rather than an "interface"). This intended phase separation may be thwarted by vigorous mixing with consequent rapid immobilization (solidification) of the mixture to "freeze-in" the nonequilibrium extent of dispersion. Such systems, although in a state of thermodynamic nonequilibrium, often have a lifetime of relative stability sufficient to promote their use as commercial materials of value.

It is common practice to add block or graft copolymers, composed of the same or similar monomers corresponding to the major polymeric components of the polymer blend, to modify the interphase regions between the domains of the major components. The blocks of similar material associate with the corresponding domain materials, thereby establishing a mechanical linkage across the phase boundaries. The polyblend is then said to be "compatibilized" by the added copolymer or other compatibilizer material.

There are many other types of multicomponent/multiphase polymer–polymer materials. These include interpenetrating polymer networks (IPNs), interpenetrating elastomer networks (IENs), and semi-IPNs, where one polymer is not cross-linked and the other one is cross-linked into a network intermixing with, but not bonded to, the first polymer. Other systems are rubber-rubber blends, bicomponent polymer fiber materials, multilayer films made by coextrusion or other methods, and various other laminated products. Polymers containing ionic groups along their chain structure (ionomers) also can be put in this category since the ionic groups form multiplets or clusters within the otherwise continuous hydrocarbon matrix.

Composite polymer materials are differentiated from polymer blends since the base polymer matrix now is modified by the incorporation of a nonpolymeric material such as carbon black, glass fibers, or a gas (cellular polymers, foams). The category could be expanded to include the use of high-tenacity synthetic polymer fibers being developed for composite reinforcement and other applications.

These fillers reinforce the polymeric matrix or increase the volume of the material with or without a reinforcing or other effect, such as increasing abrasion resistance or reducing photochemical or oxidative attack. Other uses include the addition of pigments to give color and opacity, resilient materials for sound-damping, and polymer-impregnated wood, cements, mortars, and concretes. Such materials have been, and still are, under intensive development to produce materials that are strong and have other desired properties, yet are lightweight. They are of interest especially to the transportation and construction industries.

A dispersion of a particulate filler in a polymeric matrix is usually characterized by a distribution of aggregate structures due to association of the smaller, more or less spherical, filler particles. Seldom, if ever, is a filler uniformly dispersed as individual particles. In addition, a superstructure, called agglomeration, usually forms during the sample fabrication molding process. This superstructure is destroyed or modified by subsequent mechanical deformation or extended thermal or solvent swelling treatments. This change contributes to the change in modulus and other properties associated with the Mullins' softening effect (see below).

The relative chemical compositions of the polymer matrix and the filler surface (region) largely determine the degree of adhesion. It is noteworthy that the surface composition of most materials usually is not the same as the bulk composition of the material. This is due both to migration of substances to the surface from the bulk phase and to adsorption of substances from the environment. It is common practice to coat glass fibers with an oil or other coating in order to reduce abrasion and breakage during handling and shipping. These surface features certainly may affect adhesion and adhesion-sensitive properties.

On the other hand, it is common practice to add surface active agents, such as silanes, to increase the degree of adhesion between the components. These "coupling agents" are not applied as a monolayer, but rather as finite layers in which the concentration of agent decreases with distance from the filler. The structure and properties of this interphase affect significantly the properties of the composite.

One effect is to provide a gradient of modulus from the surface into the matrix, which generally is beneficial to the material's mechanical performance.

It is noteworthy that the presence of a filler surface can affect the chain conformations and segmental mobilities of matrix polymer chains adsorbed onto or near the surface. This can range from actual chemical bonding (e.g., "bound rubber" in filled elastomers) through major to minor perturbations in chain conformations. These affect properties in much the same way that cross-linking of the polymer does. The detailed structure and arrangement of these perturbed polymer chains are subject to change by imposed mechanical deformations, leading to such effects as the "Mullins' softening effect," a decrease in modulus of the filled material after initial mechanical deformation.

The purposes of the polymeric matrix in a fiber-reinforced plastic (FRP) composite are to maintain the orientation of the fibers, separate the fibers from each other, transmit stresses to the fibers, and protect the fibers from environmental attack. The advantage for most applications is found in the mechanical properties of the fiber, its high modulus, and its strength. The polymeric binder is a necessary means for the effective utilization of those properties. The major structural features determining the properties of FRPs are the concentration (volume fraction) of fibers and their size (length) and orientation relative to the axis of loading and deformation.

Molecular Weight and Its Distribution

As mentioned previously, the primary characteristic of a polymer is its high molecular weight. The molecular weight of any polymer is not unique: a given polymer can be prepared with a wide range of molecular weights, unlike low-molecular organic compounds, which have unique molecular weights reflecting their precise molecular composition. For example, benzene (C_6H_6) always has a molecular weight of 78.11, or else it is not benzene. There is no precise molecular weight boundary differentiating between polymer and nonpolymer—it is a matter of definition and property behavior. If one needs an arbitrary guideline, let us say that any organic material of molecular weight greater than 20 000 is a "high" polymer.

The distribution of molecular weight in a polymer is initially determined by the synthesis process. The distribution can be more or less broad depending on the polymer and the method of synthesis. It can be altered by subsequent chemical reactions, including degradation, or by fractionation, either as an analytical procedure or as a consequence of the polymers exposure to appropriate solvent extraction processes.

The molecular weight of a polymer is usually expressed as an average. The four most commonly used averages are number average (obtained by colligative property measurements), weight average (from light-scattering measurements), Z-average (from ultracentrifuge measurements), and viscosity average (from dilute solution viscosity measurements). These can be individually determined by several different experimental techniques for each average. Chromatographic techniques (e.g., gel permeation and size-exclusion) are commonly used now to obtain a measure of the entire molecular weight distribution. Often the molecular weight is given in practice without indication of which average it is, in which case one must exercise caution in interpretations and predictions based on the datum.

The molecular weight and its distribution in any given polymer have great effects on polymer properties. Generally, properties such as tensile strength increase with increasing molecular weight and tend to level off at high molecular weights. Similar behavior is shown by melt viscosity, which increases as a function of M_w to the 3.4th power for sufficiently high molecular weights. This leads to difficulties in the processing of very high molecular polymers desired for their corresponding good properties in the molded product.

Typical commercial polymer molecular weights and corresponding degrees of polymerization (dp), which indicate the number of repeat units in the average polymer chain, generally cover a wide range of grades of different molecular weights. This allows the user to select materials suitable for various chemical and physical treatments. Of special importance is that different processing methods require materials of different flow properties and different viscosities, dependent on molecular weight.

A factor closely related to and dependent upon molecular weight is chain entanglement. Most high polymers exhibit some degree of chain coiling in solution and in the amorphous solid state, as we discuss below. These chains interpenetrate each other to greater or lesser extent—they become entangled. This entangling promotes the transmission of stress from one chain to another. It ties the polymer structure together.

Very slow rates of deformation allow the tangled chains to flow past one another, leading to creep flow deformation of the material. Somewhat higher rates do not allow this in the time scale of the stress rate (Newtonian region), and so the chain entanglement acts effectively as a cross-link to "increase" the molecular weight of the system. This is the reason that the melt viscosity increases so rapidly with molecular weight for molecular weights greater than a "critical" value for these rates of viscous flow. The behavior at higher rates of flow deformation is more Theologically complex.

Polymer Chain Configurations

Chain configurations are the spatial arrangements of the atoms composing the main chain of a polymer and the side groups attached to the main chain. Variations in these configurations have profound effects upon chain conformations, polymer morphologies, and consequent properties.

The configuration of a polymer molecule is established when the polymer is synthesized, and it cannot be altered without breaking and rearranging chemical bonds. Three cases are considered: (1) the sequence of copolymer repeat units along a chain; (2) the occurrence of defect structures such as head–head linkages in a homopolymer; and (3) the stereo and geometric isomerism of side-group arrangements in a polymer.

In the copolymerization of two different monomers, we obtain data that allow us to determine the reactivity ratios, r_1 and r_2, which describe the relative reactivites of a propagating radical of monomer 1 to the addition of monomer 1 and monomer 2 ($r_1 = k_{11}/k_{12}$) and of a radical of monomer 2 to monomer 2 and monomer 1 ($r_2 = k_{22}/k_{21}$). These constants determine the sequence distribution for a given monomer composition of the chain. This composition, however, is itself an average due to composition fluctuations. Confirmation or determination of the sequence distributions in polymers can be carried out by high-resolution nuclear magnetic resonance (NMR) and other techniques.

The occurrence of chain defect structures, such as head-head linkages

$$\sim\!-CH_2-CH-CH_2-CH-CH_2-CH-CH-CH_2-CH_2-CH-CH_2-CH-\!\sim$$
$$\quad\quad \underset{\varnothing}{|} \quad\quad \underset{\varnothing}{|} \quad\quad \underset{\varnothing}{|}\; \underset{\varnothing}{|} \quad\quad\quad \underset{\varnothing}{|} \quad\quad \underset{\varnothing}{|}$$

is due to entropic variations during the polymerization process. Their importance lies in the fact that they constitute weak linkages in the polymer chain due to the greater degree of steric hindrance and polar repulsion. This facilitates their decomposition at relatively low temperatures, leading to premature degradation of the polymer chain. Since their concentrations are small, their analysis is difficult.

A more important type of configurational structure is stereoisomerism, which arises from differences in symmetry between substituted carbon atoms along the chain. The major types are atactic, isotactic, and syndiotactic. The isotactic form has the pendent side group in the same configuration ("same side of the main chain") as we go down the chain, repeat unit by repeat unit:

$$\overset{\displaystyle CH_3}{\underset{\displaystyle H}{}}\quad \cdots$$

$$\sim\!-CH_2-\overset{CH_3}{\underset{H}{C}}-CH_2-\overset{CH_3}{\underset{H}{C}}-CH_2-\overset{CH_3}{\underset{H}{C}}-CH_2-\overset{CH_3}{\underset{H}{C}}-CH_2-\overset{CH_3}{\underset{H}{C}}-CH_2-\overset{CH_3}{\underset{H}{C}}-CH_2-\!\sim$$

In the syndiotactic form, the configuration alternates from one side to the other:

$$\sim\!-CH_2-\overset{R}{\underset{H}{C}}-CH_2-\overset{H}{\underset{R}{C}}-CH_2-\overset{R}{\underset{H}{C}}-CH_2-\overset{H}{\underset{R}{C}}-CH_2-\overset{R}{\underset{H}{C}}-CH_2-\overset{H}{\underset{R}{C}}-CH_2-\!\sim$$

The atactic form is a random mixture of the isotactic and syndiotactic triad forms. In more complicated main chain structures there are other forms. The stereoregular chain structures are obtained by coordination polymerization, as described in the section on coordination polymerization.

Another type of isomerism is the cis-trans geometrical isomerism about double bonds, as found in polyisoprene:

$$cis: \quad \sim\!(-CH_2-\overset{CH_3}{C}=\overset{H}{C}-CH_2-)\!\sim$$

$$trans: \quad \sim\!(-CH_2-\overset{CH_3}{C}=C-CH_2-)\!\sim$$
$$\underset{H}{|}$$

The importance of these forms of isomerism is that they affect the regularity of substitution and the conformation of the chain structure. This determines the ability of the polymer chains to pack efficiently and, hence, to crystallize. For example, isotactic polypropylene, illustrated above, is a semicrystalline material, whereas atactic polypropylene is an amorphous gum. Also, *trans*-1,4-polyisoprene is a semicrystalline material, whereas *cis*-1,4-polyisoprene is amorphous at rest but crystallizes upon stretching. Thus, the properties of the polymeric material are strongly affected by the configurational state, determined in the synthesis process.

Another aspect of the configurational nature of a polymer is the linearity of the polymer chain structure. Thus far, we have inferred that the polymer chain is completely linear. There are commercial polymers that have that configuration, such as high-density polyethylene and many condensation polymers. However, many important commercial polymers, usually made by addition polymerization, contain branches along their chain structure.

The branches along the chain may be either, or both, short chain or long chain branches. Short chain branches may result either from intramolecular chain transfer between a terminal free radical and a hydrogen atom back in the chain or, more intentionally, by copolymerization of two comonomers, one of which gives a short chain branch as a normal reaction product. An example of the latter is the production of linear low-density polyethylene (LLDPE) by the coordination polymerization of ethylene and, say, butene-1. That results in a more or less random sequence of ethyl side-chains along an otherwise linear chain.

Long chain branching is caused by intermolecular chain transfer or by chain transfer to polymer. It is comparable to graft copolymerization, where an active (radical) site is formed along a polymer chain by the abstraction of a proton by a propagating polymer radical. The secondary radical thus formed initiates the propagation of a new growing chain that is "grafted" onto the existing polymer chain. Since it is the same polymeric material, it constitutes a branch rather than a graft in our nomenclature.

The final topic concerned with polymer configurations is cross-linking. Many important polymeric materials are cross-linked for use; they are thermosets. This includes such materials as elastomers, phenol formaldehyde resins, and many adhesives such as epoxies.

In elastomeric materials, the cross-link density is about one cross-link per 100–120 repeat units. In thermoset polymers, like phenol-formaldehyde resins, the densities are 10–50 times higher. A range is found in the wide spectrum of materials that are cross-linked to reduce flow and creep or otherwise modify the properties of the polymeric materials.

Polymer Chain Conformations

Consider the two limiting extremes of a fully ordered crystal and the amorphous melt, respectively. In the former, the chains are fully extended to their maximum length, thus proportional to the dp of the individual chain molecules. If a chain has a regular configuration, it is possible to extend and align locally sections of several chains to pack into ordered regions. Such conformations are predominantly stabilized by the internal energy. This, in turn, is determined by inter- and intramolecular van der Waals interactions. Special interactions, such as H-bonds, can result in the formation and packing of helical regions.

On the other hand, in the melt, the rubbery state (a cross-linked melt), or in solution, chains assume coiled conformations. This occurs in the absence of sufficiently intense energetic interactions, stabilizing extended conformations (the rigid chain). This situation is illustrated, for example, by poly(amino acids) below their denaturation temperature. In the commercial polymers discussed here, the former situation of coil formation prevails which is regulated primarily by conformational entropy.

Consider the classical example of a long skeleton of C–C single bonds. The energetically unrestricted internal rotations around these bonds for a large number of such bonds result in a *statistical* distribution of molecular shapes. This is what is implied by the term "flexibility." As an important consequence, the *average* dimensions (end–end distance and radius of gyration) are proportional to the square root (in contrast to the first power) of the dp. The attractions and repulsions between the substituents in real chains change the proportionality factor, but leave the dp-dependence unaltered in bulk or slightly increased in good solvents.

These considerations, we note in passing, are especially important for an interpretation of the thermoelastic properties of rubbers. Elastic deformation involves a resistance to the entropy reduction, due to the extension of coiled conformations by the applied stress.

Polymer Solid-State Structure, Morphology, and Transitions

Polymer chain conformations and the rate of changes in conformations at a given temperature largely determine polymeric structure and associated properties. As the temperature is changed, most polymeric materials exhibit characteristic changes in their conformational state—for example, melting or crystallization—or in the rate of change of noncrystalline, coiled conformations—for example, the glass and other transitions.

The crystalline state is a low-energy, extended chain conformational structure in which vibrational and rotational motions about lattice sites increase with temperature to the point at which translational motions occur and the crystal melts with a discontinuous decrease in density. For those polymers that can crystallize, the melting point, T_m is the highest temperature at which crystallization and melting can take place at atmospheric pressure; it is the temperature at which the most perfect crystalline regions melt.

Bulk synthetic polymeric materials are never completely crystalline but contain lattice vacancies and defects and, usually, substantial amounts of coiled conformation material that cannot enter into a crystal lattice, due either to their nonregular configurations or to steric restraints imposed by their involvement in two different crystalline domains. Consequently, a semicrystalline polymer has a range of melting below T_m, where the more imperfect crystalline regions progressively melt as the temperature is raised. This is in marked contrast with the single sharp T_m of most low-molecular materials. The range of melting, and density change, in a polymer has pronounced effects on its properties.

Details of the process of crystallization of a given polymer are of great importance in determining the ultimate structure and properties of the material. The resultant semicrystalline structure (texture, morphology) is effectively permanent, as are the corresponding properties, although slow changes due to secondary crystallization and structural relaxation continue over very long periods of time. The integrity and the strength of a semicrystalline material are functions of the molecular weight of the polymer, which determines the probability of a given chain molecule participating in more than one crystalline domain region, thereby serving to "tie" the system together to transmit stress, etc. The effectiveness of such "tie molecules" depends on the nature and conditions of the crystallization process.

As the melt cools, the amount of supercooling below T_m introduces a thermodynamic driving force for crystallization, as well as increasing melt viscosity. Small degrees of supercooling lead to the formation of relatively few nuclei that grow into large spherulitic structures. Large degrees of supercooling produce a profusion of nuclei to give many small spherulites. In either case, the individual spherulitic domains grow in the primary crystallization process until they impinge upon each other. The subsequent secondary crystallization process involves crystallization of the material trapped within the spherulitic structure. The rate of crystallization is determined by the competition between the magnitude of the thermodynamic driving force (degree of supercooling) and the kinetic restraint on segmental mobility due to the increase in viscosity with decreasing temperature. The maximum rate of crystallization generally occurs at about $0.97\,T_m$ Kelvin.

Also influencing semicrystalline material performance and application is deformation, because it alters crystalline morphology to yield oriented structures such as fibers or biaxially oriented films. In fiber formation, the spherulitic morphology is converted into microfibrils. During deformation, small blocks of the chain-folded lamellar structure break away and align themselves transverse to the microfibril axis (and the draw direction).

Single crystals of many polymers may be obtained by crystallization from very dilute solutions. Extensive studies of those lamellar materials have shown, at the time of discovery, surprisingly, that the chain is oriented transverse to the plane of the thin platelet lamellae. This means that the individual polymer chains fold back on themselves or their neighbors in a very regular pattern. The nature of the crystal fold surface and the relationship of single crystals to spherulitic bulk crystalline structures are still subjects of discussion. It does appear that crystallization in spherulites is by chain-folded lamellae fanning out from a common nucleus as twisting and branching ribbons. The resultant material is polycrystalline with uncrystallized polymer trapped within the crystalline matrix.

An interesting example of the effects of chain conformational factors on materials structure and properties is liquid crystalline polymers (LCPs), the subjects of intense research and development efforts at this time. These materials are prepared from monomers which are rodlike (e.g., ones containing para-linked benzene rings) with chain-extending linkages such as amide or ester. The rodlike group may be in the main chain or as a side chain. Main chain polyesters based only on such molecules as hydroquinone, para-hydroxybenzoic acid, and terephthalic acid have very high melting points and cannot be processed successfully. Incorporation of flexible spacer groups (e.g., ethylene glycol) or side-group aromatic substituents or nonlinear rigid links or copolymerization of two liquid crystalline monomers reduces T_m and facilitates processing.

The LCPs associate with one another in smectic or nematic crystalline arrays as randomly oriented domains in a melt at rest. When placed under shear they become highly oriented along the shear axis. Since the amount of chain entanglement is low, the melt viscosity is relatively low. The oriented structure is preserved upon cooling. The materials are characterized by excellent flexural and tensile strengths and a high modulus along the orientation (flow) axis, but significantly lower values along the transverse axis. Impact strength is high, as are the heat distortion and continuous-use temperatures. Incorporation of fibrous fillers further enhances the properties.

In amorphous polymers and in the noncrystalline, disordered regions of semi-crystalline polymers, another characteristic temperature region is reached upon cooling. It signals the transition from a rubbery or a melt state (in amorphous systems) to a brittle glass, characterized by an increase of elastic modulus by several orders of magnitude and significant changes in such properties as thermal expansivity, compressibility, and heat capacity. This, it should be noted, is a kinetic rather than a thermodynamic effect. Under usual procedures, the experimental time scale is sufficiently long relative to the time scale of molecular responses to an imposed perturbation in the melt such as a temperature or pressure change. This situation is reversed on approaching the glass transition region because large-scale molecular motions are slowed down. Indeed, an enlargement of the experimental time scale, such as a slower cooling rate, shifts the transition to lower temperatures. An important consequence of this nonequilibrium state is a drive towards equilibrium observable in long-time experimentation. This implies a slow change in physical properties (physical aging), with possibly significant practical consequences.

The glass transition (T_g) is the most important transition and results in the most significant property changes. Additional characteristic temperatures below T_g (sub-glass relaxations) and above T_g, observable by mechanical or dielectric dynamic measurements, dilatometry, and other methods, generate quantitatively less significant changes. Below T_g they arise from the restriction of small-scale segmental motions in the chain or by substituent sidegroups and can be of importance for low-temperature properties, such as ductility. Attempts to describe the underlying molecular mobility changes in terms of structural characteristics—for example, so-called free volume quantities—have been undertaken.

To summarize, the melting temperature, T_m, and the glass temperature, T_g, are primary material parameters characterizing major changes in a great many different properties of the material. Their mutual dependence on chain flexibility, as related to conformational changes, is reflected by the observed simple linear relations between them for many polymers. Depending on symmetry, T_g is approximately one-half or two-thirds of T_m.

Polymer Surface Structure

The structure of surfaces of polymeric materials, and the associated properties, almost always differs from that in the corresponding substrate bulk phase. This is due to the thermodynamic driving force for a system to achieve the lowest possible surface-free energy and to the lack of a solid phase above the surface at the interfacial discontinuity, which limits the conformational freedom of a chain in the surface domain. This latter factor is usually considered in the context of "unbalanced forces"; the surface region is subjected to intermolecular forces from only below the interface (with equal forces in the plane of the interface) with effectively nil force from above the interface (except for usually minor vapor phase interactions; but see below).

This inequality of interactions with its neighboring molecular environment leads to the phenomenon of "surface tension," denoted as γ. This is defined as the work spent in forming a unit area of surface (also called the surface free energy). It must be distinguished from the surface stress, σ, (also called the surface energy), which is the change in the Helmholtz free energy, dA^σ, of the surface associated with a unit increase of surface area. The relation between σ and γ is established as follows: the force required to extend the surface of an isotropic solid by $d\Omega$ is $\gamma d\Omega$, and the work required must equal the increase in total surface energy dA^σ:

$$\gamma d\Omega = dA^\sigma = d(\Omega\sigma)$$

where $\gamma = d(\Omega\sigma)/d\Omega$ and $\gamma = \sigma + \Omega(d\sigma/d\Omega)$.

For a liquid, any extension of the surface usually will result in molecular flow from the underlying bulk regions into the surface that results in no change in composition; that is, $(d\sigma/d\Omega) = 0$. Thus, the surface tension γ equals the surface energy σ. For a solid, the surface is deformed by stretching, causing a change in surface density (and orientation) so that $(d\sigma/d\Omega)$ is not equal to 0. In general, for a solid the surface energy and the surface tension are not equal. This aspect is of particular importance in the interpretation of the results of polymer deformation studies, which generally have ignored the difference between γ and σ.

Major applications of surface properties of polymers have been related to adhesion, friction and wear, adsorption of substances onto the polymer surface, degradation and deformation, and fracture. Several other applications, including various electrical, electronic, ablative, mechanical, and chemical processes, especially the nature of chemical modifications, are of interest. Appreciation of the nature of polymer surfaces is essential for proper interpretation of those properties and applications and effective conduct of procedures involving those aspects.

The drive to achieve the lowest possible surface free energy often leads to segregation of components comprising a polymeric system. One type of example is given by the behavior of polymer-blend systems. For example, the polymer blend of polymethylmethacrylate (PMMA) and poly(vinylidene fluoride) (PVDF) is miscible in the range of zero to about 35% PVDF at room temperature. Surface characterization of these miscible blends showed that the surface was enriched in the component with the lower surface tension, PVDF. This enrichment was at the expense of the immediate subsurface region which was found to have a decreased concentration of PVDF. This led to pronounced effects on subsequent chemical modifications of the material and on the physical properties of both the surface and bulk material.[1] Similar results have been noted in many other multicomponent systems, including graft and block copolymers, plasticized polymeric materials, and fiber- and particle filler-reinforced polymeric composite materials.

An aspect of particular importance is the migration of lower-molecular-weight material to the surface region. This lower-molecular-weight material is composed of both the low end of the molecular weight distribution of the base polymer and any additives and impurities in the material, especially plasticizer. These substances have relatively high mobilities within the polymeric matrix due to their small size. In addition, they usually plasticize the polymeric matrix by increasing the free volume of the material, thus increasing the segmental mobility of the polymer chain segments. Indeed, this is precisely why plasticizer is added to polymers such as poly(vinyl chloride-*co*-vinyl acetate): the plasticizer increases the segmental motions and decreases T_g, thereby making the material more flexible.

The presence of low-molecular-weight material on the surface profoundly affects properties such as adhesion and friction. Adhesive strength is often drastically decreased since the lower-molecular-weight material has a correspondingly lower cohesive strength. It constitutes a "weak boundary layer" in terms of adhesion technology. The bonding of an adhesive to the boundary layer material, per se, may be very strong, but the system fails at a low stress due to cohesive failure within the weak boundary layer. This general mechanism is a primary cause of adhesive failure in practice.

A related phenomenon is the adsorption of substances on the surface from the ambient environment. These may be gases, vapors, and solids ("dust") deposited from the gas phase and low- to high-molecular-weight substances deposited from contact with liquids and solids. Such substances are often tenaciously retained on the surface or they may migrate into the surface regions to swell and plasticize the material. They modify the surface composition and structure to affect such properties as adhesion and friction. Their detection is a primary objective of surface characterization procedures.

The surface structure and morphology of semicrystalline polymers is very sensitive to the cause of crystal nucleation during the crystallization process. Surface-induced nucleation by contact with various surfaces, for example, a metal mold surface as compared with an air environment during crystallization, leads to the development of a transcrystalline surface domain in which spherulitic growth occurs only into the body of the material from the surface. There usually are a great many nuclei, so that the half-spherulite size is small and the crystalline domains contain many defect structures. The transcrystalline domains show features of orientation and partial porosity normal to the surface. Comparable effects are noted in LCP materials such that the properties are very dependent on processing procedures and conditions.

Since for both crystalline and amorphous polymers the chain conformations in the surface regions are perturbed relative to those in the bulk of the polymer material, it is expected that there will be differences in the transition behavior of surface and bulk regions. This effect will generally lead to a slight increase in T_g, due to the additional constraint on segmental motion for chains in the surface domain. However, a decreased density in the surface region will tend to counteract this effect. Similar considerations apply to melting and other transitions. In any case, the behavior is very dependent upon the history of the sample since it determines the composition/structure/morphology of the surface region.

In conclusion, it is to be reemphasized that the surface region of almost all materials is different from the underlying bulk material. In many materials this comprises a "skin and core" morphology, which are distinct structures. This is true for many fibers where the skin region comprises a significant proportion of the total fiber and therefore is a major factor determining fiber properties. The same is true for many film and membrane materials. Modification and control of the surface composition and structure of separation membranes is a major factor affecting the utility of such materials, a technology which is in rapid growth at this time. Bulk materials such as composites, liquid crystalline and semicrystalline polymers, and other multicomponent polymer systems are significantly affected by the formation and presence of a surface skin-effect. Detection and control of such surface features is essential for optimization of processing and application of most polymeric materials.

1.3 Polymer Synthesis

We consider here three major types of polymerization reactions that are widely used for the commercial production of polymeric materials: (1) chain reaction polymerization, (2) coordination polymerization, and (3) step reaction polymerization. Other types of reactions can yield polymeric substances, but their use is limited and the details are complicated. Many excellent textbooks and monographs consider the organic chemistry and kinetics of polymerization reactions; a selected few are listed in the Supplementary Reading section.

Chain Reaction Polymerization

In this type of reaction, also called *addition polymerization*, polymer molecules are the only products of the reaction. A large number of monomers are susceptible to this type of reaction, primarily those compounds with an asymmetrically substituted vinyl (CH_2=CHR) group or diene (C=C–C=C) group. These include such common polymers as polyethylene [~(–CH_2–CH_2–)$_n$~], polypropylene [~(–CH_2–$CHCH_3$–)$_n$~], polystyrene [~(–CH_2–CH∅–)$_n$~], poly(vinyl chloride) [~(–CH_2–CHCl–)$_n$~], butyl rubber [~(–CH_2–$C(CH_3)_2$–)$_n$~] and styrene-butadiene rubber (SBR), a random copolymer of [~(–CH_2–CH∅–)$_n$~] and [~(–CH_2–CH=CH–CH_2–)$_m$~] repeat groups. Symmetrically substituted vinyl groups usually do not polymerize, with a few major exceptions such as polyethylene and polytetrafluoroethylene (Teflon) [~(–CF_2–CF_2–)$_n$~].

The polymerization reaction involves the following elementary reactions, for a free radical initiator:

$I \rightarrow 2R\cdot$ initiator decomposition

$R\cdot + M \rightarrow RM\cdot$ initiation

$RM\cdot + M \rightarrow RMM\cdot$ propagation

$RM_n\cdot + M \rightarrow RM_{n+1}\cdot$ propagation

$RM_n\cdot + RM_m\cdot \rightarrow RM_{n+m}R$ termination (combination)

$RM_n\cdot + RM_m\cdot \rightarrow RM_n + RM_m$ termination (disproportionation)

Other elementary reactions also may occur, such as

$RM_n\cdot + X \rightarrow RM_n + X\cdot$ chain transfer

The reacting molecule, X, may be any molecular specie in the reaction medium. This includes initiator, monomer, polymer, solvent, impurity, added inhibitor or stabilizer, chain transfer agent, or other additives. In any case, the previously propagating chain ($RM_n\cdot$) is limited in the molecular weight that it could have obtained in the absence of the chain transfer reaction.

The new reactive center, $X\cdot$, also may react with any of the species present in the medium. In the case of monomer (and other species of comparable reactivity, such as added chain transfer agents), this reestablishes the kinetic chain reaction with little, if any, change in overall rate of propagation:

$X\cdot + M \rightarrow XM\cdot$ initiation/propagation

Chain transfer agents, such as mercaptans, often are added to control the molecular weight of materials such as synthetic elastomers.

In the case of chain transfer to polymer, this leads to the formation of chain branching, as in the high pressure and temperature production of low-density polyethylene, where X· is a secondary main chain carbon atom free radical:

$$\sim(-CH_2-\underset{\cdot}{C}H-CH_2-CH_2-)\sim + M \longrightarrow \sim(-CH_2-CH-CH_2-CH_2-)\sim$$
$$\overset{|}{\underset{CH_2-CH_2-CH_2-CH_2\cdot)}{}}$$

If X is a very reactive molecular specie, the product, X·, will be very unreactive, thereby slowing or effectively stopping the kinetic chain reaction. In that case, X is called a retarder or an inhibitor (stabilizer). Such materials are commonly added to monomers to improve shelf-life performance and to polymers to improve their resistance to degradation.

Free radical polymerization initiators usually are peroxides such as benzoyl peroxide [Ø–CO–OO–OC–Ø], azo compounds such as azobisisobutyronitrile [$(CH_3)_2C(CN)N=NC(CN)(CH_3)_2$], or redox agents such as persulfates plus reducing agents. Radiation of various kinds (thermal, visible, UV, ionizing) also can be used. UV radiation (and to a lesser extent, electron beams) is of special interest since it can promote very fast polymerizations of certain monomer systems (urethanes, acrylics, vinyl ethers, epoxies, etc.) of interest to the adhesives and coatings industries. Plasma polymerization and chemical vapor deposition (CVD) are valuable methods for surface modifications and the formation of uniform, thin films (for example, CVD growth of polyimide films[2]). Electrolytic electron transfer is another method that is the subject of considerable study.

It is noteworthy that the initiator (and chain transfer) fragment is incorporated into the polymer chain. Each initiating specie is responsible for the formation of one polymer chain. This often has consequences for the properties of the polymeric material. Such groups may serve as chromophores to aid in the photodegradation process or as labels to aid in a characterization procedure.

The chain propagation process in free radical systems is very rapid, with a concentration of about 10^{-8} mol/L of growing chains in a typical reaction. The average molecular weight in all addition polymerizations depends on the relative probabilities of propagation to termination. In general, the temperature dependence of termination reactions is greater than that of propagation, so that an increase in temperature results in an increase in reaction yield but a decrease in molecular weight of the product. As reaction proceeds at a given temperature, the amount of polymer chains of nearly the same molecular weight increases with time until the monomer is nearly depleted.

In concentrated polymerization media, when the viscosity becomes high and, with the low thermal conductivity characteristic of organic compounds, the propagation reaction accelerates due to the immobility of propagating chain ends, the relatively high mobility of monomer to those chain ends and the local rapid increase in temperature. The result is an often drastic acceleration in polymerization rate (Norrish–Trommsdorff or gel autoacceleration effect) which can lead to an explosion. The diffusion-controlled reaction, with suppressed termination reactions, leads to an increase in molecular weight. The effect is particularly pronounced with methyl methacrylate, methyl acrylate, and acrylic acid.

Other reactive centers for chain polymerizations are ionic species, both anionic and cationic. The polymerization rates in ionic systems are more rapid than in free radical systems. Ionic polymerizations are usually conducted at subzero temperatures (e.g., −78 °C) to obtain useful, but safe, rates of reaction and to control the molecular weights of the product. In general, ionic initiators are very reactive species that must be guarded from exposure to atmospheric water, oxygen, and other impurities, as also must the polymerizing system.

Typical *anionic polymerization* initiators are Grignard reagents, sodium (e.g., in THF or liquid NH_3), lithium, and their organometallic derivatives such as *n*-butyl lithium. Electron transfer (radical anion) systems, such as sodium naphthalenide in an active ether such as THF or dioxane, have been widely used, especially to prepare "living polymers." In living polymers, the termination step is avoided due to the electrostatic repulsion between propagating anions and the lack of any chain transfer reaction in the absence of Lewis acid impurities. These systems can produce polymers with very narrow molecular weight distributions, nearly monodisperse. The method also can be used to produce block copolymers by the sequential addition of different monomers to the reaction medium.

Only a limited number of monomers are suitable for anionic polymerization, and generally these contain electron-withdrawing groups. Typical monomers include styrene, methyl methacrylate, and acrylonitrile. Chain transfer and chain branching usually do not occur in anionic polymerizations. Termination is by reaction of the propagating anion with a Lewis acid, present either as an impurity or deliberately added.

A newer method of obtaining living polymers is *group-transfer polymerization*, involving the repeated addition of monomer to a propagating chain end with a reactive silyl ketene group. During the addition, the silyl group transfers to the incoming monomer, regenerating a new acetal ketene group for reaction with the next monomer. The reactions proceed rapidly at room temperature (a marked advantage over anionic polymerization at subzero temperatures), giving polymers in quantitative yield and a narrow molecular weight distribution. The molecular weight is easily controlled by the monomer/silyl ketene initiator ratio. A particular advantage is that the method can provide polymers with side groups that would preferentially cross-link during normal free radical polymerizations, for example, allyl ester side groups. A disadvantage is that the method is limited to monomers with C=O or C≡N side groups.

Initiators for *cationic polymerizations* include strong acids such as H_2SO_4 and $HClO_4$ and Lewis acids and their complexes such as BF_3, $BF_3 \colon O(C_2H_5)_2$, and $SnCl_4$. The Lewis acids seem to require a cocatalyst, such as water, in equal or less concentration. Monomers suitable for cationic polymerization are olefins containing electron-rich groups; typical monomers are isobutylene, vinyl ethers, aldehydes, and para-substituted styrenes.

The initiation process includes two steps—the generation of a proton and its addition to the monomer. Chain transfer reactions increase as the reaction temperature increases, becoming of major significance near room temperature. Under conditions where termination is more important than chain transfer, the rate of cationic polymerization increases as the temperature decreases, in contrast with the behavior of free radical polymerizations, in which the reaction rate decreases exponentially with decreasing temperature. Since the activation energies for termination and chain transfer are usually greater than the activation energy for propagation, the molecular weight also increases with decreasing polymerization temperature. Control of initiator concentration and temperature provides control over the product molecular weight.

Coordination Polymerization

Coordination polymerization can be considered a special case of chain reaction polymerization in which the initiating specie is a catalytic surface. This type of polymerization can produce polyolefins, and some other types of polymers, at polymerization conditions of atmospheric pressure and room temperature. These polymers have stereoregular, linear configurations that can crystallize to higher degrees of crystallization than can the atactic or branched polyolefins produced by noncatalytic initiator-initiated chain reaction polymerizations. Commercially important polymers made by this general method include high-density polyethylene, linear low-density polyethylene, isotactic polypropylene, and stereoregular polydienes such as polybutadiene and polyisoprene.

Coordination polymerization catalysts (now generically termed "Ziegler–Natta" catalysts after the discoverers of the general catalyst system and its use for polymerizations) are made from two components: a transition metal compound from periodic table groups IVB to VIIIB (a halide or oxyhalide of Ti, V, Cr, Mo, or Zr) and an organometallic compound from groups IA to IIIA (alkyl, aryl, or hydride of Al, Li, Mg, or Zn). The best known systems are the ones based on $TiCl_4$ or $TiCl_3$ and an aluminum trialkyl such as triethylaluminum. The catalyst systems are usually formed in a high boiling hydrocarbon media, often followed by a period of "aging" at a high temperature (e.g., 180 °C for 30 min), which affects both the composition and crystal structure of the catalyst (the nature of these materials is still a subject of debate). A reaction solvent (e.g., heptane, cyclohexane) is then added to the cooled system, and the monomer (e.g., propylene gas) is bubbled through with external cooling to control the evolved heat of polymerization.

Several mechanisms have been proposed for the polymerization leading to stereoregular linear chain polymers. In all cases, the incoming monomer is adsorbed onto the catalyst surface by coordination (π-complex), either with a vacant site on the transition metal or by interaction with both the transition metal and the cocatalyst (e.g., Al compound). The adsorption process requires that side groups, such as the methyl group in propylene, be in the same stereoregular configuration (isotactic) before and after polymerization. A related mechanism leads to precisely alternating configurations (syndiotactic). Entropic perturbations and catalyst surface defects produce an amount of atactic material which must be largely separated (solvent extraction, etc.) from the polymer product before its use. The lack of chain transfer reactions leads to the formation of linear chains without long chain branching due to chain transfer to polymer. The addition of a comonomer such as butene-1 to ethylene leads to a copolymer with a controlled amount of short chain branching with desirable properties, linear low-density polyethylene.

There are a great many variations in the catalyst systems, the monomer or monomers that are used, and the reaction conditions that lead to a wide range of polymeric materials. Changes in the catalyst system lead to variations in polymer yield, molecular weight, and stereoregularity. Some catalyst systems are heterogeneous (Ti-based) and some are homogeneous (V-based). The addition of Lewis bases and some other additives alters the stereoregularity of the product polymer. In all cases, an important factor is the chemical reduction of the transition metal to a low-valence state, with unfilled ligand sites serving as the active catalytic specie.

Step Reaction Polymerization

This major class of polymerization reactions also is known (historically) as *condensation polymerization*. This arises from the fact that in most of the polymerizations of this type the products include not just polymer but also a low-molecular-weight side product which must be "condensed" out of the reaction mixture. An example is the reaction of exactly stoichiometrically equal portions of diacid and dihydroxyl molecules for the formation of a polyester and water:

$$n\text{HOOC} - \text{R} - \text{COOH} + n\text{HO} - \text{R}' - \text{OH} \longrightarrow$$
$$\sim (-\text{OC} - \text{R} - \text{COO} - \text{R}' - \text{O}-)_n\sim + n\text{H}_2\text{O}$$

There are several polymers of this type in which there is not actually a condensation product given off. A prime example is the formation of Nylon-6 from the polymerization of ε-caprolactam, a cyclic amide with six carbon atoms. A molecule of water enters into the reaction to hydrolyze the lactam into a transient state linear molecule with acid and amine functionalities, which then "condenses" with another of the same structure to eliminate a water molecule to form the polyamide.

$$
\begin{array}{c}
\text{CH}_2 - \text{C} = \text{O} \\
\text{CH}_2 \qquad\qquad \text{NH} \xrightarrow{\text{H}_2\text{O}} [\text{H}_2\text{N} - (\text{CH}_2)_5 - \text{COOH}] \xrightarrow{-\text{H}_2\text{O}} \sim (-\text{NH} - (\text{CH}_2)_5 - \text{CO}-)\sim \\
\text{CH}_2 \\
\text{CH}_2 - \text{CH}_2
\end{array}
$$

This reaction also is an example of another important type of polymerization class, *ring-opening polymerizations*. Cyclic organic compounds that can be polymerized include cyclic ethers (trioxane, THF, epoxides, etc.), lactones (cyclic esters), anhydrides, imines (cyclic amines), and lactams (cyclic amides, as shown above).

These two examples illustrate a crucial factor in step polymerizations—the need for exact equivalence of the two different reacting functionalities. Any, even minor, nonequivalence lowers the final molecular weight due to the nonreactivity of chain ends of the same kind. This problem is overcome in the second example above by the inherent equivalence in the cyclic monomer. In other cases, the formation of acid–base salts (e.g., diacid–diamine salts) with subsequent purification before polymerization by heating to eliminate water, or other condensate, is one common method to achieve the necessary molecular functional group stoichiometry.

Step reaction polymerizations do not require an initiator, although they usually do benefit from acid or base catalysis. The inherent reactivities of the components mean that all reactive functional groups have an equal chance to react at any time. This leads to a progressive increase in the molecular weight of the reaction mixture as a function of reaction time. Useful molecular weights in step polymerizations are only achieved at extents of reaction very close to 100%, in contrast to addition polymerizations, where high-molecular-weight polymer is produced from the very start with the yield of such polymer increasing with reaction time.

Step polymerization reactions in which the reactants have an average functionality greater than two lead to the formation of initially branched structures, then to random three-dimensional networks. As the reaction nears completion, the network extends throughout the body of the reacting sample, converting it into one giant molecule. This network structure may swell in an imbibed solvent, but it will not dissolve, nor will it melt and flow. These *thermoset* materials, once produced, cannot be returned to a thermoplastic state in which they can flow. Examples are phenolic and amino resins. In the case of phenolics, the classic example is the reaction of phenol (active hydrogens at the para and two ortho positions, thus tri-functional) with formaldehyde (bi-functional); the average functionality is 2.5.

Thermoset polymers also may be formed by the cross-linking ("curing") of thermoplastic materials. The prime example is the vulcanization of rubber using sulfur to form cross-links of various sulfur atom content between neighboring rubber chain molecules. A number of important adhesive and coatings polymers are thermosets cured after application (e.g., epoxies, urethanes). A typical epoxy compound is the reaction of bisphenol A with epichlorhydrin, which cross-links through the terminal epoxy groups with a primary amine cross-linking agent. Structural cured thermosets include unsaturated polyesters with styrene and a free radical initiator. These materials are used extensively in many applications as sheet molding compound (SMC) or bulk molding compound (BMC).

1.4 Summary

The foregoing discussion illustrates the immense variety of polymer chemical compositions, structures, and molecular characteristics that can be generated by a variety of preparation methods. As a consequence, a broad spectrum of properties and thus of materials applications is offered both by polymers, per se, and their combinations with other materials, such as ceramics.

Supplementary Reading

Allcock, H. R., and F. W. Lampe. *Contemporary Polymer Chemistry*. 2nd ed., Prentice Hall, Englewood Cliffs, NJ, 1990.

Billmeyer, Jr., F. W. *Textbook of Polymer Science*. 3rd ed., Wiley, New York, 1984.

Boyer, R. F. "Transitions and Relaxations in Amorphous and Semicrystalline Organic Polymers and Copolymers." In *Encycl. Polym. Sci. Tech*. Suppl. Vol. 2., Wiley, New York, 1977, pp. 745–839.

Boyer, R. F., and R. Simha. "Secondary Transitions in Polymers." In *Encyclopedia of Materials Science and Engineering*. (M. B. Bever, Ed.) Pergamon, New York, 1986, pp. 4327–4329.

Brydson, J. A. *Plastics Materials*. 5th ed., Butterworths, London, 1989.

Comprehensive Polymer Science. (G. Allen, S. L. Aggarwal, J. C. Bevington, J. C. Booth, G. C. Eastmond, A. Ledwith, C. Price, S. Russo, and P. Sigwalt, Eds.) 7 vols., Pergamon, Oxford, 1989–90.

Cowie, J. M. G. *Polymers: Chemistry and Physics of Modern Materials*. 2nd ed., Chapman and Hall, New York, 1991.

Encyclopedia of Polymer Science and Engineering 2nd ed. (H. F. Mark, N. M. Bikales, C. G. Overberger, C. G. Menges and J. I. Kroschwitz, Eds.) Vols. 1–15 and suppl., Wiley, New York, 1985–90.

Flory, P. J. *Principles of Polymer Chemistry*. Cornell Univ. Press, Ithaca, NY, 1953.

Hall, C. *Polymer Materials: An Introduction for Technologists and Scientists*. 2nd ed., Halsted Press, Wiley, New York, 1989.

Hopfinger, A. J. *Conformational Properties of Macromolecules*. Academic Press, New York, 1973.

Odian, G. *Principles of Polymerization*. 2nd ed., McGraw-Hill, New York, 1981.

The Physics of Glassy Polymers. (R. N. Haward, Ed.) Halsted Press, Wiley, New York, 1973.

Polymer Handbook. 2nd ed. (J. Brandrup and E. H. Immergut, Eds.) Wiley, New York, 1975.

References

1   T. Yang and C. E. Rogers. *AIChE Symp. Ser*. **85** (272), 11, 1990.

2   S. P. Kowalczyk, C. D. Dimitrakopoulos, and S. E. Molis. *Proceedings*. Mat. Res. Soc. Symp. **227**, 55, 1991.

2

Polymer Fabrication Techniques

L. J. MATIENZO, D. W. WANG, and F. EMMI

Contents

2.1 Introduction

Polymers are substances composed of molecules containing repeating units of one or more low-molecular-weight monomers.[1] Polymers have a high molecular weight and can be linear or branched or have cross-linked networks. A general classification for these materials includes both natural (e.g., proteins and cellulose) and synthetic compounds (e.g., rubber and polyethylene). Essential to life, complex macromolecules are routinely synthesized by nature. Recent advances in biotechnology have made it feasible to modify and produce many important biopolymers.[2] The first synthetic polymers were developed during the last century, and one of the first commercial processes to be used was the vulcanization of rubber. As a result of extensive research and development, polymers have become the fastest growing group of materials since World War II. Polyethylene, polypropylene, polystyrene, and polyvinyl chloride are the four polymers with the highest production volume. During 1989, the United States shipped $37 billion of polymer resins to foreign markets, and the domestic production volume was estimated to be 62 billion pounds (28 million kilograms) in 1990.[3]

Polymeric materials are generally grouped into thermoplastics, thermosets, elastomers, and composites. Their applications range from films for packaging (polyethylene and polypropylene) and electrical insulation (polyimides) to resins for photoresists (acrylates and polyisoprenes) and to advanced composites (epoxies and aramids). Light weight and ease of processing are two main advantages of polymers when compared with conventional materials such as wood, metal, or ceramics. Polymers that exhibit excellent physical properties and can replace conventional materials in some applications are known as engineering polymers. At the present time, their production is about one-tenth the volume of synthetic resins.[4–7] Table 2.1 lists common commercial polymers and some of their typical applications.

In this chapter, we focus first on some common processing techniques used in the fabrication of synthetic polymers. Since there are many polymers, we emphasize some materials selected from a variety of industrial applications. It is also important to remind the reader that the properties of the finished product are a consequence of the fabrication process selected. In general, molecular segregation, molecular orientation, thermal degradation, and cross-link density are properties that can vary from the surface to the bulk of the same film, depending on the fabrication method chosen to produce a material. The second part of this chapter presents some of the fabrication methods listed in Table 2.1 and describes how a material with a given composition can vary in chemical and physical properties as a function of a fabrication sequence. Unique and extremely specialized approaches are required by some fabrication schemes; these techniques are not discussed here (the reader is referred to more specialized textbooks). Specific methods developed for plastics recycling and reprocessing, important for environmental reasons, are not included in this discussion.

2.2 Processing Techniques

Under the influence of heat, pressure, and mechanical stress, polymers can be readily fabricated. Typically, additives are required to alter their properties and make them more resistant to degradation or less flammable. Fillers are usually added to resins to improve the resulting properties and also to reduce polymer volume and lower fabrication costs. Other additives used in industrial fabrication include lubricants, plasticizers, stabilizers, flame retardants, antioxidants, antistatic agents, and colorants. Additional information on these materials can be found elsewhere.[8–12]

Polymer fabrication is usually performed with the use of a die/mold, by melt processes, or by solution casting. As expected, surface properties, such as smoothness and texture, are influenced by the physical contact of the polymer with the tool. Annealing, curing, and heat treatment also affect surface topography, morphology at the molecular level, and chemical characteristics of the fabricated item, since many processing aids might remain on surfaces. In order for residual levels of surface contaminants to be minimized, fabrication processes may require additional steps such as solvent washing, chemical etching, or physical abrasion. If the final application of the fabricated object requires bonding or attachment to another material, other methods for surface modification such as ion implantation, radiation hardening, plasma or corona discharges need to be performed accordingly.[13]

Table 2.2 gives a general description of commonly used fabrication techniques. Several of these methods are briefly described.

Polymer	Applications
Polyethylene	Films, flexible molded parts, high modulus fibers
Polypropylene	Films, carpet yarns, bottles
Polybutylene	Films and piping
Polymethylpentene	Food packaging and containers
Polyvinyl chloride	Flexible film, hoses, home siding
Polyvinyl acetate	Coatings and adhesives
Polyvinyl butyral	Adhesive for glass windshields
Polystyrene	Sheet stock, cellular foam
Polyacrylonitrile	Fibers, barrier material, carbon fiber precursor
Polymethyl methacrylate	Lenses, signs, skylights, optical fibers
Polyhydroxyethylmethacrylate	Hydrogels, soft contact lenses
Polyacrylamide	Flocculating agents, paper sizing
Acrylonitrile-butadiene-styrene	Appliance and business machine housings, pipes, fittings
Methylcellulose	Food-thickening agents
Polyglycolides	Bioabsorbable sutures, surgical staples, composites
Polyamides	Fibers, bearings, gas tanks
Aramids	Fibers, composites, structural honeycomb reinforcements
Polyethylencoxide	Water thickeners
Polyacetals	Plumbing fixtures and fittings
Polyalkyleneterephthalates	Fibers, films, bottles, implants
Polyarylesters	Electrical connectors, chip carriers
Polycarbonates	Lenses, windshields, housings
Sulfone polymers	Microwave ware, pumps, printed circuit boards
Polyimides	Flexible circuits, cable insulation, passivation layers
Polyetherimides	Electrical connectors, housings
Polyamideimides	Gears, motors, fittings
Polyphenyleneoxide/PS	Car wheel covers, valves, pumps
Polyetheretherketone	Wire coverings, composites
Fluoropolymers	Coatings, release agents, seals
Polyphenylenesulfide	Valves, pumps, coatings
Parylene	Conformal coatings
Phenol-formaldehyde resins	Adhesives, electrical components
Furane resins	Cement, coatings, molds
Amino resins	Decorative laminates, dinnerware
Alkyd resins	Coatings, moldings
Allyl resins	Electrical connectors
Unsaturated polyesters	Molding compounds, coatings
Epoxy resins	Adhesives, encapsulants, coatings
Cyanate ester resins	Advanced composites, laminates
Polyurethanes	Foams, fibers, elastomers
Siloxane resins	Encapsulants, elastomers, coatings
Polyisoprenes	Tires, gaskets, elastomers
Polybutadiene and copolymers	Tires, footware, adhesives, paints

Table 2.1   **Some commercial polymers and their applications.**

Technique	Description
Foam processing	Reactive foam forming, foam spraying, pellet formation
Film-forming methods	Blowing, casting, Langmuir–Blodgett films, skiving
Composite material processing	Molding, filament winding, continuous and batch lamination, pultrusion, rigidized thermoforming, centrifugal casting
Extrusion	Conventional extrusion and extrusion covering
Molding	Blow molding, rotational molding, injection molding, reactive injection molding, compression molding
Coatings	Planar, contour coatings, spin casting, spray coatings
Other methods	Calendering, sheet thermoforming, fiber spinning

Table 2.2   Polymer fabrication techniques.

Foam Processings[14–17]

Plastic foams have found many applications as thermal and electrical insulators and structural materials. During polymer processing, various amounts of gas can be incorporated in the form of voids and cells, resulting in lightweight products. Hollow fillers like microspheres can be mixed with epoxy or phenolic thermosets to yield syntactic foams. Other methods for foam production use chemical and physical blowing agents. A gas can also be generated by a side reaction of thermoset resins such as isocyanates. In other cases, expanded polyurethane shell filling requires the addition of small levels of low boiling temperature fluorocarbons. Heat generated by the curing reaction is used to vaporize the fluorocarbon and induce voids in the coating. Expanded polytetrafluoroethylene (PTFE) is also made by incorporating a blowing agent and mechanical stretching. With a reduced dielectric constant and an improved heat insulation factor, this material has found applications in microelectronic fabrication as well as in the garment industry.[18]

Film-Forming Processes[5, 19, 20]

Thermoplastic thin films can be made by film blowing. Resins are continuously extruded in a tube form and subsequently inflated to several times their diameter. Film thicknesses from 0.3 to 8 mils (0.008 to 0.2 mm) are achievable by this method. Multiple coextruded layers containing 3–11 components can be fused together under heat and pressure for barrier packaging. Materials most often used in this process include polyethylene, polypropylene, polyamides, polyvinyl chloride, and polyvinyl alcohol.

Sheets and films up to 6 ft (1.8 m) wide can also be produced by extrusion with appropriate dies. Film forming requires orientation and stretching, and thicknesses of up to 0.25 mils (6 µm) can be obtained in this manner. A simple method for characterization of film orientation may involve birefringence measurements or polarized infrared spectroscopy. For further details, see general descriptions of these techniques in the appropriate volume of this series. Solution casting involves the deposition of a solution of a polymer onto a belt. Subsequently, the solvent must be removed or the polymer cured by the application of heat. Cellulose and polyimide films are usually made by this process, and in some cases, both sides of the film can have different properties that affect surface roughness and solvent diffusion.[21]

Ultrathin organic films are currently being investigated in microelectronic applications since specific molecular orientation during film formation provides unique properties such as nonlinear optical properties, dichromism, redox activity, electrical conductivity, and thermal conductivity. Known as Langmuir–Blodgett (L–B) films, these are formed by spreading monolayers of amphiphilic compounds at the air–water interface. With the application of a compressive force parallel to the surface, the molecules become oriented. Deposition of L–B films onto a substrate is done by dipping the substrate perpendicular to the layer direction. Molecular orientation is maintained even after the transfer. Recently, these materials have been used as chemical sensors, electrochromic displays, barrier layers, optical modulators, and resists.[22, 23] More specific applications of these materials are discussed later in the chapter.

Another process for film fabrication is skiving. This method is used to cut a thin layer of film in a continuous fashion. Fluoropolymer films are specifically made in this manner since these polymers can be softened by heat but are not easy to process from the melt. Surface topography is thus strongly influenced by the cutting tool selected.[5] A practical application of this preparation method is that the enhanced surface roughness is conducive to better adhesion than with standard smooth films.

Composites[24–27]

Polymer composites are a fast-growing segment of engineering materials. With recent innovations in synthetic and processing technologies, polymers have found uses as matrix and reinforcement materials. For example, traditional sheet molding compounds (SMC) now contain glass fibers that allow their use as auto parts. Woven glass–epoxy composites are produced in high volume for printed circuit manufacturing, and carbon and aramid fiber composites are replacing metals in aerospace applications. Adhesion at the reinforcement–matrix interface usually is improved by using adhesion promoters. These materials, also known as coupling agents, are either organosilanes or organometallic complexes.[27] Adhesion promoters are di-functional compounds with each molecular end capable of reacting with either the matrix or the reinforcement material. In general, coupling agents also reduce moisture absorption by composites. For electronic device fabrication, aminosilanes are the material of choice since they can react with silicon or ceramic substrates at one end and through the amine end with photoresists or polyimide thin films.

Typical composite processing involves the application of a matrix to a reinforcement material and a thermal and pressure treatment to yield a product. Mechanical

Fiber	Diameter, μm	Density, g/cm^3	Tensile Strength, psi	Modulus, 10^6psi
Carbon	7	1.66	350000–450000	17–25
S-glass	7	2.50	500000–660000	13
Aramid	12	1.44	400000	10–25
Boron	100–140	2.50	510000	58
Quartz	9	2.2	500000	10
SiC fibers	10–20	2.3	400000	28
SiC whiskers	0.002	2.3	1000000	—

Table 2.3   Fiber properties (psi × 6.896 = kPa).

properties are optimized by incorporating reinforcement fibers with high aspect ratios. Some reinforcements commonly used in composite fabrication are listed in Table 2.3.[28] In contrast to other polymer products, composites consist of at least two different components, and the resulting product is strongly affected by the chosen fabrication method since the selected cure cycle, pressure range, and thermal source to enhance curing will provide unique surface and bulk properties.

Bulk molding composites (BMC) can be formulated by the addition of chopped fibers into a thick polymer paste such as polyester containing a thickener like magnesium oxide placed in a mechanical mixer. Compression, transfer, or injection molding can be used to fabricate the desired product. If sheet materials are desired, compression molding by successive application of fiber rovings and paste onto a mechanically driven belt is required.[29, 30]

Reinforced plastics can also be fabricated by spraying methods. In this approach, a fiber chopping gun and a two-component resin delivery system are used simultaneously on a mold surface. The application is usually followed by mechanical compaction to ensure good fiber wetting and air removal. Curing can proceed at room temperature or inside a vacuum bag. This process offers a relatively simple method to build large structures, especially in the automotive industry.

Pultrusion and filament winding composite fabrication involves direct impregnation of liquid resin onto continuous fiber rovings. Similar to aluminum extrusion, pultrusion offers a process to produce continuous structural profiles. The shape of the structure is formed when filaments are pulled through a die in which resin is simultaneously cured. Process speed is usually limited by curing with a large die, usually 3 ft (0.9 m) long. Depending on the resin system and cross section, speeds up to 12 ft/min (0.06 m/s) are attainable. Radio frequency curing, preheating, and postcuring are several methods commonly used to speed the process. Although the reinforcement is mainly unidirectional, many techniques exist to incorporate circumferential and random mat ply to improve mechanical properties in transverse directions.[31]

Filament winding uses rovings or tapes mechanically wrapped around a core or mandrel coated with a release agent that can inflate or collapse. Reinforcements are impregnated in a low-viscosity resin bath (epoxy, polyester thermoset, or phenolic resin) before being laid down on a mandrel. Mechanical strength can be controlled using a particular type of winding pattern. Final curing is performed by convection, infrared heating, radio frequency curing, or in autoclaves. Many large structures such as pressure vessels, tanks, cylinders, rocket motor cases, and aircraft rotor blades are made by this process.[32]

Composites can be also fabricated from prepregs, fabrics, and rovings impregnated with resins that are partially cured (B-staged). If ultimate strength is a requirement, prepregs are made from unidirectional tapes by melt impregnation. Since various types of resins are used, shelf-life of the prepreg must always be specified to ensure acceptable performance. Autoclave processing is used for parts with complex shapes such as airplane wings, radomes, or armor plate. In order to reduce weight, honeycomb cores and foam are usually incorporated in the part layup. In recent years, thermoplastics-based composites have been introduced in the aerospace industry. These prepregs have a small amount of residual solvent to facilitate layup during fabrication by autoclave/vacuum bag processing. Epoxy/E-glass circuit board laminates and melamine-formaldehyde/Kraft paper laminates are made by flatbed lamination.[33] Epoxy-based composites are typically made at 180 °C and 500 psi (3.4 MPa). Thermoplastics and more heat-resistant materials can require temperatures as high as 600 °C and pressures of 2000–3000 psi (13.8–20.7 MPa).[34]

Extrusion [35–39]

The extrusion process provides continuously formed polymer parts with high throughput. This method is usually employed for the manufacture of films, tubing, sheets, and fibers. A conventional extruder consists of an electrically driven motor connected to a single metallic screw rotating in a barrel. These machines can be equipped with elements for heating, cooling, and pressure application. Polymers—typically in pellet form—are fed through a hopper, quickly melted, and thoroughly homogenized by the rotation of the screw before they are expelled through a die. Resin flow is laminar and no significant mixing occurs. Interlayer adhesion is achieved by the melting operation. Most extrusion barrels are designed to withstand pressures up to 10 000 psi (69 MPa). The throughput of larger extruders can exceed 1000 lb/h (0.13 kg/s). Shear-sensitive materials are best handled by twin-screw extruders. Special applications may require planetary motion extruders to ensure homogenization prior to molding.

If a special application requires coextruded multilayer films or sheets, two or more extruders equipped with special dies must be used. This approach allows engineers to obtain a composite with improved properties over those of individual materials. Another method widely used to coat paper, metal, or other films for heat-sealable packages is extrusion coating. Interfacial adhesion is achieved by using mechanical or chemical treatments on the substrate prior to polymer application. Fiber yarns can also be fabricated by extrusion. When this is done, the process is known as fiber-spinning. The fabrication sequence requires extruders to push the molten polymer (melt spinning) or a solution (solution spinning) through spinneret holes. Subsequent steps such as fiber-drawing and annealing are required to achieve the desired molecular orientation and enhance ultimate tensile strength for a fiber. Extrusion covering uses an extruder to apply complete polymer covers on a continuous substrate. This is the method of choice to form protective or insulation layers on electrical wires or cables.

Molding

This description groups methods that rely on a cavity (mold) to shape polymer parts. The selection of a specific method depends on the desired physical shape as well as the starting material characteristics. It is important to mention here that these methods usually require the use of mold releasing agents to facilitate the removal of the product from the mold and surface properties may be modified by thin layers of these materials transferred during the fabrication process. This is important to remember since additional joining operations to other surfaces may be difficult if these residues are not properly removed prior to bonding.

Injection molding offers advantages for the production of parts at high speed.[40–42] Simple molding devices use a plunger to push a molten material into a heated mold coated with a mold-releasing agent. Commercial devices use reciprocating feed screws for higher efficiencies. The injection molding process involves polymer shaping, injection, and product ejection. Peak pressures during molding can range from 3000 to 30 000 psi (20 to 200 MPa). Clamping force, another term used in molding technology, can be as high as 10 000 tons (2×10^7 kg). If co-injection is required, two feed units are connected to the same mold. During injection molding, flow rate (shear rate), temperature, and the rheological behavior of a polymer melt have a strong effect on orientation, residual stresses, and shrinkage of molded parts. During mold filling, a frozen surface that is highly oriented is often formed. Thus, the surface of a molded part can have a different morphology, solvent resistance, and porosity from the bulk of the same material.

Although injection molding has been primarily used for thermoplastics, thermosetting materials such as rubbers, phenolics, and unsaturated polyesters can also be processed. These systems, however, experience vulcanization and curing reactions in the heated molds, and a short process cycle must be designed to make them economically attractive.

Reaction injection molding (RIM) was primarily developed for polyurethane resins.[43, 44] Two low-molecular-weight, fast-reacting liquids are mixed and immediately injected into a mold to allow polymerization to take place. Since low-viscosity liquids are used, the pressure requirements for RIM processes are much lower than those of conventional injection molding methods. During the process, in-mold pressure usually reaches 50–100 psi (340–690 kPa) with a moderate clamping force of 300 tons (6×10^5 kg) for large machines. Urea resins, caprolactam (Nylon-6), and aliphatic hydrocarbon resins can also be processed by this method.

Blow molding is used to produce hollow plastic parts by inflating an extruded or injection-molded soft plastic preform against a cold mold.[45] Extrusion blow molding, injection blow molding, and injection-stretch blow molding are the three most common industrial methods. A wide variety of products such as carbonated beverage bottles (polyethyleneterephthalate), canisters, storage tanks, and shipping drums (high-molecular-weight polyethylene) are fabricated by this method.

Compression molding is mostly used to process thermoset resins and rubbers.[29, 42, 46] This process involves forming materials in a confined cavity under pressure and heat. Since curing reactions are often taking place, molding time can vary from minutes to over an hour. Molding pressures ranging from several hundred to 10 000 psi (up to 70 MPa) are needed to facilitate the flow of a viscous resin for mold fill. Typically, molding temperatures are below 200 °C, but for fluoro-polymers or polyimides higher temperatures are required.

Transfer molding is a method of fabrication for thermosets and rubber products.[5] Fabrication requires the initial fusion of a low-viscosity resin in a chamber and the forcing of the material through a narrow gate into a heated mold for final cure. The pressures required are very similar to those used for compression molding. This molding method is essential for encapsulating microelectronic components with epoxies or phenolic resins.

Coatings[47–49]

Many industrial processes are used in coating applications for decorative purposes, electrical insulation, packaging, or film forming. Usually, surface activation of the substrate by such methods as plasma oxidation or mechanical roughening is required to enhance adhesion. Thin-film coatings have surfaces that are highly dependent on processing conditions because they affect the resulting film morphology. For example, in the case of polystyrene and polyimide block copolymers, the selection of coating solvent and curing process effectively provides an optimum surface condition suitable for subsequent bonding through polymer interdiffusion.[50]

Planar coatings are usually applied by an apparatus containing a roller, a knife or die, a moving belt, and an oven for solvent evaporation or polymer curing. Film thicknesses up to 4 mils (0.1 mm) can by applied at high speeds of about 1000–2000 ft/min (5–10 m/s). Painted metal, photographic films, adhesive tapes, and wallpaper are representative examples of products made by this approach. Articles with complex shapes are coated by contour coating, a technique that includes painting, dipping, and electrostatic deposition of films.

Spin-coating is extensively used in the electronics industry for uniform thin-film application. The thickness of the film, usually less than 10 μm, is achieved by dropping a solution of a polymer onto a rotating substrate. The speed of the substrate and polymer solution concentration affect film characteristics. More details on these films are presented in the applications section.

Other Methods

Calendering is most often used for making films or sheets of polyvinyl chloride resin and rubber.[4] The materials are fed between two counter-rotating heated cylinders that move at different speeds. Materials are deformed by passing through a series of rollers until the desired thickness is reached. Surface characteristics are controlled by the last roller. In addition to highly polished rollers, engraving and embossing rollers can be used in the final step of production. Calendering is usually done at speeds of 100–400 ft/min (0.5–2 m/s), with an average throughput of 1000–10000 lb/h (0.13–1.26 kg/s).

A description of fiber-spinning can be found in the section on extrusion methods; sheet thermoforming methods are included in the composites section.

2.3 Applications

In the following sections, several examples of the application of various fabrication methods for polymeric materials are described. Here, the approach also covers other areas because, in the development of a particular product, other factors such as mechanical, thermal, physical, and chemical requirements of the fabricated object are incorporated into the development schemes. Analytical methods, when applicable, are described as part of the fabrication process, and the reader is referred to the volume in this series dealing with physical and chemical methods for polymer analysis for a general background on the techniques discussed. In addition, when appropriate, descriptions of other approaches and instrumental methods currently under development in various research laboratories are incorporated to show the potential of these new ideas in current fabrication technology.

Development of Spray-On Insulation for the Thermal Protection System of the External Tank in the Space Shuttle Program

Cellular plastic coatings are widely used due to the versatility of their applications.[51] A representative family of these materials is the polyurethanes, although the resulting coatings are not made from strictly urethane monomers nor do the resulting films contain an appreciable number of urethane groups. Some applications for these materials have already been described, but they can also be used as insulation barriers, flexible layers, water-repellent coatings for fabrics, or as encapsulants in microelectronic packaging.[52] The fabrication of polyurethane coatings can yield flexible, semirigid, or rigid polyurethane foams or isocyanurate coatings. These coatings are formed either by reactive injection molding or by spray-on methods.

An extremely large area application of urethane coatings is best accomplished using spray-on methods. The external tank for the Space Shuttle missions performs

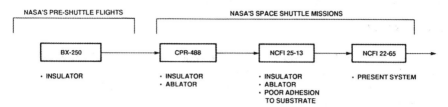

Figure 2.1 Chronological development of spray-on insulation materials for the external tank of the Space Shuttle orbiter. (Reprinted with permission from the principal author of Reference 54.)

a dual task: it must serve as a structural support to the orbiter and it must also carry and store fuel (hydrogen and oxygen in separate compartments) for the orbiter's main engines. The fuel storing tanks are covered by an external shell made of an aluminum alloy that is protected by a lightweight thermal protection system based on polyurethane technology. In areas of higher friction, an ablative layer was required in earlier flights for additional thermal protection. As the program developed, it was found that because the initial insulation system was not reliable to the temperature range between –257 and 538 °C, a new coating was required.[53] Isocyanurate foams were chosen as an improvement. These coatings develop their thermal stability from the absence of labile hydrogen atoms in ringlike structures formed during curing. Isocyanurates are brittle and friable, and other flexibilizers such as epoxies, polyimides, or urethane segments are added into these formulations.[4] In some areas, silane coupling agents are also incorporated to enhance adhesion of the insulation to the metallic substrate. Further details about the reaction chemistry of polyurethanes and isocyanurates can be found elsewhere.[51, 54]

Matienzo et al.[55] have reported the development of more advanced insulation formulations to meet the flammability, weight reduction, substrate adhesion, and performance of spray-on isocyanurate coatings. These changes have increased the cargo capability for the orbiter in more recent missions. A comparison of the various insulation requirements used in the Space Shuttle program are presented in Figure 2.1. In this diagram, the last three formulations are based on urethane/isocyanurate materials with flame-retarding capabilities. The adhesion of insulation to metallic substrate proved satisfactory using the approach previously described.

The most current formulation meets flammability and mechanical requirements and yields an isocyanurate foam from a commercial two-package system (NCFI 22-65, North Carolina Foam Industries).Table 2.4 gives a compositional breakdown of the two-part system as determined by instrumental methods of analysis.[55]

As may be noticed, two flame retardants are incorporated; however, Voranol® in flame retardant II is also used as an adhesion promoter to the substrate. A single flame-retardant system (Formulation NCFI 25-13) resulted in good flame resistance but poor adhesion. A more comprehensive selection of the appropriate formulation

Package	Component	Type	Function
A-side	Reactive molecule	Isocyanate	Foam formation
B-side	Catalysts	Amines, tin compounds	Isocyanurate formation
	Polyols	Polyol I and II	Urethane
	Flame retardant I	Fyrol PCF (tri-chloroisopropyl phosphate)	Flame retardancy
	Flame retardant II	DOW XNS 50054.020 Voranol® (a glycol) dimethyl methyl-phosphonate dibromoneopentyl-glycol	Flame retardancy and adhesion promotion
	Blowing agent	Freon 11-B	Foam expansion

Table 2.4   **NCFI 22-65 isocyanurate formulation.**

with two flame-retardant mixtures is made on the basis of bond tension testing and radiant recession rates needed to meet the engineering design criteria. In these experiments, the mixing ratios of the two reactive packages of the formulation were 2 to 1 (A-side versus B-side). Figure 2.2 shows the responses of bond tension and recession rate on flame-retardant ratios used in foam optimization.

Coating performance can be reproduced only if the reactive sides of the formulation are not affected by storage conditions and if their performances are monitored regularly. For example, the isocyanurate portion of the formulation (side A) was found to be sensitive to storage temperature and the formation of acidic groups was easily detected by Fourier transform infrared (FTIR) analysis.[55] Two other factors important for reproducible coating characteristics are the amount of blowing agent present in a new or aged formulation and the relative concentration of the two flame retardants. The blowing agent, a fluorocarbon, is used to ensure appropriate cell dimension growth during the reaction process, and the ratio of the two flame retardants determines adhesion and flame retardation specifications. These components can be measured simultaneously by high-performance liquid chromatography (HPLC).[56] Similar methods can be used to detect aging effects on the B-side of these formulations.

Another critical parameter that must be determined is the performance of the application gun, since coating quality depends on the appropriate mixing ratios of the reactive packages as well as the homogeneity of coating layering on the substrate. Correlations of foam composition uniformity at different depths and mechanical properties can be obtained by analysis of cored samples. For example, FTIR spectroscopy can be used to develop relationships between unique bands attributed to the A-side of the formulation (isocyanurate trimer band at about 1420 cm^{-1}) and the P=O stretch of a flame retardant on the B-side of the formulation (band at about 1230 cm^{-1}), as illustrated in Figure 2.3. Although in practice the A

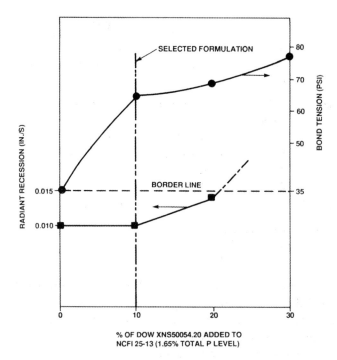

Figure 2.2   Bond tension and radiant recess rate responses as a function of flame-retardant package ratios. (Reprinted with permission from the principal author of Reference 54.)

to B ratio is formulated as 2 to 1, variations in mixing rate and reaction performance as a function of layer deposition can be easily detected by FTIR analysis and foam replacement options can be carefully evaluated.

Development of Thick, Thin, and Ultrathin Polyimide Films

Polyimides are polymers with high thermal stability and good mechanical properties. Their applications include coatings, radiation barriers, composite laminates, and dielectric layers in microelectronic circuitry.[57] Polyimides show high thermal stability up to 350–400 °C for short times, they are excellent barriers, and they can be metallized in batch or continuous processing. Some drawbacks with polyimides are their processing requirements, sensitivity to high pH solutions, solvent retention, and relatively high cost.

Polyimides and other common polymers such as epoxies are used in current electronic packaging designs that involve large-scale integration of devices with multilayered metallization components at the silicon chip and package level. Generally, these structures contain insulating (polymers) and conducting layers (mainly copper layers) with their associated interfaces. Recent industrial trends have produced flexible circuits with high packaging densities; however, the mechanical and

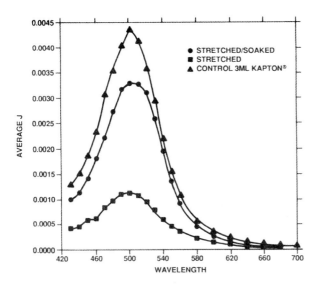

Figure 2.3 Photocurrent response for Kapton® film as a function of wave-length and film treatment. (Data reprinted with permission after Reference 68.)

physical requirements of the dielectric layers have made polyimides the material of choice.[58] The use of tape automated bonding (TAB) processes is a means for continuous fabrication and bonding of electronic circuits in a thin-roll form.[59] The most common polyimide is produced by the initial reaction of pyromellitic dianhydride (PMDA) and oxydianiline (ODA) to yield a polyamic acid. Upon additional reaction, the acid loses water and yields a polyimide known as PMDA-ODA. In industry, PMDA-ODA-based films are produced by E. I. duPont de Nemours, Inc., and sold as Kapton® films. An excellent set of physical and chemical properties for Kapton films can be found in Reference 60.

Although commercial polyimide films are sold as rolls, other applications may require films produced only as needed. The resulting surface and bulk film properties for these films can be quite different because curing and fabrication parameters affect the characteristics of the product. The goal of this section is to illustrate how a common polymer such as PMDA-ODA, when transformed into a thick, thin, or ultrathin film, can yield different surface properties as a function of the fabrication and curing conditions and further treatments employed. It is also believed that the same approach can be used to describe the variable properties of other materials such as epoxies when subjected to different fabrication methods. Since other polyimides with superior properties to PMDA-ODA films are being used in industry, comparisons are made between these materials and the standard type of PMDA-ODA films.

PMDA-ODA film casting is done commercially via a chemical imidization reaction of the polyamic acid solution in dimethyl acetamide (DMAC) as a solvent.

The resulting mixture is cast over a heated drum to yield a solvent-swollen gel that is finally thermally cured and stretched as it passes through an oven to increase tensile strength and modulus.[61] The critical selection of conditions that affect drying rates during casting has been studied by Gupta[62] for a natural convection process and for one with induced-draft conditions. The second approach is the method of choice for continuous roll fabrication. Although several types of Kapton films are known, our description in this section is limited to Kapton-H, the most commonly used film type in industry.

PMDA-ODA coatings can also be formed by spin-coating deposition methods. This type of fabrication is useful since it offers advantages in integrated circuit (IC) fabrication technology,[63] dielectric layers on ceramic substrates, or as α-particle barriers.[64] These coatings are typically applied from a solvent solution (NMP or N-methyl pyrrolidone) of a dilute polyamic acid (e.g., DuPont PI-2255) onto a rotating substrate, and the films are thermally imidized.

Kapton-H films and their spin-cast counterparts show similar bulk vibrational spectra. Differences are observed if surface analysis using X-ray photoelectron spectroscopy (XPS) or secondary ion mass spectrometry (SIMS) is performed. For example, in some cases, small amounts of organic silicon-containing compounds can be detected on roll-produced films. These contaminants are associated with lubricants or with transferred layers from silicone rubber rollers as they thermally decompose in the process ovens.[65] A general problem, this has been observed on several commercial films.[66] Contaminant detection is crucial if surface modification reactions in oxygen-rich plasmas are to be used, since silicones will yield thin layers of silicon dioxide.[67]

Molecular orientation also results from spin-cast film formation. This can clearly be shown by wide-angle X-ray diffraction.[68] Another advantage that can be exploited using spin-cast films is their enhanced photoconductivity. This approach is promising for electrostatic imaging by incorporating electron donors. Increases in crystallinity are proportional to photocurrent response.[69] Figure 2.3 presents a comparison of photocurrent response for an untreated film, 60% elongated and solvent crystallized material, as reported in Reference 69.

Structural analysis of spin-coated PMDA-ODA films has recently been refined by means of grazing incidence X-ray scattering (GIXS) using a synchrotron X-ray source.[70] This technique allows for high surface sensitivity and the detection of molecular order or crystallinity on the top 90 Å of the film. Conventional wide-angle X-ray diffraction work[68] would not be able to discriminate between surface and bulk orientations. The use of new techniques for the morphological analysis of insulators such as atomic force microscopy (AFM) can differentiate between the average surface roughness of spin-coated and roll-formed PMDA-ODA films. Figure 2.4 shows two isometric views obtained for these samples by AFM. As may be concluded, the spin-coated films may be slightly smoother than their roll-cast equivalents.[71, 72]

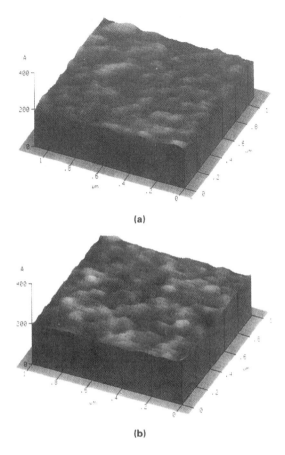

(a)

(b)

Figure 2.4   Three-dimensional views for (a) Kapton-H and (b) spin-coated PMDA-ODA surfaces obtained by AFM. (Photographs courtesy of W. N. Unertl and X. Jin. Department of Physics, University of Maine at Orono.)

Thermal stresses can develop between a substrate and a polymer film during a curing cycle; thus, it is important to match thermal expansion coefficients. Geldermans et al.[70] have demonstrated these effects during the curing of polyimides produced by the reaction of the ethyl ester of benzophenone tetracarboxylic acid and methylene dianiline (ETBA-MDA) or PMDA-ODA films on silicon wafers. Stresses measured as a function of temperature using an X-ray double crystal lattice curvature bending technique showed that increased stresses were proportional to curing temperatures and the chemical composition of the film. Stress relaxation occurred after solvent absorption, but solvent loss induced additional stresses.

During electronic packaging applications, polyimides are subjected to interactions with beams or different atmospheres to alter their surface properties. For example, the bombardment of spin-coated polyimide films on silicon wafers with Ar ions results in surfaces that are rich in graphite-like structures with enhanced conductivities.[73] Even analytical techniques can modify film surfaces, and in some cases they do not have the resolution to show film alterations. For example, Rutherford backscattering spectrometry (RBS) analysis of a series of Kapton-H films and other polyimides as a function of irradiation time are identical, but if XPS or FTIR analyses are performed on the same films, they show carbonization and imide ring bond breaking.[74, 75] These differences arise from the depth resolution of the techniques used in film characterization. Irradiation of polyimides induces the formation of graphite-like domains that are proportional to increasing optical and electrical irradiation.[76, 77] Overall, interactions of these types of polyimides with oxygen plasmas for the purpose of surface modification and adhesion improvement are very similar, leading to the formation of oxygen-rich surfaces.[66]

During a fabrication process, a film may be exposed to a variety of solvents. Pawlowski et al. have compared the relative absorption rates of various halogenated solvents into various commercial polyimide films.[78] Solvent absorption proceeded through case II non-Fickian diffusion. In addition, by a combination of gravimetric measurements and RBS analysis, a slight difference was noticed for opposite sides of the films. The rough side (casting drum side of film) produced a slightly higher solvent absorption rate than the smooth side (air side) of the films. More recent studies comparing the absorption rates of methylene chloride for spin-cast- and roll-produced biphenyltetracarboxylic dianhydride oxydianiline (BPDA-ODA) films have shown similar results but with much slower rates than those observed with PMDA-ODA films.[79] The presence of traces of chlorinated solvents on metallized polyimide films is critical since appreciable adhesion degradation can result with further processing. Matienzo et al.[80] have shown that extremely small amounts of methylene chloride can produce total adhesion loss of chromium/copper layers on Kapton-H. Corrosion at the metal–metal interface arises from thermal steps, such as encapsulant curing. Thermal decomposition ($T > 100\ °C$) of methylene chloride in the presence of a metal surface and some water yields HCl, and further exposure to a humid environment triggers metal corrosion reactions.

Some fabrication approaches using film-casting technology require the successive application of polymer layers, and the adhesion between these consecutive applications must be excellent. Kramer et al. have shown that curing parameters are important to optimize interlayer adhesion.[81] Forward recoil spectroscopy (FRS) was employed to follow the diffusion of a deuterated polyamic acid into a previously cured layer. If the curing temperature for the first layer was greater than 200 °C and the curing temperature for the polyamic acid layer was less than 200 °C, no interdiffusion occurred. Substantial interdiffusion was detected when the curing temperature for the first layer was less than 400 °C and the imidization temperature of the polyamic acid overlayer was greater than the initial curing temperature. A more recent publication has shown that adhesion measurements using

90° T-peel tests for samples prepared as indicated above substantiate the requirements for polymer–polymer interdiffusion.[82]

Recent developments in analytical techniques such as laser interferometry,[83] photothermal and photoacoustic methods,[84] and fluorescence spectroscopy[85] show applicability in the monitoring of film fabrication. Laser interferometry is an approach that can be used to follow film dissolution, solvent evaporation from films, or reversible solvent absorption. This technique is simple and inexpensive when compared with other techniques such as RBS and FRS for solvent diffusion studies. Photothermal methods of analysis are based on radiation absorption by a film and the local heating produced by the interaction. Radiation-induced thermal transients are the best-suited sources for the technique. Thermal diffusivity and temperature-dependent properties can also be measured in real-time scales. In addition, film inhomogeneities can be detected and some depth profiling can be done using photoacoustic methods. Although FTIR has been shown to be an excellent technique to monitor imidization reactions, it offers limited information on thermally cured reactions above 200 °C. An alternative method based on the molecular fluorescence enhancement with curing temperatures above 200 °C has been reported. This approach has the potential for direct monitoring of polymer webs and their cure homogeneity to ensure uniform film performance.

Vapor deposition polymerization (VDP) of PMDA-ODA has been shown to yield reproducible films with thicknesses in the range of 100 Å from the polyamic acid precursor.[86] Additional heating is required to form the fully imidized film with a well-controlled stoichiometry. The apparatus used for this fabrication method is shown in Figure 2.5. VDP films have FTIR and XPS spectra similar to those obtained with spin-coated films; however, upon curing, they produce reduced stresses at silicon–polyimide interfaces, probably because they are less anisotropic than spin-coated films. Stoichiometric films formed by vapor deposition contain some small amounts of unreacted materials since they tend to lose more weight than films developed by the spin-coating method, but they have similar decomposition temperatures. Molecular weight determinations on these materials have placed their M_w at about 13000. Developing a completely dry process in VLSI applications has been shown to be feasible. Figure 2.6 shows the fabrication scheme for this application. This approach is encouraging since it reduces harmful solvent effects on substrates, especially with copper layers known to diffuse into polyimide films.[87] An important factor to consider in the application of VPD technology is the stoichiometry of the precursors, since it has recently been reported that an excess of ODA leads to films with low molecular weights that are thermally less stable since they contain imine groups.[88]

The chemical reactivity of fully cured VDP polyimide films with copper substrates has been recently examined.[89] Stronger interaction is detected for VDP films than for spin-coated PMDA-ODA films. XPS gives evidence of an initial attack on the imide portion of the molecule with some imide ring breakage. At greater metal

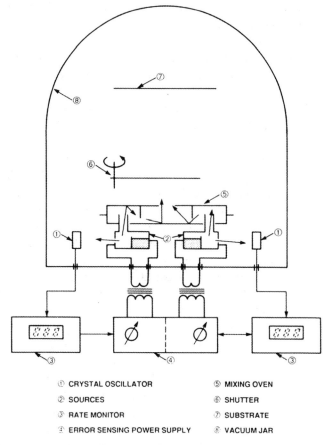

①	CRYSTAL OSCILLATOR	⑤	MIXING OVEN
②	SOURCES	⑥	SHUTTER
③	RATE MONITOR	⑦	SUBSTRATE
④	ERROR SENSING POWER SUPPLY	⑧	VACUUM JAR

Figure 2.5 **Schematic diagram of apparatus used in the vapor deposition of polyimide films. (Diagram reprinted with permission after Reference 85.)**

coverages, copper diffuses into the films and the surface becomes rougher as detected by scanning tunneling microscopy (STM) measurements. With more reactive metals, surface reactions have a different pathway. For example, it titanium is deposited onto VDP polyimide, the polymeric backbone experiences some bond breakage at the oxygen portion of the ODA segments and at the nitrogen atom in the imide ring position.[90] During the initial metal–polymer interaction, titanium is preferentially oxidized to Ti^{3+} At greater coverages, nitrides and carbides form. Metal oxidation during the initial reaction of chromium with a PMDA-ODA surface has also been reported by Clabes et al., although no attack on the ODA portion of the molecule was observed.[91] Again, VDP films have stronger interactions with the metal when compared to spin-coated films.

Ultrathin polyimide films can be obtained using L–B film technology; general details are presented in the subsection on film-forming processes. The fabrication

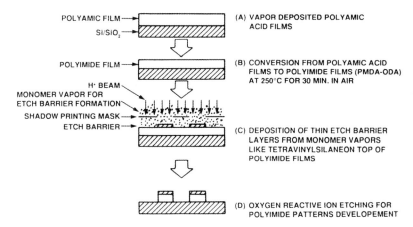

Figure 2.6   Fabrication of a polyimide film for VLSI applications using chemical vapor deposition. (Diagram reprinted with permission after Reference 85.)

of these layers with very specific properties follows the general scheme derived by Kakimoto et al.[92] Basically, polyimide L–B films are formed by spreading a solution of the desired polyamic acid after reaction with a long chain amine to yield an ammonium salt. The second step is the deposition of the L–B film onto an appropriate substrate. The last step includes the treatment of the multilayered film with a mixture of acetic anhydride and pyridine to form the polyimide layer. This method of fabrication can be used to produce films as thin as 4 Å. The advantages of this technology when compared with other fabrication methods such as spin-coating are that L–B polyimides do not require thermal treatments for cure and that polyimides not obtainable by spin-coating methods can be successfully manufactured. For instance, Nishikata et al. have demonstrated that the L–B technique can form PMDA-PDA or polyimides with large pendent aromatic moieties.[93]

A recent application of L–B films in an industrial environment has been reported by Ikeno and co-workers.[94] They have demonstrated that twisted-nematic, super-twisted nematic, and ferroelectric liquid crystal displays (LCDs) can be fabricated by using L–B polyimide films as orientation molecules for the liquid crystals. A careful evaluation of the birefringence requirements was made, and the maximum number of L–B layers needed to produce the maximum output was determined to be 5. The fabrication scheme is much simpler than conventional LCD methods currently used in industry. Additional applications for L–B films are listed in the following section.

Langmuir–Blodgett (L–B) Films

L–B films have recently found application in the bioelectronics field. Skotheim et al. have prepared a new class of conducting materials with specific electroactive sites using polypyrrole films designed for electronic communication with enzymes.[95]

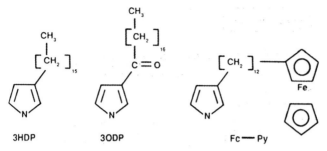

Figure 2.7 Molecular structure of alkyl pyrroles and ferrocene derivatized alkyl pyrroles. (After Reference 94.)

The chemical structure of the starting materials can influence the resulting film properties (see Figure 2.7). For example, if 3HDP was used, the surface active pyrrole could copolymerize. For 3ODP, the pyrrole could no longer polymerize and the film had behaved differently. However, Fc–Py, when mixed with 3HDP, formed stable monolayers with surfaces containing ferrocene moieties which were used to affect charge transfer to redox enzymes.[96]

Phthalocyanine can yield L–B films useful as chemical sensors when placed on electrode surfaces.[97] Stable L–B films can be made with monolayer films of metal–phthalocyanine compounds as shown in Figure 2.8. Coatings of five or more monolayers have good electrical properties, and the magnitude of the electrical charge response when exposed to ammonia gas is dependent on coating morphology and metal substitution. Film morphology was characterized by visible spectroscopy, magnetic resonance spectroscopy, vapor pressure osmometry, and X-ray diffraction. Results showed that the monolayers consisted of stacked phthalocyanine aggregates.

L–B films have also been fabricated as nonlinear optical (NLO) materials. Hoover et al.[98] have investigated a series of main chain α-cyanocinnamate polyesters having fluorinated spacer groups (Figure 2.9). Chromophore mobility was minimized through the chemical bonding of dye groups in unique configurations along the main chain. Amphiphilic characteristics were altered by varying the chromophoric and spacer groups. The glass transition temperature is influenced by the large amount of flexible fluorinated alkyl linkage between relatively small chromophores. This arrangement allows the molecule to fold in an accordion-like manner. These materials offer a possibility for enhancement of second- and third-order NLO properties. For example, films with a 10-layer Z-type film provided noncentrosymmetric dipole assemblies, as determined by second harmonic generation. This generated approximately 2% of the 532-nm-intensity light compared to a quartz standard, and an 18-layer Y-type multilayer film generated a second harmonic light with an 8% intensity.

L–B films can produce conducting polymers, or these films can be made partially conductive by ion irradiation or implantation.[99, 100] Conducting L–B films can be made from a mixed monolayer of poly(3-alkylthiophenes) and a

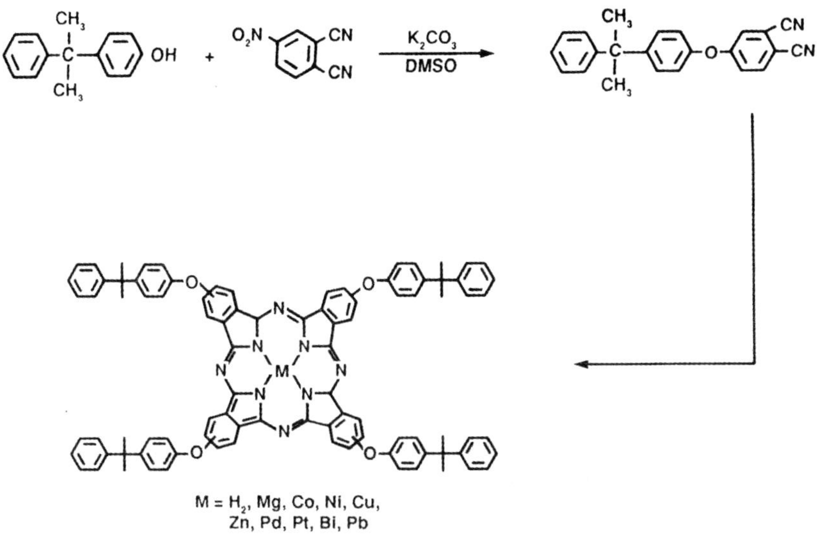

Figure 2.8 Reaction sequence for solubilizing phthalocyanine in organic solvents. (After Reference 96.)

surface active agent.[101, 102] The treated film contains domains of electroactive polymer in a mixed matrix of well-ordered surface active molecules. These systems can facilitate electronic communication between layers of different electroactive polymers. Ion irradiation has been performed on fluorinated and other polyimide L–B films.[100, 102, 103] Irradiation (150 keV Ar ions) and implantation (40 keV I or 20 keV Cs ions) effects on the polymers was monitored by FTIR. Surface conductivity of the resulting films was measured as a function of film thickness; Figure 2.10 shows the resulting surface conductance as a function of the number of deposited layers.[100]

Polymer Composites

Polymer composites are multiphase materials of two or more components in which the continuous phase or matrix is a polymer. This combination of materials results

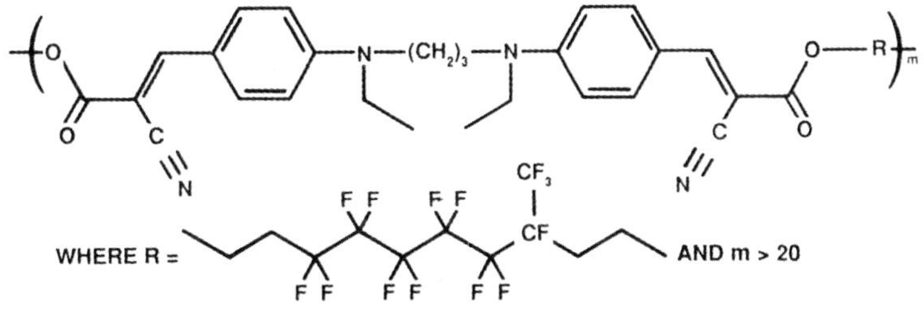

Figure 2.9 Chemical structure of an α-cyanocinnamate polyester with fluorinated spacer groups. (After Reference 97.)

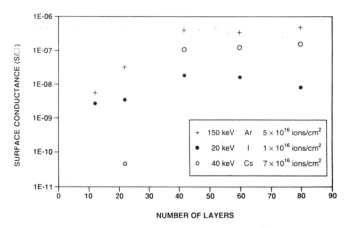

Figure 2.10 Surface conductance of ion-irradiated and ion-implanted fluorinated polyimide L–B films as a function of the number of deposited layers. (After Reference 99.)

in an optimization of chemical, physical, and mechanical properties over those found with each single component. Although polymers have some disadvantages such as low density, low mechanical strength, creep, and some low resistance to environmental effects, they can be easily fabricated into simple and complex shapes. A variety of fillers can be used in polymer composite fabrication. Among these can be listed particulates, random chopped fibers, mixed paniculate/chopped fibers, flakes, oriented short fibers, filament wound unidirectional single component or hybrid fibers, woven fibers of a single material or hybrids, laminated sheet, woven hybrid clothes, and interpenetrating networks. Composite application determines the choice of materials that meet engineering design requirements. Since a strong interaction between the filler and the matrix is required, optimization of surface properties for the reinforcement prior to fabrication must be accomplished. Due to the extensive variety of reinforcements that can be used, surface preparation methods can be quite different. Fabrication requirements can also influence the resulting surface characteristics and the success of subsequent steps to yield a finished object. In this section, several examples are given to illustrate the role of surface treatments of the reinforcement in the resulting composite properties. A description of how a fabrication cycle is optimized to yield the desired properties is also given. Since in fabrication technology there is need for new methods of characterization and fabrication, the discussion is expanded to describe some novel approaches.

Mechanical properties of composites, especially interlaminar shear strength, are critically dependent on the strength of the interface between the filler and the continuous phase.[104] Also, fabrication methods will affect other composite properties. For example, the presence or absence of release sheets or peel ply layers will affect further adhesion and bonding. Solvent evaporation during the spraying of the matrix onto the reinforcement is a critical step to consider, since solvent evaporation rate may cause surface changes in polymer density, cross-linking, or thermal resistance. Finally, one must be aware of approaches for the identification of surface treatments of reinforcements since they also can define composite performance. For example, surface-sensitive techniques such as XPS and Auger electron spectroscopy (AES) can provide contrasting results with classical methods of analysis. Even for common reinforcements such as S- and E-glass fibers, the numbers and types of surface sites are not comparable when results from classical and surface analytical techniques are used.[105] Quantitative XPS data indicate that S-glass fibers contain more magnesium than E-glass fibers. The latter fiber type, however, also contains calcium on its surface. These findings are important since they suggest that the number of silanol groups available for reaction with silane-based coupling agents may not be the same in both types of fibers. Thus, surface modification for these two fillers requires different optimization parameters.

Linear unsaturated polymer composites containing particulate fillers such as micas are used in a variety of molding operations for automotive parts. When enhanced adhesion between the filler and the continuous phase is desired, silane-based coupling agents are used. Typically, the filler is suspended in a solvent and mixed continuously while a solution of the coupling agent chemical is added. This step is followed by a dispersion of the treated material into a polymeric phase that is able to react thermally with the filler during the molding operation. For polymers containing some double bonds, the coupling agent of choice is an azido-terminated silane. Matienzo and Shah[106] have presented an analytical approach to optimize mechanical properties of these composites by monitoring the surface composition of the silane coupling agent on phlogopite micas using XPS and diffuse reflectance infrared spectroscopy (DRIFT). Both techniques have shown to be in a good agreement with regard to the relation between surface modification levels and the expected performance of the composite.

XPS has been extensively applied to identify the surface composition of carbon fibers.[107] However, due to poor chemical grouping resolution in the spectra, the results are not always completely clear. Another problem with this technique is the lack of spatial resolution below 100 μm to analyze localized areas within a fiber. Some of these disadvantages seem to be more easily solved by the application of time of flight-secondary ion mass spectroscopy (TOF-SIMS). Hearn and Briggs have recently demonstrated that the technique has a spatial resolution of 0.2 μm and that fragments dislodged from a surface can be unquestionably identified by their mass fragmentation patterns.[108] These results can also complement topographical information derived from SEM analysis. In this manner, specific fractured areas in composites can be evaluated for surface contamination or cohesive failure.

Organic fibers such as Kevlar (E. I. duPont de Nemours & Co., Inc.), which is a poly(p-phenylene terephthalamide), have been recognized for a number of years as fibrous reinforcements for composites. These fibers are relatively unreactive, and less than adequate wetting to the matrix may result in lower shear strengths. Surface modification of Kevlar via controlled lamination reactions improves surface reactivity and wetting to epoxy resins without an appreciable decrease of fiber modulus or tensile strength.[109]

Another approach to enhanced impact strength without the reduction of interlaminar shear strength between a fiber and a polymeric matrix has been studied by Rhee and Bell.[110] They used an energy-absorbing layer at the interface to increase impact strength of graphite fiber-epoxy composites. Electrochemical polymerization of glycydyl acrylate/methyl acrylate and a subsequent reaction with epoxide rings in the matrix yielded enhanced impact resistance and interlaminar shear strength by up to 30 and 14%, respectively. These results were maximized when the surface treatment layer was 0.1–0.15 µm thick. An alternative method to increase interlaminar shear strength in these composites is based on the use of nonpolymerizable and polymerizable surfactants with the matrix at the fiber–continuous phase interface.[111] In this case, shear strength was greater with nonpolymerizable surfactants (37% enhancement) than with a polymerizable treatment (10% enhancement).

Another important aspect of composite fabrication is adhesive bonding of laminates. Surface preparation and the interaction of residual process materials may influence bond strengths and composite durability. Composite surface preparation for adhesive bonding includes such methods as mechanical abrasion and the use of peel ply layers. These operations are performed to increase surface area during adhesive bonding or to minimize surface contamination during molding operations. Typical mold release agents are organic silicon compounds that are known to be detrimental in adhesive bonding. Graphite–epoxy laminates used in the aircraft industry have been critically analyzed to determine the role of surface preparation on adhesive bonding.[112] Single-lap shear results on intentionally contaminated surfaces with a silicon-based mold release agent were found to correlate with silicon levels left on the composite surface measured by XPS. The interaction of the peel ply with the composite surface was also important for contaminant elimination and the resulting composite adhesive strength depended on contaminant level as well as composite roughness. Similar work performed with carbon fiber–polyimide composites joined by an adhesive layer (LARC-TPI) has shown that the best surface preparation was obtained by oxygen plasma treatments, since they provided a reduction of surface contaminants.[113]

Poly(etheretherketone) or PEEK is a thermoplastic material that shows excellent toughness and durability. The fabrication of molded PEEK composites is carried out in metallic molds. The selection of the mold is important on the resulting surface properties of the composite since surface reactions of PEEK and copper have been shown to be thermally enhanced.[114, 115]

Processing and structural optimization of composite fabrication is an important issue to meet design criteria in any application. Lustiger et al. have given an excellent description of the requirements for the fabrication of PEEK–graphite fiber composites.[116] By a preliminary use of differential scanning calorimetry (DSC), several conditions for composite processing were selected. Samples were physically aged, annealed just below the melting point, slow-cooled, fast-cooled, or prepared under low pressure. Using a combination of microscopic techniques, isolated and graphite fiber nucleated spherulites were observed and measured. DSC analysis provided an estimation of the degree of crystallinity. At low cooling rates, larger spherulites were produced and the level of crystallinity increased. Figure 2.11 presents DSC curves measured at three cooling rates. After mechanical testing was performed using fracture toughness and impact delamination tests, the resulting performances of the composites were rated using ultrasonic C-scans. Once again, slow cooling rates yielded the highest compression strength but lowest fracture toughness. A similar evaluation with physically aged samples resulted in no degradation of mechanical properties. The results shown in this work indicate that spherulite growth as well as degree of crystallinity must be balanced by processing conditions to produce the desired mechanical properties for PEEK–graphite fiber composites.

Conventional fabrication methods have been described in earlier sections of this chapter. However, with changes in process demands and fabrication dimensions, new methods are continuously being investigated. Typical composite fabrication requires a high-pressure and thermal environment for thermoset curing—this generally limits maximum dimensions of the part to be processed. An alternative approach, which has been shown to operate at low pressures, is free of dimensional limitations such as those found in autoclave curing. Semipermeable membranes used in composite lay-up for graphite fiber–epoxy and glass or carbon fiber–polyimide composites can facilitate the transport of solvent and volatiles produced during composite curing, thus reducing the formation of voids at the fiber–polymer

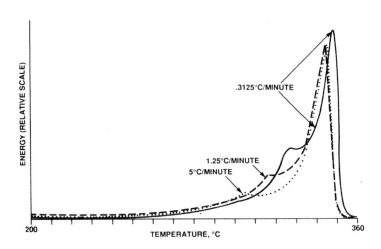

Figure 2.11 DSC scans for three different cooling rates for PEEK–graphite laminates. (Data reprinted with permission after Reference 115.)

interface.[117] The presence of these voids is responsible for reduced interlaminar shear strengths in laminates. With a combination of techniques such as DSC, thermomechanical analysis and on-line viewing of composite curing, the process optimization can be achieved. Since, as has already been mentioned, composite porosity is a detractor to performance, sensitive techniques that are capable of providing information on void formation must be developed. A technique that has potential uses is called positron annihilation spectroscopy (PAS). This method measures the lifetime of a positron in a matrix. Void trapping causes longer lifetimes and signal broadening, an indication of free volume or porosity level. This analytical method is nondestructive and it can equally be applied to polymer films or composites.[118, 119]

Because another concern in composite fabrication is process throughput, alternative methods to conventional autoclave curing have been investigated. Microwave curing is an attractive approach since a laminate can be rapidly and uniformly heated without the need of heating the surrounding container (oven) to induce curing. A disadvantage of the approach is the need to determine exactly the distribution of the electric field within the system, otherwise uneven heating may occur. Figure 2.12 gives a graphic description of a microwave curing system for composites. A recent review of microwave curing applied to polymers and composites can be found in Reference 120.

The disadvantages of autoclave processing can also be surmounted if radiation-curable composites are used, since curing can occur at room temperature in a short time and with a reduced formation of volatiles. The only disadvantage is that the prepreg used in composite fabrication must be tailored to be radiation curable. Saunders et al. have elegantly illustrated the development of radiation curable epoxy–

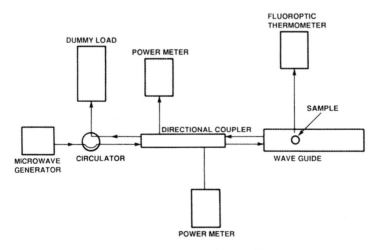

Figure 2.12 Schematic representation of a microwave curing system for composites. (Diagram reprinted with permission after Reference 119.)

graphite graphite fiber composites using a 10 MeV electron beam.[121] Their results have demonstrated that mechanical properties comparable to those obtained by conventional thermal curing methods were obtained by the new curing approach. A point of concern not mentioned in this work is the sensitivity of the composite surface to potential chemical alteration by the beam. This is a key issue to consider for further composite fabrication that may include adhesive bonding between laminates.

Molding Processes

Residual levels of mold release agents on finished surfaces can reduce adhesion of subsequent finishes such as paints, overcoats, or metallization coatings. Many commercial formulations of release agents are available; typically, mold release agents contain silicones, waxes, fluorocarbons, or inorganic materials in a variety of carriers. Since extremely thin layers of material remain on finished parts, their detection is rather difficult. Matienzo et al. have described an approach to characterize and detect general types of commercial mold release agents on molding tools and parts using gas chromatography-mass spectrometry (GC-MS), XPS, and FTIR.[122] The GC-MS technique is useful in detecting the solvent carriers in commercial mold release agents. Once solvents are evaporated, reflectance FTIR spectroscopy can easily detect the specific type of surface modifier. When XPS is combined with FTIR, the identification can be expanded to include tool surfaces and to determine if mold release transfer to simulated parts has occurred.

Some semiconductor molding compounds contain epoxy resins, phenolic hardeners, and silica glass fillers. Molding of these materials is made by batch processes. A rapid technique is always needed for quality control and compositional verification. Walker and Wehman have described an X-ray fluorescence (XRF) technique for process and product control. XRF results were compared to results of standard gravimetric or atomic absorption (AA) techniques for the analysis of bromine, arsenic, silica filler, iron, and calcium.[123] These materials are present in the formulation in the epoxy as either fillers, flame retardants, or catalysts. Results obtained were excellent for barium, antimony, and silica filler; iron and calcium analysis were reproducible but interferences due to X-ray absorption demonstrated that the technique needed further work. Overall, XRF was shown to be a quick and reproducible technique to monitor the quality of the material.

Electrochemical effects during thermoset molding have been described by Rudd et al.[124] Earlier work investigating resin transfer molding processes indicated that thermoset resins, such as hot-setting polyester, had electrolytic properties.[125–128] The electrochemical effects were investigated by probes made of dissimilar metal electrodes and a thermocouple imbedded in the resins. Changes in electromotive force (emf) and temperature were monitored between a copper electrode and a second electrode during the molding process. The emf changes during the process showed that the emf values followed the trend: Cu/Al > Cu/brass > Cu/steel > Cu/Cu. These researchers have also investigated the effects of filler (calcium carbonate) addition to the resin. Ratios of resin to filler were varied at values of 3:1,

2:1, 1:1, and 1:2. Results showed that emf changes during curing were smaller as the filler/resin ratio increased. The result was interpreted as a dilution of the electrolyte; also, the addition of filler was thought to affect reaction rates and modify surface acidity. A prototype sensor has been constructed for industrial use and shown to be useful in determining the state of cure of the resin during injection molding.

Miller et al. have discussed how NIR-FT-Raman can be applied to determine the flexural moduli of RIM polyurethanes for automotive applications.[129] This method requires minimal sample preparation. The approach was shown to work for various formulations.[130–132] Several RIM polyurethanes were prepared from 4,4′-diphenyl-methane diisocyanate (MDI), diethyl toluene diamine (DETDA), and an ethylene oxide-capped poly(propylene oxide) triol. In these mixtures, the mass percentages of MDI and DETDA, referred to as the hard-block percentage, were varied. Flexural modulus data were obtained with an Instron Tester for each sample prior to Raman analysis. The NIR-FT-Raman vibrational spectra were obtained; the vibrational structure in the spectra contained no observable fluorescence interference effects. Results showed that NIR-FT-Raman and NIR diffuse reflectance spectroscopy could be used for nondestructive analysis of RIM polyurethanes. However, the diffuse reflectance method was more accurate for predicting flexural modulus.

Fiber Drawing and Surface Modification Processes

A new two-step process for continuous high pressure drawing for polyoxymethyl-ene (POM), a crystalline polymer widely used as an engineering film, has been described by Komatsu et al.[133–135] During fiber drawing, the draw ratio and tensile modulus were determined at each step. Silicone oil was used as the pressurizing and heating medium. These POM fibers were dense ($1.45 \, \text{g} \cdot \text{cm}^{-3}$) and transparent and had good mechanical properties. On the other hand, conventionally drawn fibers are white, with an apparent tensile modulus and strength of 40 and 1.4 GPa, respectively. After the new treatment, the room-temperature tensile modulus (53 GPa) and strength (2.0 GPa) along the fiber axis improved. The increase of pressure is thought to suppress the generation of voids during drawing. The chemical and thermal resistance of fibers produced in this manner was determined by the retention of tensile modulus, tenacity, and weight after exposure to 4% NaOH (60 °C), 4% HCl (60 °C), or thermal aging using a sunshine weatherometer at 63 °C. The pressurized drawn films were transparent and were found to have better chemical resistance than conventional drawn fibers due to a denser structure and molecular orientation.[135]

Organic fibers which are used as reinforcement in polymer composites must have good interlaminar properties when used for applications requiring structural and dimensional stability. Although extended chain polyethylene (ECPE) has many good characteristics (e.g., high modulus and strength and good chemical resistance), its adhesion to organic resins is poor. In order to improve the adhesion, surface modification of the fibers must occur. Chappell et al. have demonstrated that ECPE fibers treated with ammonia and oxygen plasmas and corona discharges improve adhesion of fibers to a polymer matrix.[136] Surface analysis and mechanical testing (shear test) were used to evaluate the adhesion of the modified fibers to an epoxy matrix. XPS and dye assay techniques were used to determine the amine concentration on ammonia plasma-treated fibers; the amine concentration was proportional to the time in the plasma. The results of interlaminar adhesion indicated that the magnitude of the shear strength increased with ammonia plasma treatment, corona treatment, and oxygen plasma treatment of material, in that order. However, although the concentration of surface amine groups increased with ammonia plasma treatment time up to 20 min, the interlaminar shear strength did not increase significantly after 1 min of treatment. SEM was used to reveal the locus of failure. For example, untreated ECPE composites failed at the interface, but after plasma treatment, failure involved fiber fibrillation and internal shear fracture. The interfacial shear strength was found to be proportional to surface amine concentration until it was greater than the shear strength of the fiber. XPS was used to assess change in chemical bonding on the surface; the increase in adhesion was attributed to specific chemical interaction between the fiber surface and the resin.

Ding has tested the compatibility of antistatic polyethylene (PE) packaging materials with several surface finish materials (e.g., aluminum, stainless steel, epoxy paint, and printed wire boards) using an accelerated stress test (140 °F, 95% RH).[137] The material was placed in contact with different PE bubble packaging materials, exposed to the stress test, removed, separated, and inspected for discoloration, staining, and corrosion. Results showed that staining occurred when the PE material was in tight contact with the finished surface. The degree of staining was affected by the additive to the PE, the contact pressure, and the hardness of the metal finishing material. This example illustrates the fabrication requirements for specific applications of the desired product.

References

1   F. Millich. In *Encyclopedia of Materials Science and Engineering*, 2nd ed., Vol. 12. (H. F. Mark, N. M. Bikales, C. G. Overberger, and G. Menges, Eds.) Wiley-Interscience, New York, 1987, p. 398.

2   T. Orr. *Genetic Eng. News.* **11** (5), 3, 1991.

3   Society of Plastics Industry Reference Survey. Society of Plastics, Brookfield, CT, 1990.

4   J. M. Charrier. *Polymeric Materials and Processing: Plastics, Elastomers, and Composites.* Hanser (Oxford University Press), New York, 1990.

5   S. S. Schwartz and S. H. Goodman. *Plastics Materials and Processes.* Van Nostrand Reinhold, New York, 1982.

6   H. Saechtling. *International Plastics Handbook.* Hanser (Oxford University Press), New York, 1987.

7   *Engineering Thermoplastics.* (J. M. Margolis, Ed.) Marcel Dekker, New York, 1985.

8   *Handbook of Fillers for Plastics.* (H. S. Katz and J. V. Mileski, Eds.) Van Nostrand Reinhold, New York, 1987.

9   *Plastics Additives Handbook.* (R. Gachter and H. Muller, Eds.) Macmillan, New York, 1985.

10 E. W. Flick. *Plastics Additives, An Industrial Guide.* Noyes Data, Park Ridge, NJ, 1986.

11 *Polymer Additives.* (J. E. Kresta, Ed.) Plenum, New York, 1984.

12 *Thermoplastics Polymer Additives.* (J. T. Lutz, Ed.) Marcel Dekker, New York, 1988.

13 *Plastic Finishing and Decoration.* (D. Satas, Ed.) Van Nostrand Reinhold, New York, 1986.

14 F. A. Shutov. *Advances in Polymer Science.* **39**, 1, 1981.

15 F. Rodriguez. *Principles of Polymer Systems.* Hemisphere Publishing, New York, 1989.

16 *Foam Based Reactive Oligomers.* (A. A. Berlin, F. A. Shutov, and A. K. Shitinkin, Eds.) Technomic Press, Lancaster, PA, 1982.

17 B. C. Wendle. *Structural Foams.* Marcel Dekker, New York, 1985.

18 R. W. Gore. U.S. Patent 3,953,566, 1976.

19 C. J. Benning. *Plastic Films*, 2nd ed. Technomic Press, Lancaster, PA, 1983.

20 R. E. Kesting. *Synthetic Polymer Membranes*, 2nd ed. Wiley, New York, 1985.

21 W. M. Edward. U.S. Patent 3,179,6419, 1965.

22 C. D. Bain and G. M. Whitesides. *Angew. Chemie Adv. Mater.* **101**, 522, 1989.

23 B. Tieke. *Adv. Mater.* **5**, 222, 1990.

24 *Handbook of Composites.* (G. Lubin, Ed.) Van Nostrand Reinhold, New York, 1982.

25 *Encyclopedia of Composite Materials and Components.* (M. Greyson, Ed.) Technomic Press, Lancaster, PA, 1983.

26 *Handbook of Reinforcements for Plastics.* (J. V. Milewski and H. S. Katz, Eds.) Van Nostrand Reinhold, New York, 1987.

27 E. P. Plueddemann. *Silane Coupling Agents*, 2nd ed. Plenum, New York, 1991.

28 A. J. Klein. *Advanced Composites.* May/June, 32, 1988.

29 B. Stillhard. *Adv. Polym. Technol.* **10**, 13, 1990.

30 R. W. Meyer. *Handbook of Polyester Molding Compounds and Molding Technology*. Chapman and Hall (Methuen), New York, 1987.

31 W. B. Goldsworthy. In *Encyclopedia of Polymer Science and Engineering*, 2nd ed., Vol. 4. (H. F. Mark, N. M. Bikales, C. G. Overberger, and G. Menges, Eds.) New York, 1986, p. 1.

32 R. W. Meyer. *Handbook of Pultrusion Technology*. Chapman and Hall (Methuen), New York, 1985.

33 D. W. Wang. *Mater. Res. Soc. Symp. Proc.* **108**, 125, 1988.

34 *High Temperature Matrix Composites*. (T. T. Serafini, Ed.) Noyes Data, Park Ridge, NJ, 1987.

35 C. Rauwendaal. *Polymer Extrusion*. Macmillan, New York, 1986.

36 M. L. Steven. *Extruder Principles and Operation*. Elsevier Applied Science, New York, 1985.

37 J. R. A. Person. *Mechanics of Polymer Processing*. Elsevier Applied Science, New York, 1985.

38 S. Levy. *Plastics Extrusion Technology Handbook*. Industrial Press, New York, 1981.

39 *Plastics Extrusion Technology*. (F. Hensen, Ed.) Hanser (Oxford University Press), New York, 1988.

40 *Injection Molding Handbook*. (D. V. Rosato and D. V. Rosato, Eds.) Van Nostrand Reinhold, New York, 1985.

41 J. B. Dym. *Injection Molds and Molding*, 2nd ed. Van Nostrand Reinhold, New York, 1987.

42 *Injection and Compression Molding Fundamentals*. (A. I. Isayev, Ed.) Marcel Dekker, New York, 1987.

43 F. M. Sweeney. *Reaction Injection Molding Machinery and Processes*. Marcel Dekker, New York, 1987.

44 C. Mocosko. *Fundamentals of Reaction Injection Molding*. Hanser (Oxford University Press), New York, 1988.

45 D. V. Rosato and D. V. Rosato. *Blow Molding Handbook*. Hanser (Oxford University Press), New York, 1988.

46 D. H. Morton-Jones. *Polymer Processing*. Chapman and Hall, New York, 1989.

47 R. B. Seymour and H. F. Mark. *Handbook of Organic Coatings*. Elsevier Applied Science, New York, 1990.

48 M. W. Ranney. *Powder Coating Technology* Noyes Data, Park Ridge, NJ, 1975.

49 *Modern Coatings Technology*. (C. J. Colbert, Ed.) Noyes Data, Park Ridge, NJ, 1982.

50 P. F. Green, J. Palmström, J. W. Mayer, and E. J. Kramer. *Macromolecules*. **10**, 501, 1986.

51 *SPI Plastics Engineering Handbook*, 5th ed. (M. L. Berings, Ed.) Van Nostrand Reinhold, New York, 1991, p. 541.

52 E. Reichmainis and C. W. Wilkins, Jr. In *Microelectronic Polymers*. (M. S. Htoo, Ed.) Marcel Dekker, New York, 1989, p. 53.

53 L. Ronquillo and C. Williams. *J. Therm. Insul.* **7**, 228, 1984.

54 H. E. Reymore, Jr., P. S. Carleton, R. A. Kolakowski, and A. A. R. Sayigh. *J. Cell. Plast.* **11**, 328, 1975.

55 L. J. Matienzo, T. K. Shah, A. J. Gibbs, and S. R. Stanley. *J. Therm. Insul.* **9**, 30, 1985.

56 L. J. Matienzo, T. K. Shah, A. J. Gibbs, and S. R. Stanley. *Proceeding*. 16th SAMPE Tech. Symp., Albuquerque, 1984, p. 548.

57 G. P. Schmitt, B. K. Appelt, and J. T. Gotro. In *Principles of Electronic Packaging*. (D. P. Seraphim, R. Lasky, and C. Y. Li, Eds.) McGraw-Hill, New York, 1989, p. 351.

58 T. H. Shepler and K. L. Casson. *Electronic Materials Handbook*, Vol. 1. ASM International, Materials Park, OH, 1989, pp. 578–596.

59 P. Hoffman. *Solid State Technol.* **31**, 85, 1988.

60 C. E. Sroog. In *Polyimides*. (D. Wilson, H. D. Stenzenberger, and P. M. Hergenrother, Eds.) Blackie & Son, Glasgow, U.K., 1990, pp. 254–256.

61 The Netherlands Patent 6,604,263 to E.I. du Pont de Nemours & Co., 1968; British Patent 1,098,556 to E.I. du Pont de Nemours & Co, 1968.

62 S. K. Gupta. *J. Appl. Polym. Sci.* **32**, 4541, 1986.

63 D. S. Soane and Z. Martynenko. *Polymers in Microelectronics: Fundamentals and Applications*. Elsevier, Amsterdam, 1989, p. 153.

64 R. Iscoff. *Semiconductor International.* **10**, 116, 1984.

65 P. S. Wang, T. N. Wittberg, and J. D. Wolf. *J. Mater. Sci.* **23**, 3987, 1988.

66 L. J. Matienzo and F. D. Egitto. *Polym. Degrad. Stab.* **35**, 181, 1992.

67 J. G. Tackacs, V. Vukanovic, F. D. Egitto, L. J. Matienzo, F. Emmi, D. Tracy, and J. K. Chen. *Proceedings*. 10th International Symp. Plasma Chemistry, ISPC-10, I.4–29, Bochum, Germany, 4–9 August 1991, p. 1.

68 G. Sawa, K. Iida, T Tanimoto, S. Nakamura, and M. Ieda. *Proceeding*. 2nd International Conf. on Conduction and Breakdown in Solid Dielectrics, Erlangen, Germany, 7–10 July 1986, p. 381.

69 S. C. Freilich and K. Gardner. *Proceeding/Abstracts*. Third International Conf. on Polyimides, Soc. Plast. Eng., Mid-Hudson Section, Ellenville, NY, 2–4 Nov. 1988.

70　P. Geldermans, C. Goldsmith, and F. Bedetti. In *Polyimides: Synthesis, Characterization and Applications*, Vol. 2. (K. L. Mittal, Ed.) Plenum, New York, 1984, p. 695.

71　B. J. Factor, T. P. Russell, and M. F. Toney. *Phys. Rev. Lett.* **66**, 1181, 1991.

72　W. N. Unertl, X. Jin, and R. C. White. *Proceedings*. Second European Tech. Symp. on Polyimides and High Temp. Polym. Montpellier, France, 4–7 June 1991.

73　B. J. Bachman and M. J. Vasile. *J. Vac. Sci. Technol.* **A7**, 2709, 1989.

74　L. J. Matienzo, F. Emmi, D. C. VanHart, and T. P. Gall. *J. Vac. Sci. Technol.* **A7**, 1784, 1989.

75　L. J. Matienzo. Unpublished results.

76　T. Venkatesan, W. L. Brown, B. J. Wilkens, and C. T. Riemann. *Nucl. Instrum. Methods Phys. Res.* **B1**, 605, 1984.

77　T. Hioki, S. Noda, M. Sugiura, M. Kakeno, K. Yamada, and J. Kawamoto. *Appl. Phys. Lett.* **43**, 30, 1983.

78　W. P. Pawlowski, M. I. Jacobson, M. E. Teixeira, and K. G. Sakorafos. *Proceedings*. Mat. Res. Symp., **167**, 147, 1990.

79　L. J. Matienzo. Unpublished results.

80　L. J. Matienzo, F. Emmi, D. C. VanHart, and J. C. Lo. *J. Vac. Sci. Technol.* **A9**, 1278, 1991.

81　E. J. Kramer, W. Volksen, and T. P. Russell. *Electronic Packaging Materials Science II Symposium*. (K. A. Jackson, R. C. Pohanka, D. R. Uhlmann, and D. R. Ulrich, Eds.) Palo Alto, CA, 15–18 April 1986.

82　H. R. Brown, A. C. M. Yang, T. P. Russell, W. Volksen, and E. J. Kramer. *Polymer*. **29**, 1807, 1988.

83　K. L. Saenger and H. M. Tong. *Polym. Eng. and Science*. **31**, 432, 1991.

84　H. Coufal. *Polym. Eng. and Science*. **31**, 92, 1991.

85　E. D. Wachsman and C. W. Frank. *Polymer*. **29**, 1191, 1988.

86　J. R. Salem, F. O. Sequeda, J. Duran, and W. Y. Lee. *J. Vac. Sci. Technol.* **A4**, 369, 1986.

87　S. P. Kowalczyk, Y. H. Kim, G. F. Walker, and J. Kim. *Appl. Phys. Lett.* **52**, 375, 1988.

88　R. G. Pethe, W. N. Unertl, and C. M. Carlin. *J. Mater. Res*. Submitted for publication, 1993.

89　R. G. Mack, E. Grossman, and W. N. Unertl. *J. Vac. Sci. Technol.* **A8**, 3827, 1990.

90　W. N. Unertl. *High Performance Polymers*. **2**, 15, 1990.

91 J. G. Clabes, M. J. Goldberg, A. Viehbeck, and C. A. Kovac. *J. Vac. Sci. Technol.* **A6**, 985, 1988.

92 M. Kakimoto, M. Suzuki, T. Konishi, Y. Imai, M. Iwamoto, and T. Hino. *Chem. Lett.* 823, 1986.

93 Y. Nishikata, T. Konishi, A. Morikawa, M. Kakimoto, and Y. Imai. *Polym. J.* **20**, 269, 1988.

94 H. Ikeno, A. Oh-saki, N. Ozaki, M. Nitta, K. Nakaya, and S. Kobayashi. *SID International Symp., Digest on Technical Papers.* Anaheim, CA, 24–26 May 1988, p. 45.

95 T. A. Skotheim, H. S. Lee, X. Q. Yang, M. N. Simon, J. S. Wall, and Y. Okamoto. *Polym. Mater. Sci. Eng.* **64**, 265, 1991.

96 T. Inagaki, M. Hunter, X. Q. Yang, T. A. Skotheim, and Y. Okanoto. *J. Chem. Soc. Commun.* 126, 1988.

97 A. Snow, W. Barger, N. Jarvis, and H. Wohltjen. *Proceedings.* 16th National SAMPE Tech. Conf., Reno, NV, 9–11 Oct. 1984, p. 388.

98 J. M. Hoover, R. A. Henry, G. A. Lindsay, C. K. Lowe-Ma, M. P. Nadler, S. M. Nee, M. D. Seltzer, and J. D. Stenger-Smith. *Polym. Prepr.* **32** (1), 197, 1991.

99 J. H. Cheung, E. Punkka, M. Rikukawa, R. B. Rosner, A. T. Rubner, and M. F. Rubner. *Polym. Mater. Sci. Eng.* **64**, 263, 1991.

100 K. F. Schoch, Jr., J. Bartko, and W. F. A. Su. *Polym. Mater. Sci. Eng.* **64**, 83, 1991.

101 I. Watanabe, K. Hong, and M. F. Rubner. *Langmuir.* **6**, 1164, 1990.

102 M. Kakimoto, M. Suzuki, Y. Imai, M. Iwamoto, and T. Hino. *Polym. Mater. Sci. Eng.* **55**, 420, 1986.

103 K. F. Schoch, Jr., W. F. A. Su, and J. Bartko. *Proceedings.* 4th SAMPE Electronic Materials and Processes Conf., Albuquerque, 12–14 June 1990, p. 334.

104 B. Harris. *Engineering Composite Materials.* The Institute of Metals, North American Publications Center, Brookfield, VT, 1986, p. 50.

105 A. K. Rastogy, J. P. Rynd, and W. N. Stassen. *Proceedings.* 8th Nat'l SAMPE Technical Conference, Seattle, 12–14 Oct. 1976, p. 353.

106 L. J. Matienzo and T. K. Shah. *Surf Interface Anal.* **8**, 53, 1986.

107 Y. Xie and P. M. A. Sherwood. *Chem. Mater.* **2**, 293, 1990.

108 M. J. Hearn and D. Briggs. *Surf. Interface Anal.* **17**, 421, 1991.

109 Y. Wu and G. Tesoro. *J. Appl. Polym. Sci.* **31**, 1041, 1986.

110 H. W. Rhee and J. P. Bell. *Polym. Composites.* **12**, 213, 1991.

111 H. F. Wu, G. Bireshaw, and J. T. Lemmle. *Polym. Composites.* **12**, 281, 1991.

112 L. J. Matienzo, J. D. Venables, J. D. Fudge, and J. J. Velten. *Proceedings*. 30th Nat'l. SAMPE Symp., Anaheim, CA, 19–21 March 1985, p. 302.

113 D. J. D. Moyer and J. P. Wightman. *Surf. Interface Anal.* **17**, 457, 1991.

114 B. R. Prime. *J. Polym. Sci.* **25**, 641, 1986.

115 R. D. McElhaney, D. G. Castner, and B. D. Ratner. In *Metallization of Polymers*. (E. Sacher, J. J. Pireaux, and S. P. Kowalczyk, Eds.) ACS Symp. Series 440, American Chemical Society, Washington, DC, 1990, p. 370.

116 A. Lustiger, E S. Uralil, and G. M. Newaz. *Polym. Composites*. **11**, 65, 1990.

117 L. J. Matienzo, T. K. Shah, and J. D. Venables. *Proceedings*. 30th National SAMPE Symp., Anaheim, CA, 19–21 March 1985, p. 330.

118 J. J. Singh. *Proceedings*. 33rd. International SAMPE Symp., Anaheim, CA, 7–10 March 1988, p. 407.

119 B. Mayo and J. Pfau. *Proceedings*. 33rd International SAMPE Symp., Anaheim, CA, 7–10 March 1988, p. 1751.

120 J. Mijovic and J. Wijaya. *Polym. Composites*. **11**, 184, 1990.

121 C. A. Saunders, A. A. Carmichael, W. Kremers, V. J. Lopata, and A. Singh. *Polym. Composites*. **12**, 91, 1991.

122 L. J. Matienzo, T. K. Shah, and J. D. Venables. *Proceedings*. 15th Nat'l. SAMPE Tech. Conf., Cincinnati, 4–6 Oct. 1983.

123 R. J. Walker and T. C. Wehman. *Proceedings*. 14th Nat'l. SAMPE Tech. Conf., Atlanta, GA, 12–14 Oct. 1982, p. 150.

124 C. D. Rudd, K. F. Hutcheon, and M. J. Owen. *J. Mater. Sci.* **26**, 1259, 1991.

125 M. J. Owen, V. Middleton, K. F. Hutcheon, F. N. Scott, and C. D. Rudd. *Proceedings*. 16th Annual Reinforced Plastics Congress, Blackpool, U.K., Nov. 1988, p. 127.

126 M. J. Owen, V. Middleton, C. D. Rudd, F. N. Scott, and K. F. Hutcheon. *Proceedings*. 2nd International Conf. on Automated Composites, Leeuwenhorst, the Netherlands, Sept. 1988.

127 M. J. Owen, V. Middleton, K. F. Hutcheon, F. N. Scott, and C. D. Rudd. *Proceedings*. I. Mech. E. Des. Compos. Mater., 1989, p. 107.

128 K. F. Hutcheon. MPhil. thesis. University of Nottingham, U.K., 1989.

129 C. E. Miller, D. D. Archibald, M. L. Myrick, and S. M. Angel. *Appl. Spectrosc.* **44**, 1297, 1990.

130 F. J. Bergin and H. F. Shurvell. *Appl. Spectrosc.* **43**, 516, 1989.

131 D. D. Archibald, L T. Lin, and D. E. Honigs. *Appl. Spectrosc.* **42**, 1558, 1988.

132 M. B. Seasholtz, D. D. Archibald, A. Lorber, and B. R. Kowalski. *Appl. Spectrosc.* **43**, 1067, 1989.

133 T. Komatsu, S. Enoki, and A. Aoshima. *Polymer.* **32**, 1983, 1991.

134 T. Komatsu, S. Enoki, and A. Aoshima. *Polymer.* **32**, 1988, 1991.

135 T. Komatsu, S. Enoki, and A. Aoshima. *Polymer.* **32**, 1994, 1991.

136 P. J. C. Chappell, J. R. Brown, G. A. George, and H. A. Willis. *Surf. Interface Anal.* **17**, 143, 1991.

137 H. L. Ding. *Proceedings.* 4th International SAMPE Electronics Conf., Albuquerque, 12–14 June 1990.

Chemical Composition of Polymers

STEVEN P. KOWALCZYK

Contents

3.1 Introduction

The chemical composition of a polymer, along with its molecular weight, are the key factors that determine most of its important properties (dielectric constant, adhesion, etc). The determination of the chemical composition is important, especially since polymers are increasingly seeing application as thin films. In microelectronics, for example, thin films can be less than 1 μm thick; the question of a possible difference between the surface and bulk composition can be critical, and interfacial reaction and compound formation can be the determining factors for a successful application. New techniques of preparation now being employed need very fine control, such as vapor deposition from multiple monomers, where excess monomer can result in phase-separated aggregates at the surface. Surface chemical modification by wet or dry treatment is very important for the enhancement of many important properties, such as adhesion, and must be meticulously controlled.

This chapter briefly describes some of the main techniques used to determine chemical composition and discusses some of the advantages and disadvantages associated with them. The bias of this chapter is towards thin polymer films, their surfaces, and interfaces, because they are becoming increasingly important for current and future technologies. Thin polymer films present some unique and stringent experimental concerns not only in their preparation but also in their analysis. The next section presents some of the questions one should ask when considering chemical composition. This is followed by a brief summary of the main techniques used to determine the chemical composition of polymers, with consideration given to their advantages and disadvantages. Several of these techniques are applied to problems concerning the chemical composition of polymers as illustrative case examples.

The characterization of chemical composition is important to most of the chapters in this book. The characterization techniques are especially important to Chapter 9, Chemistry, Reactivity, and Fracture of Polymer Interfaces, and Chapter 7, Surface Modification of Polymers. It is also of particular relevance to Chapter 2, Polymer Fabrication Techniques, and Chapter 8, Adhesion. Of particular concern is the question of surface versus bulk chemical composition. Quite often one wants properties for the surface different from those for the bulk. For example, for a metallized polymer one wants good adhesive properties at the surface, but the bulk property of interest is different, such as low dielectric constant in a multilevel microelectron package or low weight and shock absorbence in an automobile bumper.

3.2 Chemical Composition: Questions to Ask

A number of questions about the following can be asked under the rubric of chemical composition:

- stoichiometry
- functionality/end groups
- phase separation
- surface segregation
- surface modification
- molecular orientation
- interfacial (surface) chemical composition
- taticity
- impurities (contaminants/additives).

The most obvious question concerns stoichiometry or elemental composition. As discussed below, stoichiometry is not as straightforward as it might seem to be. Once the stoichiometry is determined, the next question—usually the most important question—is what functionalities (that is, chemical moieties) are present. Related questions are what the end group functionalities and taticity are. These questions assume the polymer is homogeneous, which is not necessarily the case as, for instance, in a block copolymer or in polymer blends. Thus a key question concerns the possibility of phase segregation, the possibility of spinodal decomposition, or the presence of unreacted components of an incomplete reaction.

Often the surface of the polymer, especially a film, is intentionally modified to enhance or induce new properties, and thus the composition of the outermost region (from 10 Å to 1 μm) or a compositional depth profile is of interest. An important question for thin films is what the difference is between surface and bulk composition (i.e., surface segregation) and orientation/alignment (e.g., are the molecular units oriented such that one end containing a certain functional group is facing the surface?).

Another key consideration is interfacial chemical composition, since polymers are often used in a multilayer structure (i.e., metal–polymer interface or composite materials), and chemical integrity (stability) of the interface under adverse conditions such as weathering and temperature/humidity cycling, which is often the key to the choice of the particular structure for real-world applications. Finally, there is the question of impurities, which can be contaminants, additives, fillers, lubricants, dopants, catalysts, solvents, plasticizers, stabilizers, colorants, antistatic agents, release agents, adhesion promoters, or antioxidants/corrosion inhibitors.

3.3 Techniques for Determination of Chemical Composition: What to Use to Ask the Questions

In this section we discuss several techniques that are used to answer the questions posed in the previous section concerning the chemical composition of polymers. The discussion is brief because these techniques have been discussed in detail elsewhere in this series. We concentrate here on the most widely employed spectroscopic techniques and highlight concerns pertinent to polymer investigations. A bibliography of applications of these techniques to polymer applications is given at the end of the chapter. We focus on aspects pertinent to polymers, and we are concerned with the type of information obtained from these methods, the types of problems they can be applied to and their limitations. The techniques most commonly used for determining the chemical composition of polymers are

- X-ray photoelectron spectroscopy (XPS)

- near-edge X-ray absorption fine structure (NEXAFS)

- Auger electron spectroscopy (AES)

- Fourier transform infrared (FTIR) spectroscopy

- high-resolution electron energy-loss spectroscopy (HREELS)

- secondary ion mass spectrometry (SIMS)

- Rutherford backscattering spectroscopy (RBS)

- ion scattering spectroscopy (ISS)

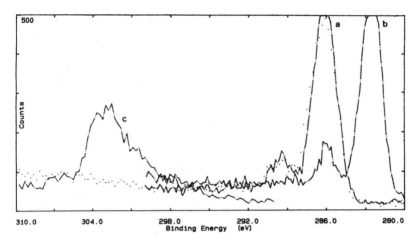

Figure 3.1 C 1*s* spectra from a 6-μm-thick polyimide film exhibiting charging effects and charge compensation. The spectra in order of accumulation are (*a*) obtained with light shining on sample, nearly normal binding energy observed, (*b*) with "zero" energy electron flood gun on, charge overcompensated and lower than normal binding energy observed, and (*c*) with no light or electron flood gun on, shift to high binding energy by nearly 10 eV and greatly distorted line shape.

- Raman spectroscopy

- nuclear magnetic resonance (NMR)

- Mössbauer spectroscopy

- energy-dispersive X-ray (EDX) analysis

- inelastic electron tunneling spectroscopy (IETS).

Of these, probably the most extensively employed technique is X-ray photoelectron spectroscopy (XPS), also known as electron spectroscopy for chemical analysis (ESCA). The power of this technique for polymer investigations was evident from the work by D. Clark and co-workers dating from the early 1970s. (A list of comprehensive reviews with emphasis on polymers is given in the bibliography.) This technique involves the use of soft X rays, usually Al Kα ($h\nu$ = 1486.6 eV) or Mg Kα ($h\nu$ = 1253.6 eV). Synchrotron radiation can be used as the photon source, providing a tunable, intense photon source. The sampling depth of the technique is on the order of 75 Å. Thus, this technique is quasi bulk sensitive, though its surface sensitivity can be greatly enhanced either by grazing exit angle analysis or by the use of photon energies to yield kinetic energies near the minimum in the electron mean free path.

One liability of the technique is possible beam-induced damage. X rays, for the most part, cause less damage than other beams such as electrons or ions, one reason this technique has become such a workhorse for polymer analysis.[1–4] The other

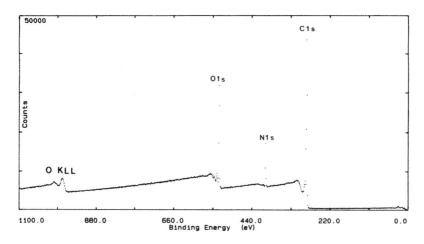

Figure 3.2 XPS survey spectrum from a polyimide film (PMDA-ODA) showing C, N, and O 1s core levels and O KLL Auger transitions.

major experimental problem in the application of this technique to polymers results from the fact that most polymers are good insulators, resulting in charging due to surface charge accumulation.[5–7] This can result in large shifts in binding energies and, in the case of inhomogeneous charging, gross distortion of line shapes, often making data analysis impossible. Figure 3.1 shows XPS spectra of a polymer under a variety of conditions. With no charge compensation, the main peak is shifted by nearly 9 eV and is greatly broadened. The use of an electron gun to flood the sample surface with low-energy electrons significantly narrows the line width and in this case overcompensates for the charging and shifts the spectra to a lower binding energy than the actual. In this particular polymer, irradiation with an external source of photons from an external lamp produces a nearly neutralized surface, as seen by the line shape and peak position. Again, this is a problem associated with all beam techniques when applied to polymers. This can be minimized by studying very thin films (<2000 Å)[8] or by charge neutralization by flooding the surface with "zero" energy electrons (again being careful of beam-induced degradation, which is minimized by the low energy of the electrons).

XPS survey spectra allow an assessment of the stoichiometry and purity of polymer samples. Figure 3.2 shows a survey spectrum from a polyimide film. The stoichiometry can be determined at the level of ~1–2% and is limited by uncertainties resulting from the many body nature of the photoemission process, in particular shake-up satellite intensity (Figure 3.3), which must be accounted for for a truly quantitative analysis.[9] The use of standards can be helpful for comparison.[†] Many impurities can routinely be detected below 0.1%.

---

† Reference materials on polymer films for surface studies are commercially available, e.g., Surface/Interface.

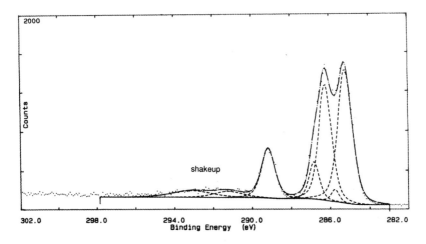

Figure 3.3   High-resolution XPS spectrum from the C 1s region of polyimide showing shake-up intensities.

A technique related to XPS is near edge X-ray absorption fine structure (NEX-AFS), which requires a synchrotron source. It has not yet been widely used in polymer studies. A key feature of NEXAFS is that it can be used to obtain information about orientation and unoccupied states.

Other related techniques are Auger electron spectroscopy (AES) and scanning Auger microscopy (SAM). However, because these techniques employ intense electron beams, their application to polymer problems is limited in comparison with the above technique due to severe beam-induced sample damage and charging problems. This technique is primarily limited to elemental rather than chemical composition information, again making it less popular than XPS for polymer characterization. A scan showing C, N, and O from a polyimide film is shown in Figure 3.4*a*. These techniques are used in conjunction with ion sputtering to obtain a depth composition profile (which can also be done but less rapidly with XPS). Figure 3.4*b* shows a compositional depth profile of a metallized polyimide structure on a silicon wafer. A key feature of this technique is its good lateral spatial resolution, ~200 Å compared to ~100 µm, which is commonly available for XPS analysis.

The other nearly ubiquitous technique for studies of polymer chemical composition is Fourier transform infrared (FTIR) spectroscopy. This technique yields detailed information about the chemical functionalities present, and polarization studies can give information about the orientation of the functionalities. Because, this technique is much more of a bulk technique than XPS, special techniques are needed to obtain surface or interface information. This technique does not suffer from the problems of beam-induced damage or charging effects inherent with the techniques involving the use of charged particle beams, nor is there the requirement of ultra-high-vacuum. These factors often make FTIR easier, faster, and cheaper

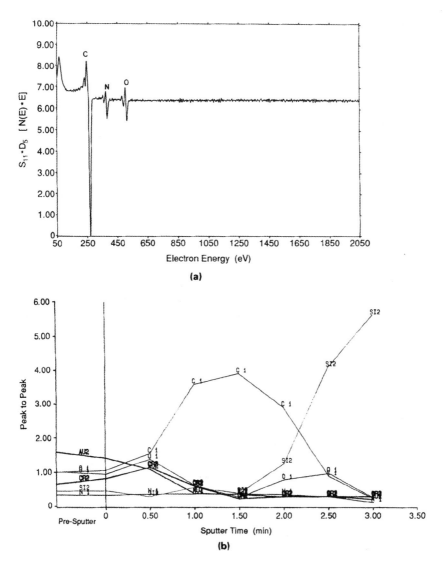

Figure 3.4   (a) Auger spectrum from polyimide showing C, N, and O transitions;
(b) Auger depth profile from a metallized polyimide structure.

than, for example, XPS. A related optical technique is Raman spectroscopy, which
can yield information similar to FTIR.

Another technique yielding information similar to that of IR techniques is high-
resolution electron energy-loss spectroscopy (HREELS). Although HREELS is much
more surface-sensitive than FTIR, it does have the problems already discussed for

XPS. This technique is usually used not as the primary analytical tool but in conjunction with other techniques, especially XPS.

Increasingly, secondary ion mass spectrometry (SIMS) has been applied to polymer studies because it yields molecular information. Especially useful are studies in the static mode with very low beam dosages that minimize (but not necessarily eliminate) beam damage; this mode is also extremely surface-sensitive. SIMS in the dynamic mode at higher doses can be used to perform depth profiling. SIMS can also be used to obtain lateral composition maps with 1-μm resolution.

Most of the techniques discussed above are surface-sensitive to some extent; Rutherford backscattering spectrometry (RBS) and energy-dispersive X-ray (EDX) analysis are more bulk-sensitive. RBS can give elemental depth profiles with about a 50-Å depth resolution. EDX can demonstrate bulk elemental composition which in conjunction with the surface-sensitive techniques can yield information about bulk versus surface differences in composition. Because EDX measurements are usually performed in a scanning electron microscope (SEM), information about surface morphology is also obtained.

A problem with many of the techniques discussed in this chapter is their inability to perform quantitative analysis of hydrogen. Direct recoil spectroscopy can complement the already mentioned techniques in this regard. The other techniques listed at the beginning of this section have had rather limited use for the polymer applications of interest here and are not discussed—the interested reader is referred to the lead volume of this series, *Encyclopedia of Materials Characterization*, or to the bibliography for further information.

Finally, it should be apparent that no one technique can solve *all* problems. Indeed, even for the solution to *a* particular problem, a multitechnique approach is required. In our bibliometric study of the above techniques as applied to chemical composition problems in polymers, we found that XPS and FTIR were by far the most popular techniques. However, the majority of studies did use two or more of the techniques, usually with XPS or FTIR playing the major role and the XPS/FTIR combination the most prevalent.

3.4 Illustrative Examples of Characterization of Chemical Composition of Polymers

In this section we present several case studies as examples to answer the questions posed in Section 3.2. We use polyimide as the canonical polymer to serve as the unifying example. In fact, this has been one of the most heavily investigated polymers in the last five years and has been studied by using nearly every technique discussed in this chapter.

Stoichiometry

Figure 3.5 is an example of following stoichiometry during a reaction, in this case the imidization of a vapor-deposited polyimide film. The upper panel shows the

XPS survey spectrum from the vapor-deposited (from the monomers PMDA and ODA) uncured film that is polyamic acid. Upon curing to 350 °C, the film imidizes with a loss of H_2O. This change can be followed by changes in the C:N:O ratios in the two spectra. The spectra also demonstrate that there are no detectable impurities from the vapor deposition process. For a discussion of vapor deposition as a fabrication process for polyimide films, see Chapter 2.

Functionality

Figure 3.6 shows core-level spectra from the C $1s$, O $1s$, and N $1s$ levels in polyimide (PMDA-ODA). Figure 3.7 shows, for comparison, the C $1s$ spectrum from another polyimide (BPDA-PDA) having a different chemical structure that has no

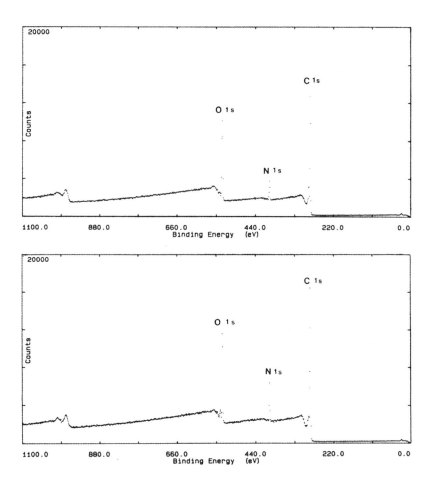

Figure 3.5   XPS survey spectra from (*upper*) a codeposited PMDA-ODA film and (*lower*) the film after it has been cured to 350 °C, demonstrating a change in stoichiometry upon imidization reaction.

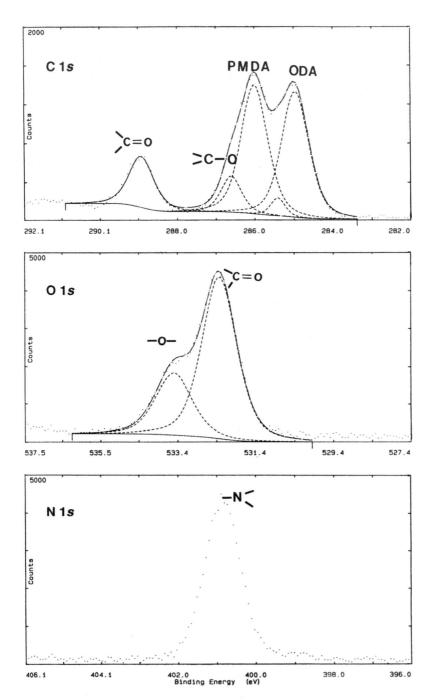

Figure 3.6   C, O, and N 1s spectra from PMDA-ODA polyimide showing different functionalities.

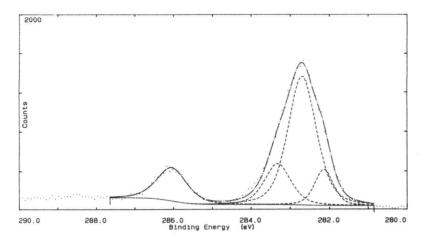

Figure 3.7 C 1s spectrum from BPDA-PDA, which is different from PMDA-ODA of Figure 3.6 due to the different chemical structure.

ether oxygen, resulting in an easily distinguishable C 1s (and O 1s) spectrum. The O 1s spectrum only consists of one peak, in contrast to that for PMDA-ODA, which shows two peaks in Figure 3.6. Figure 3.8 compares polyimide films grown by vapor deposition and spin-casting from solution. The spectra show the films to be chemically identical, and the survey spectra again show that they are stoichiometrically equivalent. Figure 3.9 shows the changes in the core level spectra related to Figure 3.5 upon curing. In particular, note the changes in the C 1s spectra in the region of about 290 eV, where initially there are two peaks, one due to the amide

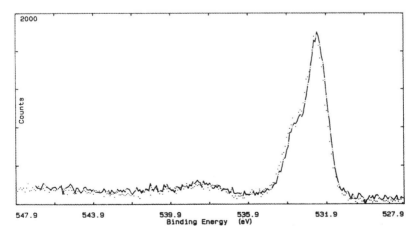

Figure 3.8 O 1s spectra from (···) vapor deposited and (−·−) spun polyimide (PMDA-ODA) films demonstrating identical chemistry for the two preparation techniques.

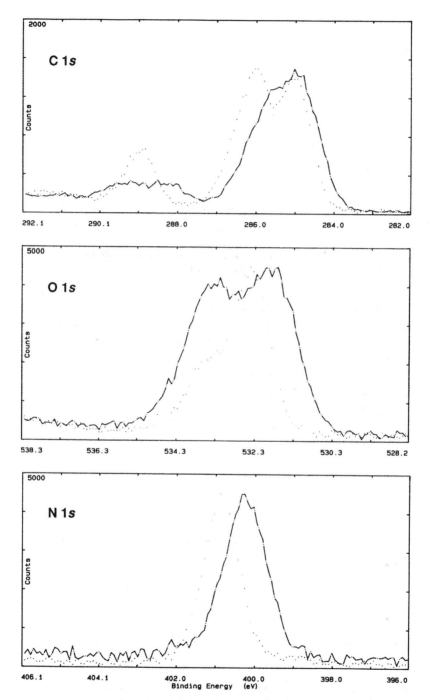

Figure 3.9   C, O, and N 1s spectra from (···) uncured and (−·−) cured polyimide
films showing chemical changes upon imidization.

CHEMICAL COMPOSITION OF POLYMERS   Chapter 3

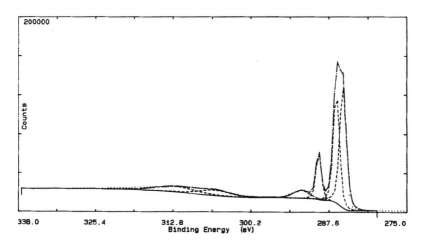

Figure 3.10 C 1s spectrum showing extent of shake-up.

carbonyl and the other due to the acid carbonyl, which upon cycloimidization becomes one peak of the imide carbonyl, as expected. Note, however, that the expected carbonyl intensity is not what is expected from stoichiometry, but this is believed due to shake-up, as discussed above. Figure 3.10 is a C 1s spectrum extending 50 eV past the primary core level, showing the structure that must be fully understood before a truly quantitative analysis is possible.

Because XPS peaks tend to be somewhat broad, changes in functionalities can be studied in more detail with FTIR. Figure 3.11 is an example of an FTIR spectrum from polyimide, with assignments given for several features. Note the asymmetric carbonyl stretch. The relative intensity of this feature can be used to study crystallinity and orientation with polarization studies.[10]

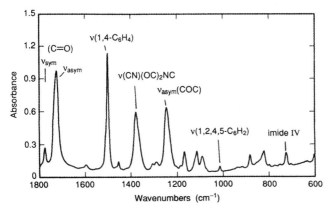

Figure 3.11 FTIR spectrum from polyimide film showing highly resolved functional groups.

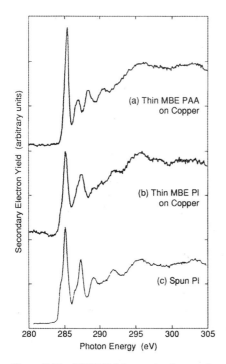

Figure 3.12   NEXAFS C 1s spectra from polyamic acid and polyimide.

NEXAFS also can be used to follow chemistry.[11] Figure 3.12 shows NEXAFS C K-edge spectra, and Table 3.1 gives some assignments for the core level to low-lying excited state transitions. The first peak in the polyamic acid spectra, for example, at ~285 eV is the C 1s to $1\pi^*$ resonance of the arene ring, whereas the peak at ~287

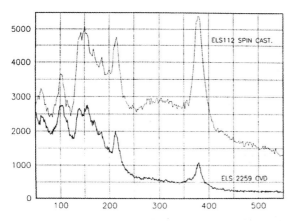

Figure 3.13   HREELS spectra from two different polyimide films.

Feature	Energy, eV	Origin	Final Orbital
1	284.8	C 1s[b]	π^*(C=C) (PMDA ring)
2	285.2	C 1s[a]	π^*(C=C) (ODA ring)
3	286.6	C 1s[b]	π^*(C=C) (ODA ring)
4	287.4	C 1s[c]	π^*(C=O)
5	289.2	C 1s[a]	π^*(C=C) (ODA second)
		C 1s[b]	π^*(C=C) (PMDA second)
6	291.9	C 1s[b]	σ^*(C–O.C–N)
7	295.4	C 1s[a]	σ^*(C=C) (ODA ring)
8	303.1	C 1s[c]	σ^*(C=O)
		C1s[a]	σ^*(C=C) (ODA second)
I.P. C 1s[a]	284.7		
I.P. C 1s[b]	285.7		
I.P. C 1s[c]	288.6		

[a] Lowest unoccupied molecular orbital.

[b] Second-lowest unbinding unoccupied molecular orbital.

[c] C 1s binding energy relative to the Fermi level.

Table 3.1   NEXAFS assignments for C K-edge transitions in polyimide.

is the resonance to the $1\pi^*$ resonance of the ODA ring. The 288 eV feature is associated with the carboxylic acid carbonyl resonance.

Figures 3.13 and 3.14 are examples of HREELS[12] and SIMS[13] spectra from polyimide. Figure 3.15 shows two other aspects of XPS spectra: valence band spectra, which contain detailed molecular structure information, and also X-ray (as opposed to standard electron) -induced Auger spectra. Neither aspect is much used, but much information can be obtained potentially from the molecular level spectra.[14]

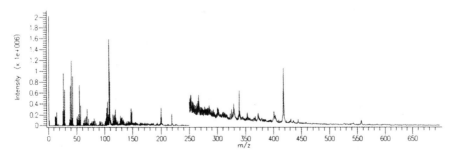

Figure 3.14   SIMS spectrum from a vapor deposited polyimide film.

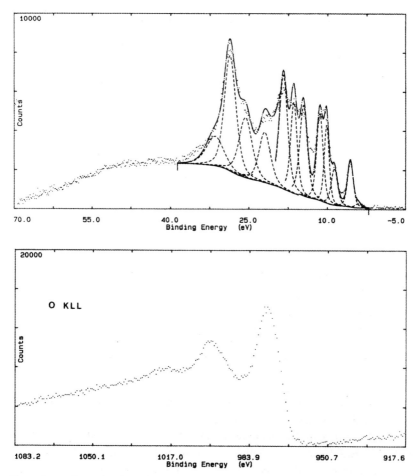

Figure 3.15 (*upper*) Valence band and (*lower*) X-ray-induced Auger spectra from a polyimide film.

Phase Separation

Phase separation can be an important processing consideration. For example, in the vapor deposition of polyimide when growing thick films, control of the flux ratio of the two monomers is very important.[10] The Figure 3.16 inset is an SEM micrograph of the surface of a film grown with an excess of PMDA, and the C 1s spectra compares two spectra grown on and off stoichiometry. FTIR microscopy was used to show that the structure was segregated excess anhydride crystallites. Another case of a separated phase is the interaction of polyamic acid with a copper surface upon curing, which produces cupric oxide precipitates within the polyimide film.[15] In this case it was demonstrated by the fact that the precipitates were large enough that in transmission electron microscopy (TEM) electron diffraction patterns could

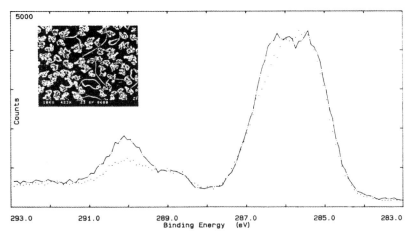

Figure 3.16 Comparison of C 1s spectra from sample grown with and without excess PMDA. Inset SEM micrograph shows surface morphology of sample with excess PMDA.

be obtained from the precipitates and be used to identify the chemical composition of the precipitates in conjunction with EDX elemental analysis.

Surface Modification

Surface modification is discussed in detail in Chapter 7; here we give a few brief examples of the techniques to characterize the chemical composition of modified surfaces. XPS is excellently suited for this application. RF sputter treatment is a common pretreatment and is known to decrease the resistivity from $\rho > 10^{18} \, \Omega \cdot$ cm to $7.9 \, \Omega \cdot$ cm.[16] Figure 3.17 shows the change in the C 1 s spectra which demonstrates that a carbonaceous layer is produced as the surface is decarboxylated and ring opening occurs. Table 3.2 compares the stoichiometry of an untreated polyimide surface with ones that have been Ar-ion-bombarded and RF Ar-sputtered. FTIR (Figure 3.18) also supports this interpretation.[17] Wet modification and its depth can also be studied.[18] Figure 3.17 shows that the surface treatment of KOH can

Element	Virgin	Ion Beam	RF Sputter
C	79	85.7	76
N	6	5.9	2
O	14	8.4	13
Cu	0.0	0.0	9
Ar	0.0	< 0.1	0.4

Table 3.2 Comparison of stoichiometry for polyimide surfaces.

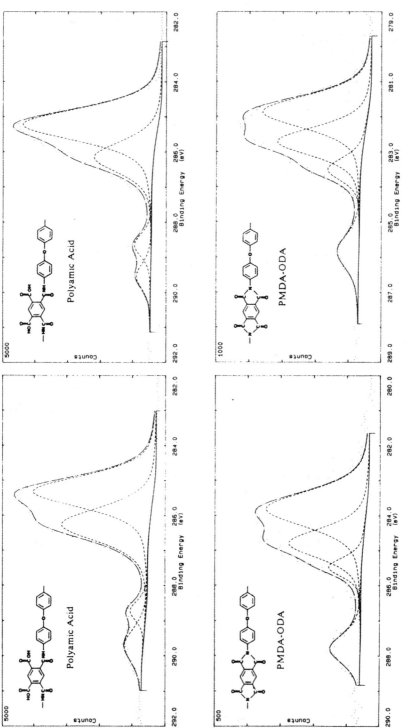

Figure 3.17 C 1s spectra from (*upper left*) a spun polyamic acid film and (*upper right*) a surface of polyimide which was converted to polyamic acid by wet chemical treatment. The lower panels are the C 1s spectra after heat treatment to form polyimide.

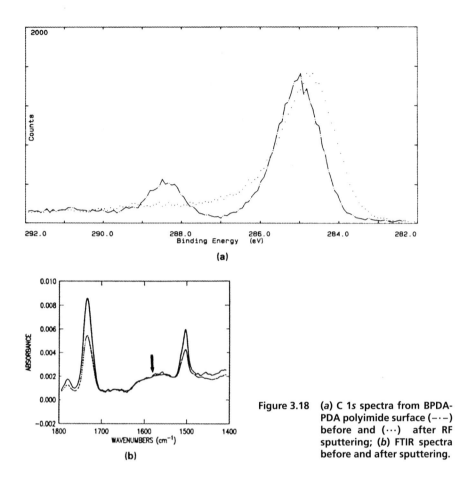

Figure 3.18 (a) C 1s spectra from BPDA-PDA polyimide surface (−·−) before and (···) after RF sputtering; (b) FTIR spectra before and after sputtering.

convert the surface from poly*amic acid* to poly*imide*, as can be seen by the changes in the carbonyl region of the C 1s spectra.

Interfacial Chemical Composition

Because this topic is discussed in detail in Chapter 9, we again give only brief examples. This is a particularly excellent application for XPS. Figure 3.19 shows XPS spectra obtained with synchrotron radiation and demonstrates the interaction of metal with the carbonyl functionality and formation of carbide at the growing interface; Figure 3.20 shows the same reaction followed by NEXAFS. An aid to studying interfacial chemistry is doing reactions with model compounds.[19]

Impurities

Impurities can have significant deleterious effects. As shown above, RF sputtering can induce chemical changes. However, another less well-known but equally serious

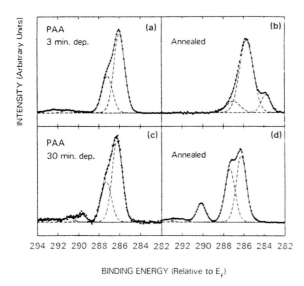

INTENSITY (Arbitrary Units)

(a) PAA 3 min. dep.

(b) Annealed

(c) PAA 30 min. dep.

(d) Annealed

294 292 290 288 286 284 282 292 290 288 286 284 282

BINDING ENERGY (Relative to E_F)

Figure 3.19   XPS spectra taken with synchrotron radiation showing C 1s interracial chemical reaction.

problem in RF sputter tools is the redeposition of metals from walls and platens. Figure 3.21 and Table 3.2 illustrate a case where a large amount of copper has been redeposited onto a polyimide surface. Figure 3.22 shows XPS spectra of a polyimide

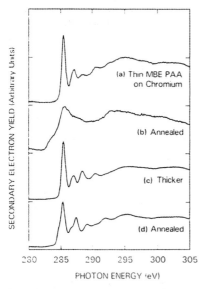

SECONDARY ELECTRON YIELD (Arbitrary Units)

(a) Thin MBE PAA on Chromium

(b) Annealed

(c) Thicker

(d) Annealed

280   285   290   295   300   305

PHOTON ENERGY (eV)

Figure 3.20   NEXAFS C 1s spectra from growing polyimide film.

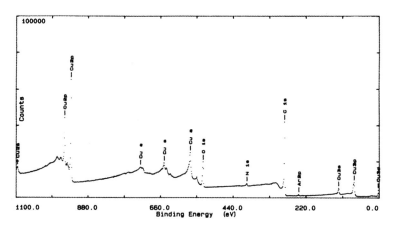

Figure 3.21   XPS survey spectrum of RF-sputtered polyimide surface.

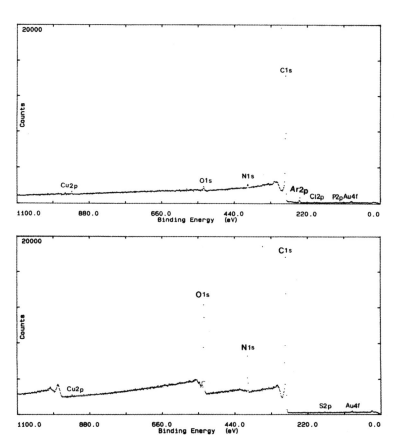

Figure 3.22   XPS survey spectra from surface of polyimide exposed to dirty CVD reactor:
(*upper*) area that had been ion-beam-sputtered and (*lower*) unsputtered area.

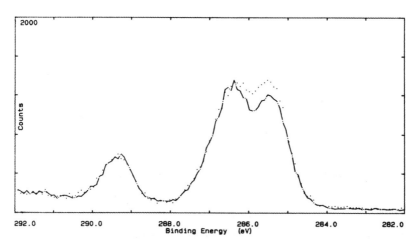

Figure 3.23 C 1*s* spectra from spun polyimide film before and after in situ UHV anneal.

film within an area that had been ion-beam-sputtered. This film then sat in a dirty chemical vapor deposition (CVD) reactor. Contamination could be found on the film, and the spot that had been sputtered was slightly different from the rest of the films, demonstrating different reactivity.

The final example shows problems of analysis, especially surface-sensitive analysis, and especially when samples are not prepared in situ—as is most often the case for polymer analysis. Polyimide films are normally spun from solution and cured in ovens; analysis is performed ex situ on such samples. Polyimide films are also known to take up H_2O. Figure 3.23 shows a comparison between a polyimide film spectrum taken "as received" and one that was heated in an ultra-high-vacuum (UHV) prior to analysis, where the change in the spectrum is obvious.

3.5 Summary: New Horizons

Most of the techniques discussed here are fairly mature and have demonstrated their usefulness in solving many important industrial and academic problems. The current trend for many of the above techniques, in particular XPS, SIMS, and FTIR, is towards better and easier microscopy and imaging.[20–24] XPS will see more use of synchrotron studies as well as use of other laboratory X-ray sources such as Ti[25] for variation of surface/bulk sensitivity. SIMS using time of flight mass spectrometry (TOF-SIMS) will become increasing important, as will NEXAFS. Further techniques to minimize beam-induced effects will continue to be important. New scanning microscopies (e.g., scanning tunneling microscopy [STM] and atomic force microscopy [AFM]) as they are increasingly applied to polymer studies will open new areas of particular relevance to polymers, especially the spectroscopic aspects of the STM.

CHEMICAL COMPOSITION OF POLYMERS   Chapter 3

Acknowledgments

I would like to acknowledge the contributions of a few of the many colleagues who have provided fruitful discussions and collaborations in various aspects of polymer characterization: Jung-ihl Kim, Steven Molis, Christos Dimitrakopoulos, Benjamin Eldridge, Ned Chou, Kang-Wook Lee, Jean Jordan-Sweet, and William Salaneck.

Bibliography

X-Ray Photoelectron Spectroscopy

Clark, D. T. *Pure Appl. Chem.* **54**, 415, 1982.

Clark, D. T., and W. J. Feast. *J. Macromol. Sci.-Rev. Macromol. Chem.* **C12**, 191, 1975.

Kowalczyk, S. P. In *Metallization of Polymers*. (E. Sacher, J.-J. Pireaux, and S. P. Kowalczyk, Eds.) American Chemical Society, Washington, DC, 1990, p. 10.

Pireaux, J.-J., J. Riga, P. Boulanger, P. Snauwaert, Y. Novis, M. Chtaib, C. Gregoire, F. Fally, E. Beelew, R. Caudano, and J. Verbist. *J. Electron Spectros.* **52**, 423, 1990.

Salaneck, W. R. *Rep. Prog. Phys.* **54**, 1215, 1991.

Near Edge X-Ray Absorption Fine Structure

Jordan-Sweet, J. L. In *Metallization of Polymers*. (E. Sacher, J.-J. Pireaux, and S. P. Kowalczyk, Eds.) American Chemical Society, Washington, DC, 1990, p. 36.

Jordan-Sweet, J. L., C. A. Kovac, M. J. Goldberg, and J. F. Morar. *J. Chem. Phys.* **89**, 2482, 1988.

Jordan-Sweet, J. L., and S. P. Kowalczyk. In *Handbook of Metallization of Polymers*. (S. P. Kowalczyk, Ed.) Marcel Dekker, New York, in press.

Ohta, T., K. Seki, T. Yokoyama, I. Morisada, and K. Edamatsu. *Physica Scripta.* **41**, 150, 1990.

Sette, F., J. Stöhr, and A. P. Hitchcock. *J. Chem. Phys.* **81**, 4906, 1984.

Stöhr, J., D. A. Outka, K. Baberschke, D. Arvantis, and J. A. Horsley. *Phys. Rev.* **B36**, 2976, 1987.

Tourillon, G., R. Garrett, N. Lazarz, M. Raynaud, C. Reynaud, C. Lecayon, and P. Viel. *J. Electrochem.* **137**, 2499–2501, 1990.

Infrared/Fourier Transform Infrared Spectroscopy

Dunn, D. S., and J. L. Grant. *J. Vac. Sci. Technol.* **A7**, 253, 1989.

Ishida, H., S. T. Wellinghoff. E. Baer, and J. L. Koenig. *Macromol.* **13**, 826, 1980.

Kelley, K. K., Y. Ishino, and H. Ishida. *Thin Solid Films.* **154**, 271, 1987.

Leu, J., and K. E Jensen. *J. Vac. Sci. Technol.* **A9**, 2948, 1991.

Linde, H. G. *J. Appl. Poly. Sci.* **40**, 2049, 1990.

Molis, S. E. In *Polyimides: Materials, Chemistry, and Characterization*. (C. Feger, M. M. Khojasteh, and J. E. McGrath, Eds.) Elsevier Scientific, Amsterdam, 1989, p. 659.

Molis, S. E., D. G. Kim, S. P. Kowalczyk, and J. Kim. *Mater. Res. Soc. Symp. Proc.* **203**, 53, 1991.

Pryde, C. A. *ACS Symp. Ser.* **407**, 57, 1989.

Pryde, C. A. *Polym. Mater. Sci. Eng.* 214, 1988.

Raman Spectroscopy

Borio, F. J., W. H. Tsai, G. Montaudo. *J. Poly. Sci. Poly. Phys.* **27**, 1017, 1989.

Siperko, L. M., W. R. Creasy, and J. T. Brenna. *J. VAc. Sci. Technol.* **A7**, 1750, 1989.

Tsai, W. H., F. J. Boerio, and K. M. Jackson. *Langmuir.* **8**, 1443, 1992.

High-Resolution Electron Energy-Loss Spectroscopy

Apai, G., and W. P. McKenna. *Langmuir.* **7**, 2266, 1991.

Becker, R., M. R. Ashton, T. S. Jones, N. V. Richardson, and H. Sotobayashi. *J. Phys. Cond. Mater.* **3A**, S29, 1991.

Jones, T. S. *Vacuum.* **43**, 177, 1992.

Pireaux, J.-J. In *Handbook of Metallization of Polymers*. (S. P. Kowalczyk, Ed.) Marcel Dekker, New York, in press.

Pireaux, J.-J., C. Gregoire, M. Vermeersch, P. A. Thiry, and R. Caudano. *Surf. Sci.* **189/190**, 903, 1987.

Pireaux, J.-J., C. Gregoire, M. Vermeerich, P. A. Thiry, M. R. Vilar, and R. Caudano. In *Metallization of Polymers*. (E. Sacher, J.-J. Pireaux, and S. P. Kowalczyk, Eds.) American Chemical Society, Washington, DC, 1990, p. 47.

Pireaux, J.-J., P. A. Thiry, R. Sporken, and R. Caudano. *Surf. Int. Anal.* **15**, 189, 1990.

Secondary Ion Mass Spectrometry

Benninghoven, A., F. G. Rudenauer, and H. W. Werner. *Secondary Ion Mass Spectrometry.* Wiley, New York, 1987.

Briggs, D. *Brit. Polym. J.* **21**, 3, 1989.

Briggs, D. *Polymers.* **25**, 1379, 1984.

Briggs, D. *Surf. Inter. Anal.* **9**, 391, 1986.

Eldridge, B. N., C. Feger, M. J. Goldberg, W. Reuter, and G. J. Scilla. *Macromol.* **24**, 3209, 1992.

Gardella, J. A. In *Handbook of Metallization of Polymers*. (S. P. Kowalczyk, Ed.) Marcel Dekker, New York, in press.

Hook, K. J., T. J. Hook, J. H. Wandass, and J. A. Gardella. *Appl. Surf. Sci* **44**, 29, 1990.

Linton, R. W., C. L. Judy, S. S. Maybury, and S. F. Corcoran. *Mater. Res. Soc. Symp. Proc.* **123**, 1989.

Mawn, M. P., R. W. Linton, S. R. Bryan, B. Hagenhoff, U. Jürgens, and A. Benninghoven. *J. Vac. Sci. Technol.* **A9**, 1307, 1991.

Russell, T. P., V. R. Deline, V. S. Wakharkar. *MRS Bulletin.* 33, Oct. 1989.

Stevie, F. A. *Surf. Int. Anal.* **18**, 81, 1992.

van der Wel, H., A. H. M. Sondag, L. Postma, and R. C. C. Roberts. *Vacuum.* 41, 1651, 1990.

Ion Scattering Spectroscopy

Baun, W. L. *Pure Appl. Chem.* **54**, 323, 1982.

de Puydt, Y., D. Leonard, and P. Bertrand. In *Metallization of Polymers.* (E. Sacher, J.-J. Pireaux, and S. P. Kowalczyk, Eds.) American Chemical Society, Washington, DC, 1990, p. 210.

Sparrow, G. R. and H. E. Mismash. In *Quantitative Surface Analysis of Materials.* (N. S. McIntrye, Ed.) ASTM, Philadelphia, 1978, p. 164.

Direct Recoil Spectroscopy

Schmidt, H. K., J. A. Schultz, S. Tachi, S. Contarini, S. Adura, J. W. Rabalais, and J. L. Margrave. *J. Vac. Sci. Technol.* **A5**, 2961, 1987.

Schultz, J. A., Y. S. Jo, S. Tachi, and J. W. Rabalais. *Nucl. Instrum. Methods.* **B15**, 134, 1986.

Rutherford Backscattering Spectroscopy

Chauvin, C., E. Sacher, A. Yelon, R. Groleau, and S. Gujrathi. In *Surface and Colloid Science in Computer Technology.* (K. L. Mittal, Ed.) Plenum, New York, 1985, p. 267.

Green, P. F., C. J. Palmstrom, J. W. Mayer, and E. J. Kramer. *Macromol.* **18**, 510, 1985.

Matienzo, L. J., and F. Emmi. In *Polyimides: Materials, Chemistry, and Characterization.* (C. Feger, M. M. Khojasteh, and J. E. McGrath, Eds.) Elsevier Scientific, Amsterdam, 1989, p. 643.

Matienzo, L. J., F. Emmi, D. C. VanHart, and T. P. Gall. *J. Vac. Sci. Technol.* **A7**, 1784, 1989.

Nuclear Magnetic Resonance

Ando, I., T. Yamanobe, and T. Asakura. *Prog. Nucl. Magn. Reson. Spectrosc.* **22**, 349, 1990.

Blümich, B., A. Hagenmeyer, D. Schaefer, K. Schmidt-Rohr, and H. W. Spiess. *Adv. Mater.* **2**, 72, 1990.

Kaplan, S., and A. Dilks. *J. Appl. Poly. Sci.* **38**, 105, 1984.

Konstadinidis, K., B. Thakkar, A. Chakraborty, L. W. Potts, R. Tannenbaum, M. Tirrell, and J. F. Evans. *Langmuir.* **8**, 1307, 1992.

Spiess, W., *Ann. Rev. Mater. Sci.* **21**, 131, 1991.

Mössbauer Spectroscopy

Leideiser, Jr., H., A. Vertes, J. E. Roberts, and R. Turoscy. *Hyperfine Interactions.* **57**, 1955, 1990.

Auger Electron Spectroscopy

Burrell, M. C., and J. J. Keane. *Surf. Int. Anal.* **11**, 487, 1988.

Ezzell, S. A., T. A. Furtsch, E. Khor, and L. T. Taylor. *J. Poly. Sci. Poly. Chem. Ed.* **21**, 865, 1983.

Furman, B. K., S. Purushothaman, E. Castellani, S. Renick, and D. Neugroshl. In *Metallization of Polymers.* (E. Sacher, J.-J. Pireaux, and S. P. Kowalczyk, Eds.) American Chemical Society, Washington, DC, 1990, p. 297.

Energy-Dispersive X-Ray Analysis

Chen, K.-M., S.-M. Ho, T.-H. Wang, J.-S. King, W.-C. Chang, R. F. Cheng, and A. Hung. *J. Appl. Phys.* **45**, 947, 1992.

Wang, P. S., T. N. Wittberg, and J. D. Wolf. *J. Mater. Sci.* **23**, 3987, 1988.

Inelastic Electron Tunneling Spectroscopy

Mallik, R. R., R. G. Pritchard, C. C. Harley, and J. Comyn. *Polymer.* **26**, 551, 1985.

References

1   R. Chaney and G. Barth. In *Polymer Surface Dynamics.* (J. D. Andrade, Ed.) Plenum, New York, 1988, p. 171.

2   D. R. Wheeler and S. V. Pepper. *J. Vac. Sci. Technol.* **20**, 226, 1982.

3   G. J. Leggett and J. C. Vickerman. *Appl. Surf. Sci.* **55**, 105, 1992.

4   F. Emmi, L. J. Matienzo, D. C. VanHart, and J. J. Kaufman. In *Metallization of Polymers.* (E. Sacher, J.-J. Pireaux, and S. P. Kowalczyk, Eds.) American Chemical Society, Washington, DC, 1990, p. 196.

5   C. E. Bryson. *Surf. Sci* **189/190**, 50, 1987.

6   G. Barth, R. Linder, and C. Bryson. *Surf. Int. Anal* **11**, 307, 1988.

7   B. D. Ratner, D. G. Castner, T. A. Horbett, T. J. Lenk, K. B. Lewis, and R. J. Rapozo. *J. Vac. Sci. Technol.* **A8**, 2306, 1990.

8   S. P. Kowalczyk. In *Metallization of Polymers.* (E. Sacher, J.-J. Pireaux, and S. P. Kowalczyk, Eds.) American Chemical Society, Washington, DC, 1990, p. 10.

9   P. S. Bagus, D. Coolbaugh, S. P. Kowalczyk, G. Pacchioni, and F. Parmigiani. *J. Electron Spectrosc.* **51**, 69, 1990.

10 S. P. Kowalczyk, C. D. Dimitrakopoulos, and S. E. Molis. *Mat. Res. Soc. Symp. Proc.* **227**, 55, 1991.

11 S. P. Kowalczyk and J. L. Jordan-Sweet. *Chem. Mater.* **1**, 592, 1989.

12 N. J. Chou, L Li, C. D. Dimitrakopoulos, R. Purtell, and S. P. Kowalczyk. Unpublished results.

13 C. D. Dimitrakopoulos, B. N. Eldridge, and S. P. Kowalczyk. Unpublished results.

14 S. P. Kowalczyk, S. Staftröm, J.-L. Brédas, W. R. Salaneck, and J. L. Jordan-Sweet. *Phys. Rev.* **B41**, 1645, 1990.

15 Y.-H. Kim, J. Kim, G. F. Walker, C. Feger, and S. P. Kowalczyk. *J. Adhesion Sci. Technol.* **2**, 95, 1988.

16 D.-G. Kim, S. E. Molis, T.-S. Oh, S. P. Kowalczyk, and J. Kim. *J. Adhesion Sci. Technol.* **5**, 509, 1991.

17 S. E. Molis, D.-G. Kim, S. P. Kowalczyk, and J. Kim. *Mat. Res. Soc. Symp. Proc.* **203**, 53, 1991.

18 K.-W. Lee, S. P. Kowalczyk, and J. M. Shaw. *Macromol.* **23**, 2097, 1990; *Langmuir.* **7**, 2450, 1991.

19 W. R. Salaneck, S. Stafström, J. L. Brédas, S. Andersson, P. Bodö, S. P. Kowalczyk, and J. J. Ritsko. *J. Vac. Set. Technol.* **A6**, 3134, 1988.

20 U. Gelius, B. Wannberg, P. Baltzer, H. Fellner-Feldegg, G. Carlsson, C.-G. Johansson, J. Larsson, P. Münger, and G. Vegerfors. *J. Electron Spectrosc.* **52**, 747, 1990.

21 P. Coxon, J. Krizek, M. Humpherson, and I. R. M. Wardell. *J. Electron Spectros.* **52**, 821, 1990.

22 I. W. Drummond, L. P. Ogden, and F. J. Street. *J. Vac. Sci. Technol.* **A9**, 1434, 1991.

23 P. Pianetta, P. L King, A. Borg, C. Kim, I. Lindau, G. Knapp, M. Keelyside, and R. Browning. *J. Electron Spectros.* **52**, 797, 1990.

24 H. Ade, J. Kirz, S. Hulbert, E. Johnson, E. Anderson, and D. Kern. *J. Vac. Sci. Technol.* **A9**, 1902, 1991.

25 T. G. Vargo and J. A. Gardella. *J. Vac. Sci. Technol.* **A7**, 1733, 1989.

Characterization of the Morphology of Polymer Surfaces, Interfaces, and Thin Films by Microscopy Techniques

DWIGHT W. SCHWARK and EDWIN L. THOMAS

Contents

4.1 Overview of Polymer Interfaces and Thin Films

Polymeric materials are employed in a wide variety of applications, a significant number of which depend upon the surface, interfacial, and/or thin-film properties. Specific examples include polymeric materials as adhesives, as composite structures with other types of materials, as membranes, or as thin-film coatings in the microelectronics industry. In order for one to fundamentally understand and further improve the surface, interfacial, or thin-film properties of polymers, a complete morphological characterization of surface and interfacial regions is required.

The importance of surface characterization is immediately apparent if one considers the influence of processing conditions on polymeric materials. For example, following the extrusion or molding of polymers, surface characterization commonly reveals the presence of a skin/core effect. Morphological and/or chemical composition differences occur in the surface region and can drastically influence the properties of the polymeric material for the chosen application.

92

This chapter discusses several morphological characterization techniques for polymer surfaces, interfaces, and thin films. Microscopy techniques are specifically dealt with in this chapter, and scattering techniques are discussed in Chapter 5. In this discussion we first include a brief description of the chemical and physical principles of the various microscopy techniques. Examples of the application of the various microscopy techniques are given for block copolymers, liquid crystalline polymers, semicrystalline polymers, and polymer–metal interfaces. For further information regarding the characterization of polymer–polymer interfaces, sec Chapter 10.

Before discussing the different microscopy techniques, we briefly discuss some of the key morphological descriptors for characterizing the interfacial regions of a solid polymer. When characterizing polymer interfaces, the most immediately obvious descriptor is the interfacial roughness. Characterization of the surface roughness has been greatly simplified by the recent introduction of the scanning probe microscopy techniques discussed in Section 4.6. A second key morphological descriptor is the composition profile or density gradient. It may be necessary to measure in directions both parallel and perpendicular to the interface. The configurations of polymer molecules are significantly perturbed in the interfacial region. The ordering or packing of chains to form microdomains or crystallites in the interfacial regions must also be considered for block copolymers and semicrystalline polymer systems. Preferential nucleation of polymer crystallization is quite common, due to both temperature gradients and heterogeneous nuclei at the surface region. Although similar morphological descriptors are required for the interfacial characterization of other materials, because we are dealing with polymeric materials, the length scale over which these descriptors are important is at least several radii of gyration (R_g) for polymeric materials. This entails probes sensitive to low atomic number materials which can assess structure and composition, not only at the uppermost surface region (~1 nm), but inwards to depths of up to 50 nm while maintaining good spatial resolution (~1 nm).

4.2 Introduction to Microscopy Techniques

Although many interfacial and thin-film analysis techniques can provide direct information on chemical composition, microscopy techniques offer the optimum in x, y, and z spatial resolution. Four different types of microscopy techniques are discussed in this chapter optical microscopy (OM), scanning electron microscopy (SEM), transmission electron microscopy (TEM), and the more recently developed scanning probe microscopy (SPM) techniques, including scanning tunneling and atomic force microscopy (STM and AFM). The order of discussion of the various microscopy techniques follows that of increasing spatial resolution of the techniques. A more in-depth discussion of the fundamentals of the OM, SEM, and TEM techniques, as well as a discussion of the specimen preparation procedures for the various techniques, may be found in Sawyer and Grubb's book[1] on the microscopy of polymers.

4.3 Optical Microscopy

Of the microscopy techniques, the oldest and most widely used is OM. Although this technique lacks the spatial resolution offered by electron microscopes and the new scanning probe microscopes, the optical microscope is commonly the first structural technique employed to investigate the sample. Recent improvements in OM have greatly improved the resolution and the quantitation of images. The digitization of images using area detectors, such as vidicons and charge coupled devices (CCDs), has permitted electronic image enhancement, digital image analysis, and real-time recording of dynamic phenomena. The use of finely focused beams and/or microfiber optic detectors, which are scanned over the sample, has increased resolution by better that a factor of 10× over the diffraction limitation of conventional optics. The observation of chin films or cross sections of bulk materials in the transmission mode is the most routinely used application of OM for polymeric materials. To investigate the surfaces of polymers, one may use the OM in either transmission or reflection modes. As concerns sample preparation, the easier of the two methods is the reflection mode.

Reflection techniques can provide information on the surface topography of polymeric materials. However, because of their relatively low reflectivity, polymers present a problem for reflectivity techniques. Due to their transparency there is substantial penetration of incident light through the surface and into the material where this light is scattered, and a portion may obscure the reflected image, resulting in a degradation of the "surface" image. Imaging of thin transparent polymer films may also be affected by reflections from the polymer–substrate interface. To overcome both of these problems, the surface of the polymer can be coated with a thin layer of metal. This can either be done by sputter coating or evaporation coating. It is critical that the amount of metal deposited does not obscure the morphological features of interest. Moreover, precautions should be taken to avoid excessive heating of the surface during deposition. Techniques which are commonly used in the reflection mode to analyze quantitatively the surface topography of polymers include interference microscopy methods. Further information regarding differential interference contrast (DIC) microscopy and the various types of interferometers available may be found in Reference 2.

In order to examine polymer surface morphology with transmission methods, one must use cross sections with thicknesses of less than 2 μm. Such cross-sectional analysis of the sample involves microtomy. A review of the techniques used to prepare thin sections for transmission OM can be found in Chapter 1 of Reference 2. Depending upon the nature of the polymeric sample, different forms of transmission OM may be employed. For example, polarized light OM may be used for semicrystalline polymers. Phase contrast OM may also be used for these materials or for noncrystalline multiphase polymer systems such as high-impact poly(styrene) (HIPS) and acrylonitrile-butadiene-styrene (ABS) terpolymers. The choice of the technique depends upon the index of refraction differences, thickness differences, and optical anisotropies present in the material. Although most polymers do not absorb in the visible spectrum, staining techniques can alter the index of refraction and enhance contrast between polymers in a blend and allow the use of bright-field microscopy techniques.

As an example of the use of OM to characterize the surface morphology of polymeric materials, we consider the case of spin-cast thin films of symmetric poly(styrene-*b*-methylmethacrylate) (SMMA) diblock copolymers. The typical lamellar microdomain size in microphase separated symmetric diblock copolymers is on the order of a few tens of nanometers. Since the resolution of the OM is on the micron scale, lamellar block copolymer microdomains themselves cannot be directly imaged. However, OM in the reflection mode shows an unusual surface topography for annealed block copolymer specimens. As shown in Figure 4.1*a*, a nonplanar "island–hole" structure develops from an initially flat, homogeneous as-spin-cast film upon annealing above the glass transition temperatures (T_g) of the material. The quantization of film thickness to give the "island–hole" structure

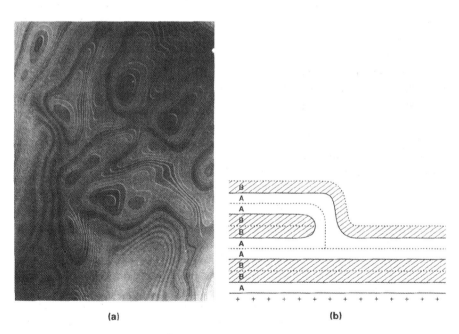

(a)   (b)

Figure 4.1   (a) Color DIC image[4] of the nonplanar "island-hole" structure of SMMA diblock copolymers, (b) Cross-sectional schematic proposed by Ausserre et al.[3] to account for the variation in copolymer film thickness of annealed thin films of symmetric SMMA diblock copolymers spin-cast onto a silicon substrate (+ + +). B denotes the PS block and A denotes the PMMA block.

occurs due to the incommensurate nature of the initial film thickness with the thickness of a set of alternating polystyrene (PS) and poly(methylmethacrylate) (PMMA) lamellar domains.[3] Direct chemical compositional analysis techniques for investigating symmetric SMMA diblocks, such as secondary ion mass spectrometry (SIMS)[4] and neutron reflectivity (NR),[5] where either the PS or PMMA block was perdeuterated, have suggested the preferential alignment of the lamellar microdomains parallel to the external surface and substrate interfaces for film thicknesses less than 500 nm. In addition, PS was found to exhibit a preferential affinity for the external surface or the interface with gold, whereas PMMA was preferentially located at the interface with a silicon substrate. A native oxide present upon the silicon substrate is thought to drive the polar PMMA block to the interface. The PS layer at the external surface or the PS layer at the Au substrate had a thickness of one-half that of the bulk PS lamellae thickness, while the ratio of the thickness of the PMMA layer at the silicon interface to that of the bulk PMMA lamellae was also one-half. On silicon substrates the overall film thickness is given by $(n + \frac{1}{2})L$, and on gold substrates the thickness is given by nL, where L is the lamellar period and n is the number of periods, typically less than 10 in these spin-cast films, depending on solution concentration and spinning speed. The combination of the SIMS and NR results with the OM results has allowed Ausserre et al.[3] to propose the schematic seen in Figure 4.1b to account for the morphology of thin films of symmetric SMMA diblock copolymers spin-cast onto a silicon substrate. The ordering of the lamellar microdomains parallel to the two interfacial constraints, and hence the oscillating compositional gradient in the film thickness direction, is thought to arise from the interfacial segregation of the lower interfacial tension block or blocks coupled with the connectivity and immiscibility of the two blocks.

A second example of the usefulness of OM for the characterization of polymer thin films concerns the defect structures of liquid crystals. Thermotropic liquid crystalline polymers (TLCPs) are attractive since they combine complementary properties from polymers (e.g., mechanical strength) and liquid crystals (e.g., optical switching). Transmission, polarized-light OM, in combination with a specimen hot stage, is used to determine transition temperatures and to identify the type of mesophase or mesophases present through their respective characteristic textures. The schlieren texture of nematic liquid crystals arises from the presence of rotational defects, termed disclinations, which are easily visible under crossed polarized transmitted light. Figure 4.2a shows a schlieren texture in a polyester TLCP material viewed under crossed polars. A thin film about 10 μm thick of the polymer was prepared by blade shearing in the melt state. Using a hot stage, we first heated the sample into the isotropic state and then quenched it into the nematic state to create a large number of disclination defects. The defects cause abrupt changes in the molecular director field; the nature of the disclinations may be determined using a first-order red plate and a rotation stage. When a first-order red plate is employed with the DIC mode, image contrast increases and yellow and blue regions will be present, depending on the local molecular orientation and film thickness. The color

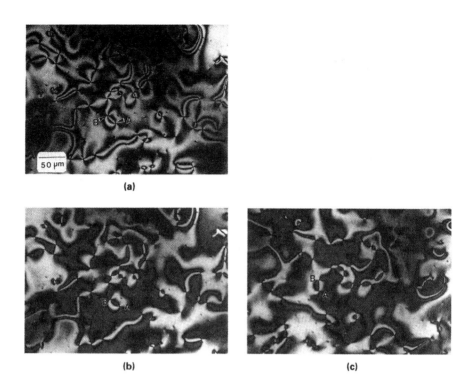

(a)

(b) **(c)**

Figure 4.2 (*a*) The schlieren texture of a polyester TLCP under cross polars. (*b*) The schlieren texture with first-order red plate. (*c*) The schlieren texture with first-order red plate, with the sample stage rotated 45° from (*b*). (Photo credit: D.-K. Ding.)

images can be used to determine the approximate molecular orientation about the defect from birefringent interference figures.

If the fast component (smaller index of refraction) in the sample and first-order red plate are parallel, the total optical path difference is increased and the interference color goes down scale to blue. If the fast component of the sample is parallel to the slow component of the plate, the interference color goes up to yellow. The fast component of the transmitted light is along the main chain axis of the TLCP polyester, so it is easy to find the approximate orientation of the molecules about a disclination. Figures 4.2*b* and 4.2*c* are DIC images related by a 45° stage rotation. Defect A is identified as a plus one defect, since the director field is symmetrical about the defect as the interference color does not change upon rotation. Defect B is identified as a minus one defect, since the blue colors of the first and third quadrants change to yellow upon 45° rotation.[6, 7]

OM offers the investigator a relatively inexpensive and less laborious means to investigate polymers without damage to the sample by the incident beam, but there are also some disadvantages to the technique. Most obvious is the relatively low resolution (200 nm resolution) of the conventional OM. Since many morphological features in polymeric materials, such as the microdomains in block copolymers and the crystallite regions in semicrystalline polymers, are smaller than this, the imaging capabilities of the traditional OM are somewhat limited.

The recent development of a near-field scanning optical microscope offers the potential to overcome the resolution limitations of the conventional OM. Bringing either a light source or a detector, with an exit or entry size smaller than the wavelength of light, very close to the specimen has been shown to increase dramatically the resolution of the OM.[8] The improvement in resolution beyond the $\lambda/3$ diffraction limit of conventional microscopy arises from the absence of an image-forming lens in the optical system. A scanning probe with a 50 nm diameter entrance aperture has yielded images with resolutions of 30–50 nm. The resolution is expected to be improved to the 1–10 nm range with finer micropipette light sources. In addition to the improvement in the resolution of the OM, the depth-profiling capabilities of the OM have also recently been more fully developed. Although the Confocal Scanning OM was invented in the 1960s, only during the last few years has there been widespread interest in this depth-profiling technique.[9] Depth profiling is possible with the use of illumination and detection apertures to limit the focal plane to an extremely thin (~0.25 μm) layer. Both of these new developments suggest that OM will indeed remain a principal investigative technique.

4.4 Scanning Electron Microscopy

The size of many polymer microstructural features is in the 2–20 nm range, making the SEM ideally suited for surface structural investigations. Modern SEM instruments feature a resolution of 5 nm or better, great depth of focus, relatively simple image interpretation, and ease of sample preparation. A striking feature of SEM images, resulting from the greater depth of focus, is the three-dimensionality of the features of the surface being investigated.

As seen in Figure 4.3, a high-energy (1–40 keV) incident (primary) electron beam impinging on the surface of the sample, with a diameter d_{pe}, causes a large number of low-energy (0–50 eV) secondary electrons (SEs) to be produced by removing bound electrons from the sample. These SEs can be attracted to a scintillator detector held at a positive potential (~200 V). Because they are of low energy, the SEs can only escape from the near-surface region (e.g., the SE escape depth, X_{se}, is ~1 nm for Au).[10] SEs produced by the finely focused primary beam provide a high-resolution image.

A second type of signal arises from the interaction of the incident electron with the nucleus of an atom in the specimen. Of the incident electrons that undergo Rutherford scattering with the nuclei of atoms, a fraction will scatter at angles sufficiently large to return to the surface of the specimen and escape with energies close to that of the incident beam. The trajectories of these high-energy backscattered electrons (BEs) are only slightly deviated by the positive potential of the SE detector, and therefore only BEs that leave the specimen traveling directly at the BE detector will contribute to the signal being measured. To maximize the efficiency of collection, the BE detector should subtend a large solid angle from the specimen. A small negative bias is applied to the BE detector to repel the low energy SEs. The yield of BEs increases monotonically with atomic number and therefore can provide compositional contrast. For samples having laterally homogeneous surface compositions, the use of the difference signal between the two halves of the solid-state BE detector permits determination of the surface topography, because then the signal collection depends only on the line of sight from the surface features to the BE detector.

Besides the SEs produced by the incident electron beam, SEs are produced by BEs. However, these BE-SEs degrade the resolution of the SE image because they may come from a region with a diameter, $d_{(be-se)}$, which as Figure 4.3 indicates is considerably larger than the diameter of the primary incident electron beam impinging on the sample.

In order to obtain an SEM image, from either the SE or the BE signal, the finely focused beam is scanned over the specimen surface. The signal emanating from the position of the probe is detected, amplified, and used to modulate the intensity of a second electron beam on a cathode ray tube. Synchronization by using the same scan generator ensures spatial correspondence between the object and the image.

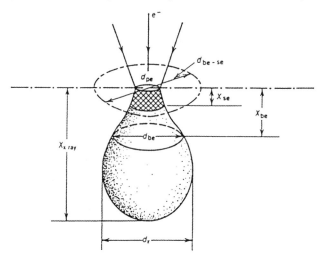

Figure 4.3 **Electron beam-sample interaction. The primary beam electrons impinge on the sample with diameter d_{pe}. The secondary electron imaging (SEI) signal arises from a region of diameter $d_{(be-se)}$, and from a specimen depth down to X_{se}. The backscattered electron imaging (BEI) signal arises from a region with a maximum diameter of d_{be} and from a specimen depth down to X_{be}. Similarly, the characteristic X-ray signals come from a region of maximum diameter d_X and down to a depth $X_{X\,ray}$.**

SEM sample preparation is relatively easy and usually involves only mounting on a specimen-stub; however, for nonconducting specimens a conductive coating is applied to the surface to prevent charging. This coating process is acceptable provided the coating does not cover the morphological features of interest. Unfortunately, the electron beam can damage the polymer specimen. Types of beam damage include cross-linking and dimensional shrinkage, loss of crystallinity, or, in certain radiation-sensitive polymers such as electron beam resists, chain scission and mass loss.[11] Further details concerning the theory and practice of conventional. SEM are given in References 10 and 12–16. Rather than discussing well-established uses of conventional SEM in the investigation of surfaces and interfaces of polymeric materials (for example, fractography for polymer deformation studies,[17] investigation of polymer wear,[17] and environmental surface damage studies[18,19]), we instead focus on new advances in SEM technology and applications of these new techniques.

Low-Voltage High-Resolution Scanning Electron Microscopy

A notable development with the SEM in the last few years involves the improvement in resolution of the instrument. By using a high-brightness field emission electron gun, investigators have noted resolutions of 0.5 nm at 30 kV.[20] Besides the type of gun employed, resolution improvements in SE imaging in the SEM have also resulted by altering the position of the sample and the SE detector. In conventional SEM, the sample and the SE detector are placed beneath the lower pole piece of the objective probe-forming lens. However, in the new models featuring high-resolution SEM (HRSEM), the sample is placed in the middle of the objective lens and the SE detector lies above the (immersion) objective lens. Such a design eliminates some of the undesired sources of SEs that create noise in the image and also facilitates the production of a smaller, but very intense, incident electron beam.

Modern HRSEMs also offer a greater opportunity to adjust the accelerating voltage of the incident electron beam. This feature is particularly useful for the study of polymer surfaces. Specimens examined in the SEM receive electrons from the incident electron beam and lose electrons as SEs or BEs. Depending upon whether there is a net loss or gain of electrons, the specimen will charge up positively or negatively, unless the balance is restored by some other means. For an electrically conducting specimen, a constant zero potential can be maintained by electron flow through the mounting stub and grounded microscope. However, polymer specimens which are typically electrical insulators either need to be coated with a conductive surface film or the electron gun voltage needs to be reduced (typically to about 0.8–1.5 kV), so that there is no net gain or loss of electrons during examination.

The reduction of the accelerating voltage, to reduce charging of uncoated polymer samples, has a somewhat detrimental effect on the resolution of the HRSEM. In a conventional SEM, as the accelerating voltage is reduced from 30 to 1 kV the resolution is greatly reduced from approximately 20 to 140 nm.[16] The decrease in resolution is primarily due to the significantly decreased flux in the electron beam that arises when the electron gun is operated at a lower voltage. For the newest generation of field emission HRSEMs, dropping the voltage from 30 to 1 kV results in a decrease in resolution from 0.7 to 4.0 nm.[21] However, a 4-nm-resolution at 1 kV is still very useful for the surface analysis of polymers. In addition to the reduction in charging that accompanies the lowering of the accelerating voltage, at lower voltages the SE signal only arises from the near-surface region because of the much reduced beam-sample interaction volume. Thus, there is currently considerable interest in low-voltage HRSEM (LVHRSEM) to analyze polymer surfaces. In our own research group we are using minimal thickness conductive coatings, low voltages, and a field emission gun for surface studies of a variety of polymeric systems.

One recent application of LVHRSEM was to investigate directly the director texture in bulk samples of thermotropic liquid crystalline polyesters[22] aligned in a magnetic field. When a liquid crystalline sample containing disclination defects is placed in a strong magnetic field, the molecules attempt to align with the applied field direction. However, in regions with disclinations, the director pattern is forced to accommodate these topological singularities by the introduction of π inversion walls. Such walls must either terminate at half-integer line defects or form closed loops.

Unlike OM, which can probe the sample texture via the optical anisotropy, LVHRSEM depends on topographic features to produce suitable SE image contrast. The so-called lamellar decoration technique of Thomas and Wood[6] employs the fact that the polyester molecules can crystallize into 30-nm-thick lamellae after quenching from the nematic state and annealing above T_g and below the crystal-to-nematic transition temperature. The lamellae grow with their long axes orthogonal to the local molecular director and decorate the director field throughout the specimen. To provide secondary electron image contrast, one may etch the glassy nematic regions between the crystalline lamellae, resulting in a surface topographic profile.

A film of a thermotropic liquid crystalline polyester oriented in a 13.5 T magnetic field was quenched, annealed, and etched with methylamine solution in water and then ion sputter-coated with Au/Pd for 15 s. Figure 4.4a shows a LVHRSEM image of an inversion wall imaged at 1 kV. The inversion walls are actually three-dimensional, as is evident from stereo-imaging in the SEM, or, as is discussed in Section 4.6, from AFM (see Figure 4.4b). Such observations enable detailed study of the director field at a resolution not possible by OM and provide information on several open questions concerning liquid crystal defect textures. For example, the detailed director field pattern about a disclination in a magnetic field has not yet been solved theoretically.

A second application of LVHRSEM involves the investigation of the surface morphology of symmetric poly(styrene-b-butadiene) (SB) diblock copolymers.[23] Using LVHRSEM, one can easily examine the orientation of the lamellar microdomains, in the near-surface region, over large (a few square microns) areas. To provide contrast between the PS and poly (butadiene) (PB) lamellae, we stained the near-surface region with osmium-tetroxide (OsO_4), a preferential heavy-atom stain

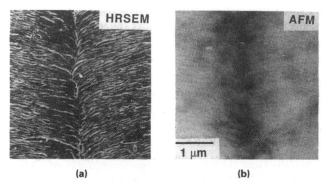

(a)   (b)

Figure 4.4   (a) LVHRSEM image of an inversion wall in a TLCP aligned by a magnetic field. (b) AFM image of a similar wall, showing that the wall
center is a valley-like feature on the specimen surface. (Photo credit:
D. Vezie).

for the PB, for 24 h. Light sputter-coating (approximately a monolayer) of the stained
surface with Au/Pd was performed to eliminate charging. Figure 4.5 shows alternating light and dark regions with a repeat spacing similar to the bulk lamellar spacing
of 59 nm, independently determined by small-angle X-ray scattering (SAXS).[24] In
the LVHRSEM micrograph, the PB phase is the light phase because of the enhanced
SE yield due to the presence of high atomic number osmium in the PB phase.

In LVHRSEM images of symmetric SB diblock copolymers, maximum contrast
between the PS and PB lamellae is obtained when the incident electron beam is
perpendicular to the lamellar normal, whereas, if the lamellae were oriented parallel to the external surface, the contrast would be uniform. Loss of contrast between
alternating lamellae when the incident beam is no longer perpendicular to the
lamellar normal can be seen in the LVHRSEM images shown in Figure 4.6. When
the area in Figure 4.6a is tilted 40° about the axis indicated, changes in the apparent size of the PS and PB lamellae, as well as a reduction in the contrast between the

Figure 4.5   LVHRSEM image of the surface of an
OsO$_4$ stained and Au/Pd sputter coated
SB 40/40 sample, indicating the non-
parallel surface orientation of the lamellar microdomains.

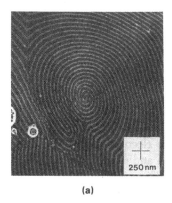
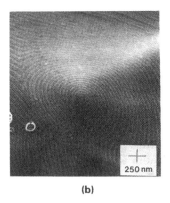

(a)   (b)

Figure 4.6   **LVHRSEM images of the surface of an OsO$_4$ stained and Au/Pd sputter-coated SB 40/40 sample, indicating the loss of image contrast and change in the apparent size of the PS (*dark*) and PB (*light*) lamellae when the region in (*a*) is tilted by 40° about the axis shown to yield the image in (*b*).**

adjacent domains are noted, as in Figure 4.6*b*. The reason for this is the fixed escape depth of the SE. When the incident electron beam is no longer perpendicular to the lamellar normal, SEs escape through regions which include PS and preferentially OsO$_4$ stained PB material. The loss of contrast and variation in microdomain size is similar to that which occurs in tilted TEM projections.

For samples which have large, flat surfaces, the incident electron beam direction can easily be oriented along the surface normal. In such a case, a determination of the orientation of the lamellae in the near-surface region is possible from a measurement of the lamellar repeat along the surface, d_s. As seen in Figure 4.7, measurement of d_s by LVHRSEM, combined with a knowledge of the lamellar repeat, L, from TEM or SAXS, allows for the determination of Θ, since Θ is given by arcsin (L/d_s). However, for samples with rough surfaces (i.e., the angle between the incident probe and the local surface normal is not always zero) deductions of Θ from changes in image periodicity become problematic.

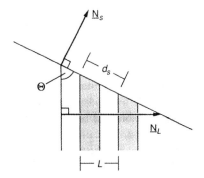

Figure 4.7   **Schematic showing the definition of the near-surface lamellar orientation, Θ, in terms of the normal to the alternating lamellae, $\underline{N}_L$, and the local surface normal, $\underline{N}_S$.**

Environmental Scanning Electron Microscopy

Electron microscopes generally require a vacuum of 10^{-4} torr or better to prevent strong scattering of the electrons. Moreover, materials such as polymers are often insulators, so that conductive surface coatings are needed for SEM investigations. These two facts would seem to rule out any SEM observations of wet or molten samples. Recently a new type of secondary electron detector which can function in vacuums of upwards of 100 torr has been successfully employed to image wet insulating specimens and liquids.[25-29] The so-called Environmental SEM (ESEM) employs differential pumping and a gaseous secondary electron detector.[25] A high- (10^{-7}) torr vacuum is maintained at the electron gun and throughout the optical column while the sample chamber can be maintained at pressures up to 100 torr, depending on the beam voltage, and inversely on the electron beam path length through the high-pressure region. The SE detector is integrated into the objective lens final aperture, which provides both a minimum working distance and high collector efficiency for improved image resolution. Details of the detector physics and the interplay of gas pressure, working distance, and collection voltage are discussed in the papers of Danilatos.[25-29] In addition to SE imaging, the ESEM is also capable of backscattered electron imaging, cathodoluminescence, and X-ray microanalysis.

A key difference of the ESEM as compared with a conventional high-vacuum SEM is that, in the former, the primary electron beam impinges on the sample in a small probe containing an outer skirt of electrons multiply scattered from passage through the short high-pressure path to the specimen. Specimen charging is suppressed by neutralization of the sample from the surrounding ionized ambient gas. Gas leaking from the pressure aperture is replaced from a reservoir. Signal amplification also occurs from SE–gas interactions, and the SEs generated by the specimen are accelerated towards the biased ionization detector and, along their path, ionize gas molecules, resulting in a cascade process and good signal amplification. The resolution of the ESEM is limited by the inner diameter of the (unscattered) probe. A rule of thumb for practical resolution is given by pressure × working distance ≈ 1 Pa·m (for a 10 keV incident beam and N_2 gas).[29] For uncoated polymers, beam penetration into the sample will be large at the higher beam voltages employed to achieve a small, high-flux probe so that the attainable resolution will be considerably less than that demonstrated for samples with a lower interaction volume, such as gold particles examined at high chamber pressures.

We have employed an ESEM to follow the dissolution of a starch-based packaging foam. The foam is produced by the extrusion of starch, whereupon exiting the high-pressure region of the extruder the superheated water undergoes a phase change to a vapor, resulting in a low-density foam suitable for packaging applications. The thermoplastic starch material is both water-soluble and biodegradable. Shown in Figure 4.8 are a series of micrographs of a portion of an uncoated packaging peanut at 200× magnification undergoing dissolution. The ESEM was operated at 20 kV, using a tungsten filament with a working distance of approximately

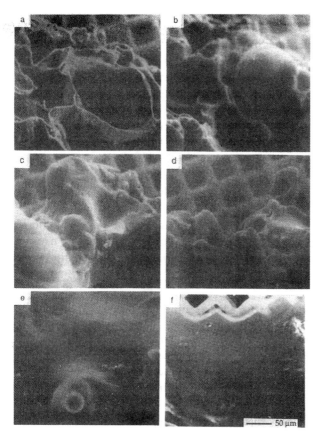

Figure 4.8 Environmental SEM micrographs displaying the effect of water condensation on a starch-based foam (sample is sitting on a 400-mesh copper TEM grid): (*a*) T = 15 °C, P = 10.0 torr; (*b*) T = 10 °C, P = 10.1 torr, 1 min after temperature change; (*c*) T = 10 °C, P = 10.1 torr, 5 min; (*d*) T = 10 °C, P = 11 torr, 10 min; (*e*) T = 10 °C, P = 12.0 torr, 15 min; (*f*) T = 15 °C, P = 4.9 torr, 25 min. (Photo credit: S. Simmons.)

9 mm. In Figure 4.8*a*, the Peltier stage was set at 15 °C and the chamber pressure maintained at 10 torr, just below the saturation pressure of water at 15 °C. The cellular structure of the starch team is evident, with cell diameters ranging from 50 to 200 μm. The stage temperature was then reduced to 10 °C, in order to condense water on the sample and to observe the changes in the sample morphology. Figures 4.8*b*–4.8*e* are a series of micrographs taken at various times after the temperature change. The cell framework begins to collapse as water droplets form on the sample. After 15 min, the cellular structure has almost completely dissolved into the liquid (Figure 4.8*e*). Setting the Peltier stage back to 15 °C results in the evaporation of

surface water from the sample stage, and the now nearly featureless, gel-like sample mass is shown in Figure 4.8f.

Although the above series of ESEM micrographs is remarkable, it should be stressed that present technology does not permit full control of the sample environment at intermediate moisture levels. The sample experiences both pressure and temperature excursions from insertion and initial pump-down of the sample chamber. Controlled environment transfer systems and more attention to attainment of variable sample chamber relative humidity will hopefully be available in the near future, not only for water but for other vapors as well. A further concern is electron beam damage, since water molecules dissociate into free radicals when irradiated with an electron beam. These radicals, as well as the primary beam, can cause significant changes in the specimen and may significantly alter the dynamics of various in situ experiments.

Environmental SEM clearly offers polymer surface science new opportunities to explore the dynamics of many surface phenomena, since they occur under "natural" conditions and will be an important tool for in situ surface characterization using both reactive gases and liquids.

4.5 Transmission Electron Microscopy

SEM is an excellent tool for providing information on surface topography and on lateral or depth concentration gradients at polymer surfaces, but TEM has greater resolution capabilities for the determination of the surface morphology of polymers. Although the illumination system in a TEM is similar to that in an SEM, the accelerating voltage is typically much higher (80–400 kV), so that the high-energy incident electrons pass completely through the thin polymer sample. TEM images are therefore two-dimensional projections along the electron beam direction. In order to infer the three-dimensional structure in the film, one must perform careful tilting to obtain a series of related two-dimensional images. Details on the theory and practice of TEM may be found in References 1, 10, and 16. A brief description of TEM contrast mechanisms relevant to polymer studies is presented here.

After the incident electrons have interacted with the thin specimen film, the amplitude and phase of their wave functions contain information about the specimen. Use of the phase portion of the electron wave function allows one to produce phase contrast with resultant changes in image intensity. More commonly, the square of the amplitude of the wave function is utilized to produce amplitude contrast. Examples of amplitude contrast include mass–thickness contrast (amorphous specimen) and diffraction contrast (semicrystalline specimen).

In amplitude contrast, image contrast arises from scattering of the electrons outside of the objective aperture. Regions of higher density, increased thickness, or crystals at the Bragg condition will scatter more and, hence, appear dark in a bright-field image.[10] In most polymer specimens there is only weak mass–thickness contrast and, hence, preparation techniques such as selective heavy-atom staining and metal decoration or shadowing are necessary to improve the contrast. It must be noted that these preparation procedures will, however, place a limit on the attainable resolution in the images. As with SEM, one must also be aware of electron beam radiation damage, which can rapidly alter the morphology of the sample during observation.[11]

In order for the incident electrons to pass through the polymer sample, the specimen thickness must be on the order of 200 nm or less, depending upon factors such as the energy of the incident electron beam and the density of the polymeric material. To prepare such thin films, one can use a number of techniques. Casting or drawing of films from dilute solutions or melts can be performed.[1] Alternatively, preparation of thin films on holey carbon grids by blotting techniques may also be used.[30] If the surface or interfacial morphology, rather than the thin-film morphology, is of interest, preparation of thin cross sections by ultramicrotomy techniques or preparation of surface replicas is necessary. Details concerning these sample preparation procedures may be found in Reference 1.

Surface replicas provide mass–thickness contrast of the replicated surface topography by employing low-angle shadowing of the replica surface with an evaporated metal. Figure 4.9 shows a bright-field image of a replica of a half-integer disclination in a semicrystalline TLCP. The pattern shows the change in the lamellar orientation about the singular core of the defect. The surface replica was prepared by

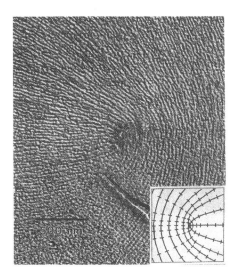

Figure 4.9 **A plus one-half disclination of a rigid polymer which has more splay at the core. Because the lamellae are perpendicular to the director, the lamellae appear to bend more at the core. The disclination center is marked with an arrow. Inset is a sketch of a disclination having more splay at the core. The director field is shown by the solid lines. The lamellae, perpendicular to the director, are drawn as hash marks. (Photo credit: S. Hudson.)**

briefly etching the surface of the polymer to produce relief via preferential attack of the noncrystalline, interlamellar regions by etchant. Approximately 50 Å of Pt/C was deposited onto the surface at a low (15°) angle, followed by a 90° coating of several hundred angstroms of carbon to provide durability for the replica. Next, a concentrated solution of polyacrylic acid (PAA) was applied and allowed to harden upon evaporation of the solvent (water). The thick layer of PAA enables stripping of the Pt/C and C layers from the TLCP material. The PAA layer is then dissolved in water and the replica mounted on a fine mesh copper grid for viewing in TEM. Regions with Pt/C exhibit strong mass–thickness contrast compared to the shadowed (uncoated) regions. The image appears to the eye as though illuminated from the deposition direction, where the print contrast is reversed.

Another type of amplitude contrast can occur for crystalline polymer specimens. Crystals appear dark when oriented so as to excite strong Bragg scattering of the electrons. An example of diffraction contrast in a bright-field image is seen in Figure 4.10. Here, an oriented semicrystalline film of polybutene-1 has been partially coated by evaporated tin. As the inserted electron diffraction pattern indicates, both polymer and tin share a common axis of orientation.[31] Indexation of the pattern allows one to know the relative crystallographic orientation of each type of crystalline lattice. Epitaxy, where the underlying crystalline substrate causes a specific orientation and location of the crystalline overlayer, is of increasing interest in the microelectronics field, where dissimilar materials often come into contact. The detailed structure of metal-on-polymer (MOP) and polymer-on-metal (POM)

Figure 4.10   (a) Bright-field image of Sn crystals deposited onto a uniaxial oriented polybutene-1 film. Deposition was carried out at 23 °C at a rate of 2 nm/s and the total Sn deposited was approximately 50 nm. (b) Selected area electron diffraction pattern of the Sn–PB-1 film shown in (a). Both the Sn and polymer reflections are highly oriented, demonstrating a type of epitaxy in which the two materials share a common axis of orientation. (Photo credit: J. Reffner.)

interfaces are crucial to the performance of solid-state devices. The combination of bright-field, dark-field and electron diffraction with a goniometer stage to facilitate systematic tilt of the crystals with respect to the incident beam direction is a powerful means for determining the structure of MOP and POM interfaces.[32]

The influence of two interfacial constraints in close proximity, that is, a thin-film constraint, on the morphology of block copolymers may also be investigated with TEM by the preparation of thin films. Such films are readily prepared by dilute solution casting or spin-coating onto a variety of substrates. The microdomain morphology may then be viewed in projection along the film thickness direction or alternatively in cross section (normal to the film thickness direction) using ultramicrotomy techniques.

In the block copolymer examples shown, the enhancement of mass–thickness contrast with OsO_4 staining has been employed. In poly (styrene-b-butadiene) (SB) or poly(styrene-b-isoprene) (SI) diblock copolymers, OSO_4 reacts preferentially with the double bonds present in the PB or poly(isoprene) (PI) domains. Because of the much higher atomic number for osmium, now present preferentially in the PB or PI phase, than for carbon or hydrogen, the PB or PI phase will scatter more electrons at larger scattering angles than the PS phase. Hence, in a bright-field TEM micrograph, the Os stained phase in a styrene-diene diblock copolymer will appear dark. It should, however, be noted that the OsO_4 staining process can perturb the dimensions, although not the shapes, of the microdomain morphologies.[33]

Figure 4.11a shows a TEM image of the projection along the film thickness direction in a thin film of a lamellar forming symmetric SB diblock copolymer prepared by dilute solution casting onto a carbon substrate.[34] Alternating dark (preferentially OsO_4 stained PB) and light (PS) lamellar microdomains can be seen in the top portion of the TEM micrograph and are associated with the case in which the lamellar normal lies in the plane of the film. An area with uniform intensity in

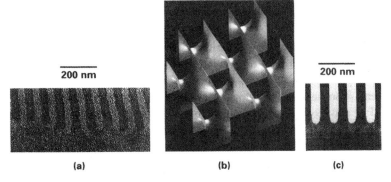

(a)   (b)   (c)

Figure 4.11 (a) Bright-field TEM image of the lamellar microdomain morphology of a PS/PB block copolymer. (b) Computer-generated image of Scherk's first minimal surface (by J. T. Hoffman). (c) Computer-simulated protection of a structure whose PS/PB interface is (a).

projection, seen in the bottom half of the micrograph, is indicative of the case in which the lamellar normal is parallel to the film thickness and electron beam direction. Besides the microdomain morphology, size, and interfacial orientation, the TEM micrograph provides information on the grain boundary structure in block copolymers. The solution to the problem of joining two sets of evenly spaced planes that meet orthogonally is the minimal surface, seen in Figure 4.11*b*, which is known as Scherk's First Surface.[35, 36] When seen in projection along one of the lamellar normals (see Figure 4.11*c*) the projection simulation of this surface bears a striking similarity to the grain boundary in the TEM micrograph seen in Figure 4.11*a*. The boundary interface therefore approximates Scherk's minimal surface and affords optimal PB and PS phase continuity while minimizing unfavorable PB–PS phase contact.

For one to investigate the upper and lower near-surface regions of a thin film, cross-sectional ultramicrotomy is necessary. This is done by embedding the thin film in an epoxy matrix and sectioning in a direction normal to the thin-film surface. In order to increase the yield of useful data, one can stack a series of thin-film samples with thin layers of epoxy between adjacent films. A cross-sectional TEM micrograph of a symmetric SB diblock copolymer thin film spin-cast from a dilute toluene solution onto a carbon-coated glass slide or a sodium chloride substrate can be seen in Figure 4.12.[24] In this case, cross-sectional TEM provides direct information on the thin-film morphology and the morphology at the polymer–air and polymer–sodium chloride interfaces. Although interfacial segregation of the lower critical surface tension PB (dark) phase is noted at both the upper and lower surfaces after annealing for 14 days, a chaotic bicontinuous microdomain morphology rather than the expected two-dimensionally continuous alternating lamellar microdomain morphology is found in the film interior. In this example, TEM also provides information on how the solvent evaporation rate controls the micro-domain morphology in thin films of symmetric SB diblock copolymers.

The microphase separation of block copolymers results in internal interface structures in the bulk polymer. TEM affords the possibility of examining the microdomain morphology of ABC triblock copolymers by viewing the structure along directions of high symmetry and corresponding simple projection.[24] Figures 4.13*a* and Figures 4.13*b* demonstrate the use of OsO_4 and methyl-iodide (CH_3I) to selectively stain the PI and poly(2-vinylpyridine) (P2VP) phases, respectively, in ABC triblock copolymers of poly(styrene-*b*-isoprene-*b*-2-vinylpyridine) (SI2VP). Sequential staining with OsO_4 and CH_3I and the location of the preferentially stained PI block in the SI2VP triblock copolymer allows one to deduce the identity of the PI (darkest phase) as the phase surrounding the circular P2VP phase (intermediate contrast phase) in the matrix of the PS phase (lightest phase). Based on the hexagonally packed P2VP surrounded by PI microdomains seen in the projection in Figure 4.13*a* and the layered projection seen in Figure 4.13*b*, a hexagonally packed concentric P2VP and PI cylindrical microdomain morphology is proposed as shown schematically in Figure 4.13*c*. An interesting feature of the proposed morphology is the perturbation

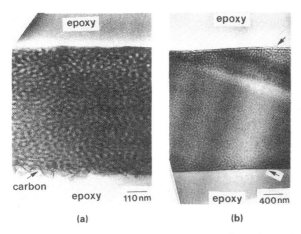

Figure 4.12   Cross-section TEM micrographs of thin films of an SB 40/40 diblock copolymer sample: (*a*) an unannealed film spin-cast onto a carbon-coated sodium chloride substrate; (*b*) a thin film spin-cast onto a glass slide substrate and annealed for 14 days at 115 °C. Continuous PB interfacial layers for the annealed film only are marked by arrows in (*b*).

of the PI–PS interface from a constant mean curvature cylindrical surface to what appears as a rounded hexagonal prism interface.[37]

4.6 Scanning Probe Microscopy

Newly developed SPM techniques are being actively applied in the investigation of polymer surfaces. Because of the extreme sensitivity of SPM to structural variations in the outermost surface, data obtained from these techniques are having a large impact on the fundamental understanding of polymer surfaces. STM and AFM have optimum resolutions of approximately 0.1 nm lateral and better than 0.1 nm depth.[38]

STM was pioneered in the laboratory of Binnig and Rohrer at IBM-Zurich in 1982.[39] The physical principle of STM relies on the tunneling current observed between two closely spaced electrodes when a voltage is applied. A conducting specimen surface and a conducting sharp probe (tip) can act as these two electrodes and a small tunneling current, which varies exponentially with the specimen surface–tip distance, can be created by applying a suitable voltage. It is this extremely strong dependence of the signal with distance that yields the subatomic height discrimination. A piezoelectric drive allows x, y, and z translations of the tip (or, in some models, of the sample) and, therefore, atomic resolution in the lateral directions. Usually, the tip is scanned across the sample and the height is varied to keep the tunneling current constant (Constant Current Mode). The vertical, z, displacement

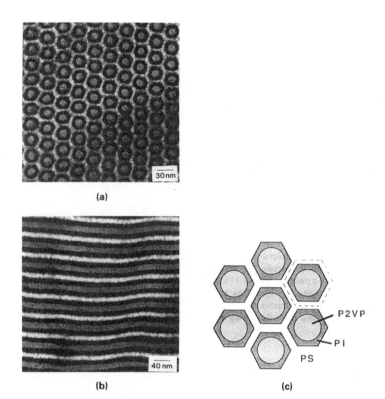

(a)

(b)

(c)

P2VP

PI

PS

Figure 4.13 TEM micrographs of the bulk microdomain morphology in an SI2VP 15/13/15 triblock copolymer sample stained with OsO_4 for 4 h and CH_3I for 12 h, showing: (*a*) hexagonally packed gray and dark concentric circular microdomains in a light matrix, for an axial (001) projection and (*b*) layers of light, dark, and gray microdomains, for a transverse (100) projection. (*c*) Schematic showing the two-dimensional (001) projection of the Wigner–Seitz cell for the concentric P2VP and PI cylindrical microdomain morphology in the SI2VP 15/13/15 triblock copolymer sample. Six diametrically opposed, flattened sides of the PI–PS IMDS, within the hexagonal projection of the Wigner–Seitz cell, are seen in the schematic.

of the tip to maintain a constant tunneling current yields an x–y–z map of the surface. In STM, the tunneling current is ideally between one atom on the tip and one on the sample surface. Therefore, the image is due to the local surface density of states rather than to simple surface topography. A change in the bias voltage will also result in a change in the contrast of the image. Reproducibility of the data, especially with respect to scan direction, and critical comparison of the data with Fourier filtered data should always be done to verify features evident in the processed image. A linear gray scale is used to relate the vertical displacement to the intensity.

STM may be performed in an ultra-high vacuum (UHV) or with the tip immersed in a fluid and hence can follow chemical and physical processes of the sample surface in situ. The main limitation of STM is its requirement of a conductive substrate for tunneling. Therefore, most polymers must have a thin (less than 10 nm), uniform layer of conductive material (usual choices are Pt/C or Au/Pd) vapor deposited onto them. This metal coating broadens the surface features and severely limits the interpretable resolution to the grain size (~3–5 nm). For this reason, investigations have also been attempted using STM on polymer surfaces without a conductive coating. Single-crystal graphite substrates have been used, and images of individual molecules and absorbed monolayers have been collected.[40] It is not clear why atomic resolution features are sometimes visible in STM on 10-nm-thick insulating lamellar polymer crystals mounted on graphite substrates, since the height image is the modulation of the tunneling current from the graphite substrate by the monolayer to the tungsten tip. A novel application is to use STM for nanofabrication: the placement of material at the nanoscale level using the tip to move single atoms into desired patterns.[41]

Binnig et al.[42] were also the first to develop the technique of AFM. Rather than employing a tunneling current, this technique utilizes the atomic forces of attraction or repulsion between a sharp tip and the sample surface. No sample preparation besides simple mounting on a stub is necessary. The insulating tip, usually Si_3N_4, is mounted on a cantilevered beam, and the deflection of the tip is monitored by optical methods as the sample is rastered underneath the fixed tip by a piezoelectric crystal. Contrast in AFM images is therefore related to the relative height of the external surface. The AFM may either be operated in the deflection mode, where the cantilever deflects according to the force, or the constant-force mode (typically 10^{-9} N), where the height of the tip is adjusted to maintain a fixed, either attractive or repulsive, force. The z resolution is comparable to that in STM.

In AFM studies of relatively soft materials such as polymers, the magnitude of the force needs to be carefully considered, since deformation of the sample is quite possible because the force is concentrated over an extremely small area. The repulsive-force mode of imaging, which has higher resolution capabilities, can cause elastic or plastic deformation of the specimen.[43] Finally, if the experiment is performed in air, care must be taken since the sample surface–tip force interaction can be perturbed by an adsorbed layer of water. In this case, the force associated with the surface tension of the water molecules will influence the force measurement. An estimate of the radius of contact a of a spherical tip and a flat substrate is given by a theoretical treatment of Johnson, Kendall, and Roberts[44] as

$$a^3 = \frac{3R}{4E}\left\{ W + 6\pi\Delta\gamma R + \left[12\pi\Delta\gamma RW + (6\pi\Delta\gamma R)^2 \right]^{1/2} \right\}$$

where E is Young's modulus, R is the tip radius, W is the load, and $\Delta\gamma$ is the difference in surface energy per unit area between tip and sample. Unertl and Jin[45] estimate that for polyimide ($E \approx 3$ GPa) imaged with a Si_3N_4 tip of $R = 40$ nm, the lateral resolution in AFM should be about 16 nm, comparable to the smallest features resolved in their images.

Recent papers have claimed molecular resolution of the parallel chains in extended chain polyethylene fracture surfaces.[46, 47] Stocker et al.[47] also investigated the surface structure of lamellar crystals of a number of paraffinic compounds and have demonstrated AFM imaging of the methyl groups at the upper crystal surface in normal alkanes [C_{32}, C_{36}]. It remains to be seen whether the fold surface of polymer single-crystal lamellae will be imaged with high fidelity and under what crystallization conditions. Although the recognition of order and quantification of spacings and height variations is possible, it is problematic to interpret disordered surface regions at atomic resolution. It is likely that researchers will need to extend and develop additional appropriate descriptors (e.g., directionally and scale-dependent surface roughness) to characterize noncrystalline polymer surfaces adequately.

Many investigations using AFM (and STM) display atomic resolution in the data, but reliable atomic resolution of polymer surfaces using AFM is much more difficult to achieve. Fortunately, there is a wealth of interesting structural detail at resolutions of a factor of 10× or more. Figure 4.4*b* is an AFM image of the inversion wall structure of the liquid crystalline polymer previously imaged by LVHRSEM (Figure 4.4*a*). The three-dimensional nature of the π inversion wall is immediately apparent, and a cross-sectional scan shows the depth of the valley region to be approximately 100 nm. The combination of LVHRSEM and AFM imaging is a powerful characterization approach for polymer surfaces.

Another example of AFM imaging of polymer surfaces, shown in Figure 4.14, is of a lamellar forming SB diblock copolymer.[23, 48] Typical heights of the one-dimensional external surface corrugations varied from 1.5 to 4.0 nm. These values are somewhat lower than the typical 6–12 nm values determined for the same block copolymer by cross-sectional TEM.[24] The differences may be due to a coupling of the lamellar tilt angle at the surface and the corrugation height or due to the convolutional smearing of the surface profile because of the relatively large radius

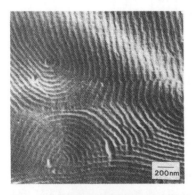

200 nm

Figure 4.14 AFM image of the surface off an OsO$_4$ stained SB 40/40 sample, showing one-dimensional corrugation of the external surface. The black regions correspond to the peaks; the white regions represent the valleys.

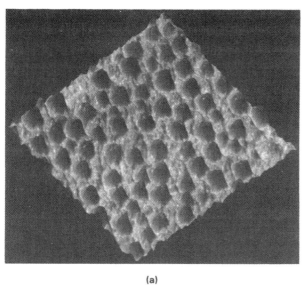

(a)

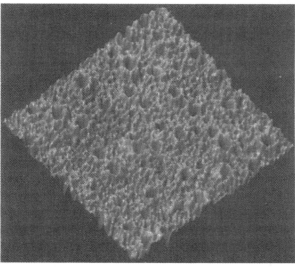

(b)

Figure 4.15   (a) 4000 × 4000 nm AFM image of PMDA-ODA polyimide cured to 400 °C. The polyimide was made from a blend of 20% precursor 1 and 80% precursor 2. The height of the hills formed from the imidization of precursor 1 that are pushed-up on subsequent imidization of precursor 2 are ~5.0 nm. The base diameter of the hills is 360 nm. (b) x–y range as (a) for PMDA-ODA cured under identical conditions as (a), but with precursor composition 80% precursor 1 and 20% precursor 2. The depth of the holes formed by shrinkage of the soft-inclusions due to the imidization of precursor 2 is ~11 nm and the lateral dimension is ~270 nm.

of curvature (~60 nm) of the silicon–nitride tip, which suggests that the tip may not probe the entire corrugation depth.

An additional feature of AFM studies of block copolymers is the ability to examine unstained samples. Since staining is known to modify characteristics of the micro-domain morphology, such as the lamellar period, it is necessary to ensure that the surface corrugation observed by cross-sectional TEM and AFM is not an artifact of the OsO_4 staining process. AFM images of both stained and unstained SB samples exhibited surface corrugation, indicating that this morphological feature is indeed real. The information provided by AFM is therefore seen to complement that pro-vided by LVHRSEM and cross-sectional TEM on SB diblock copolymer samples.

Control of surface roughness is of major interest for both microelectronics and aero-space applications[49]. Roughening of polyimide (PI) surfaces at the nanoscale level to improve adhesion has been accomplished by plasma and ion sputtering techniques. However, these techniques are difficult to control and also alter the chemical nature of the surface.[50] Recently, Saraf et al.[51] have developed a novel method to produce and control the nanoscale surface roughness by blending difference precursors of the same PI to induce spontaneous roughening of the surface subsequent to phase separation, via the different cure kinetics and shrinkage characteristics of the respective procursive poly-mers. AFM shows that this approach leads to a variety of surface features, ranging from hills to holes (see Figure 4.15). The number density is controlled by the composition of the precursor blend and by the spin casting and curing schedule. Up to 300% enhance-ment for PI/Cu adhesion compared to existing methods has been demonstrated.[51]

4.7 New Microscopies

Extension of the early SPM techniques based on tunneling current and atomic force interactions between the tip and surface has resulted in a new generation of imaging devices which rely upon other interactions between the tip and specimen surface. These new SPM techniques, known as the "Children of STM,"[52] currently include friction force microscopy, magnetic force microscopy, electrostatic force microscopy, and attractive mode force microscopy. Further information concerning the type of sample surface-tip interactions and the applicability of these new SPM techniques can be found in Reference 52. The development of these SPM techniques, especially in conjunction with special controlled sample environments, will continue to pro-vide many new opportunities for the characterization of the morphology of polymer surfaces, interfaces, and thin films.

Acknowledgments

The work reported by our group in this chapter was supported by grants from the National Science Foundation, Polymers Program, and the Air Force Office of Scientific Research.

References

1   L. C. Sawyer and D. T. Grubb. *Polymer Microscopy*. Chapman and Hall, New York, 1987.

2   *Applied Polymer Light Microscopy*. (D. A. Hemsley, Ed.) Elsevier Applied Science, New York, 1989.

3   D. Ausserre, D. Chatenay, G. Coulon, and B. Collin. "Growth of Two-Dimensional Domains in Copolymer Thin Films." *J. Phys. France*. **51**, 2571–2580, 1990.

4   G. Coulon, T. P. Russell, V. R. Deline, and P. F. Green. "Surface-Induced Orientation of Symmetric, Diblock Copolymers: A Secondary Ion Mass Spectrometry Study." *Macromolecules*. **22**, 2581–2589, 1989.

5   S. H. Anastasiadis, T. P. Russell, S. K. Satija, and C. F. Majkrzak. "The Morphology of Symmetric Diblock Copolymers as Revealed by Neutron Reflectivity." *J. Chem. Phys*. **92**, 5677–5691, 1990.

6   E. L. Thomas and B. A. Wood. "Mesophase Texture and Defects in Thermotropic Liquid Crystalline Polymers." *Faraday Discussion Chem*. **79**, 229–240, 1985.

7   D. K. Ding and E. Thomas. "Structures of Point Integer Disclinations and Their Annihilation Behavior in Thermotropic Liquid Crystal Polyesters." *Mol. Cryst. and Liquid Cryst*. Submitted 1992.

8   E. Betzig, M. Isaacson, and A. Lewis. "Collection Mode Near-Field Optical Microscopy." *Appl. Phys. Lett*. **51**, 2088–2090, 1987.

9   A. Boyde. "Confocal Optical Microscopy." In *Modern Microscopies*. Plenum, New York, 1990.

10 E. L. Thomas. "Electron Microscopy." In *Encyclopedia of Polymer Science and Engineering*, Vol. 5. Wiley, New York, 1986.

11 D. T. Grubb. "Review: Radiation Damage and Electron Microscopy of Organic Polymers." *J. Mat. Sci*. **9**, 1715–1736, 1974.

12 J. R. White and E. L. Thomas. "Advances in SEM of Polymers." *Rubb. Chem. Tech. Reviews*. **57**, 457–506, 1984.

13 J. W. S. Hearle, J. T. Sparrow, and P. M. Cross. *The Use of the Scanning Electron Microscope*. Pergamon, Oxford, 1972.

14 O. C. Wells. *Scanning Electron Microscopy*. McGraw-Hill, New York, 1974.

15 *Practical Scanning Electron Microscopy*. (J. I. Goldstein and H. Yakowitz, Eds.) Plenum, New York, 1975.

16 I. M. Watt. *The Principles and Practice of Electron Microscopy*. Cambridge University Press, New York, 1985.

17   L. Engel, H. Klingele, G. W. Ehrenstien, and H. Schafer. *Rasterelektronenmikrosckopische Untersuchungen von Kunststoffschaden.* Carl Hanser Verlag, Munich, 1978 (English ed.: *An Atlas of Polymer Damage.* [M. S. Welling, trans.] Wolfe Publishing, London, 1981).

18   O. L. Carter, A. T. Schindler, and E. E. Wormster. "Scanning Electron Microscopy for Evaluation of Paint Film Weatherability." *Appl. Polym. Symp.* **23**, 13, 1974.

19   L. H. Princen, F. L. Baker, and J. A. Stolp, "Monitoring Coatings Performance upon Exterior Exposure." *Appl. Poly. Symp.* **23**, 27, 1974.

20   T. Nagatani, M. Sato, and M. Osumi. "Development of an Ultra-High-Resolution Low Voltage (LV) SEM with an Optimized 'In-Lens' Design." *Proceedings.* 12th Internat'l Cong, for Electron Microscopy. 388–389, 1990.

21   J. B. Pawley. "LVSEM: Past, Present and Future." *Proceedings.* 12th Internat'l Cong, for Electron Microscopy, 364–365, 1990.

22   S. D. Hudson, D. L. Vezie, and E. L. Thomas. "Director Textures in Bulk Samples of Liquid-Crystal Polymers." *Makromol. Chem., Rapid Commun.* 11, 657–662, 1990.

23   D. W. Schwark, D. L. Vezie, J. R. Reffner, E. L. Thomas, and B. K. Annis. "Characterization of the Surface Morphology of Diblock Copolymers Via Low-Voltage, High-Resolution Scanning Electron Microscopy and Atomic Force Microscopy." *J. Mat. Sci. Let.* **11**, 352–355, 1992.

24   D. W. Schwark. "Influence of Interfacial Constraints on the Microdomain Morphology of Block Copolymers." Ph.D. dissertation, University of Massachusetts, Amherst, 1992.

25   G. D. Danilatos. "Foundations of Environmental Scanning Electron Microscopy." *Adv. in Electronics and Electron Physics.* **71**, 109, 1988.

26   G. D. Danilatos. "Design and Construction of an Atmospheric or Environmental SEM, Part 1." *Scanning.* **4**, 9, 1981.

27   G. D. Danilatos. "Design and Construction of an Atmospheric or Environmental SEM, Part 3." *Scanning.* **7**, 26, 1985.

28   G. D. Danilatos. "Design and Construction of an Atmospheric or Environmental SEM, Part 4." *Scanning.* **12**, 23, 1990.

29   G. D. Danilatos. "Review and Outline of Environmental Scanning Electron Microscopy at Present." *J. Microscopy.* **162**, 391, 1991.

30   J. R. Bellare, H. T. Davis, L. E. Scriven, and Y. Talmon. "Controlled Environment Vitrification System: An Improved Sample Preparation Technique." *J. Electron Microsc. Technique.* **10**, 87–111, 1988.

31   J. Reffner. "The Influence of Surfaces on Structure Formation: 1. Artificial Epitaxy of Metals on Polymers. 2. Phase Separation of Block Copolymers

and Polymer Blends under Nonplanar Surface Constraints." Ph.D. dissertation, University of Massachusetts, Amherst, 1992.

32 "Interfaces Between Polymers, Metals and Ceramics." *Proceedings*. (B. M. DeKoven, A. J. Gellman, and R. Rosenberg, Eds.) Materials Research Society, **153**, 1989.

33 C. V. Berney, R. E. Cohen, and F. S. Bates. "Sphere Sizes in Diblock Copolymers: Discrepancy between Electron Microscopy and Small-Angle Scattering Results." *Polymer*. **23**, 1222–1226, 1982.

34 C. S. Henkee, E. L. Thomas, and L. J. Fetters. "The Effect of Surface Constraints on the Ordering of Block Copolymer Domains." *J. Mat. Sci.* **23**, 1685–1694, 1988.

35 E. L. Thomas, D. M. Anderson, C. S. Henkee, and D. Hoffman. "Periodic Area-Minimizing Surfaces in Block Copolymers." *Nature*. **334**, 598–601, 1988.

36 E. L. Thomas, J. R. Reffner, and J. Bellare. "A Menagerie of Interface Structures in Copolymer Systems." *Colloque de Physique*. **C7**, 363–374, 1990.

37 S. Gido, D. W. Schwark, E. L. Thomas, and M. Gonçalves. "Observation of a Non-Constant Mean Curvature Interface in an ABC Triblock Copolymer." *Macromolecules Notes*. In press.

38 E. Occhiello, G. Marra, and F. Garbassi. "STM and AFM Microscopies: New Techniques for the Study of Polymer Surfaces." *Polymer News*. **14**, 198–200, 1989.

39 G. Binnig, H. Rohrer, C. Gerber, and E. Weibel. "Surface Studies by Scanning Tunneling Microscopy." *Phys. Rev. Lett.* **49**, 57–61, 1982.

40 J. P. Rabe and S. Buchholz. "Direct Observation of Molecular Structure and Dynamics at the Interface Between a Solid Wall and an Organic Solution by Scanning Tunneling Microscopy." *Phys. Rev. Lett.* **66**, 2096, 1991.

41 C. F. Quate. *Nature*. **352**, 571, 1991.

42 G. Binnig, C. F. Quate, and C. Gerber. "Atomic Force Microscope." *Phys. Rev. Lett.* **56**, 930–933, 1986.

43 W. N. Unetl and X. Jin. "Thin Films: Stresses and Mechanical Properties III." *Proceeding*. (W. D. Nix, J. C. Bravman, E. Artz, and L. B. Freund, Eds.) Materials Research Society, **239**, 1992.

44 K. Johnson, K. Kendall, and A. Roberts. "Surface Energy and the Contact of Elastic Solids." *Proc. Roy. Soc. (London)*. **A324**, 301, 1971.

45 X. Jin and W. N. Unertl. "Submicrometer Modification of Polymer Surfaces with a Surface Force Microscope." *Appl. Phys. Lett.* **61** (6), 657–659, 1992.

46 B. K. Annis, D. W. Noid, B. G. Sumpter, J. R. Reffner, and B. Wunderlich. "Application of Atomic Force Microscopy (AFM) to a Block Copolymer and

an Extended Chain Polyethylene." *Makromol. Chem., Rapid Commun.* **13**, 169–172, 1992.

47 W. Stocker, G. Bar, M. Kunz, M. Möller, S. Magonov, and H. Cantow. "Atomic Force Microscopy on Polymers and Polymer Related Compounds, 2: Monocrystals of Normal and Cyclic Alkanes $C_{33}H_{68}$, $C_{36}H_{79}$, $(CH_2)_{48}$, $(CH_2)_{72}$." *Polymer Bull* **26**, 2, 215, 1991.

48 B. K. Annis, D. W. Schwark, J. R. Reffner, E. L. Thomas, and B. Wunderlich. "Determination of Surface Morphology of Diblock Copolymers of Styrene and Butadiene by Atomic Force Microscopy." *Makromol. Chem.* **193**, 2589–2604, 1992.

49 *Polyimides.* (D. Wilson, H. Stenzenberger, P. Hergenrother, Eds.) Blackie and Sons, Glasgow, 1990.

50 R. C. White and P. Ho. In *Polyimides: Materials, Chemistry and Characterization.* (C. Ferger, M. Khojasteh, and J. McGrath, Eds.) Elsevier Applied Science, New York, 1989.

51 R. Saraf, T. Derderian, and J. Rolden. Unpublished results. IBM, 1992.

52 R. Pool. "The Children of the STM." *Science.* **247**, 634–636, 1990

5

Structure and Morphology of Interfaces and Thin Films by Scattering Techniques

RAVI F. SARAF

Contents

5.1 Introduction

The characterization of the physical structure and morphology of polymer thin films and interfacial structures is crucial to the understanding and, subsequently, to the tailoring and optimization of their adhesive strength, interfacial stresses, thermal expansivity, optical and dielectric properties, gas permeability, tribological behavior, friction and wear-tear nature, etc. In this chapter, "structure" refers to the physical structure of the material (unless otherwise specified). The measurement of the structural parameters of the interface or thin films—such as molecular orientation, optical anisotropy, density, extent of order (i.e., percent crystallinity), and surface roughness— may also serve as a quality control method during manufacturing.

The length scale of interest for the structural characterization of a surface or interface and thin film depends on the application and industry. Consider surface topography (i.e., roughness) as an example: The characteristic lateral length scale (parallel to the surface) of interest for measuring surface roughness to improve adhesion via interface mechanical interlocking is 0.1–1 µm; to reduce surface scatter in polymer wave guides it is <0.1 µm; (comparable to the wavelength of the optical signal); and to improve lubrication characteristics of thin organic films for magnetic disk applications it is <100 nm. Similarly, the characteristic thickness of a thin film and interfacial layer will depend on the industrial application. Phospholipid membranes for biomedical applications have thicknesses of <100 nm. Thin-film dielectrics for multilayer packaging in microelectronics range from 100 to 0.5 μm in thickness. The thickness of thin-film coatings for magnetic disk applications to improve tribology is 10–100 nm.

In the next section, the general structural features of interfaces and thin films are briefly described to provide a motivation for structure and morphology characterization. Although thin films such as coatings are not discussed, their interfacial and size effects (where thickness is comparable to the molecular size) may be measured by methods similar to those for Langmuir–Blodgett (L–B) films. The section does not review in detail structural properties, but touches on certain aspects of the subject to give the reader an appreciation of this rapidly developing area. The selected references are intended to help the reader study the subject of interest in more detail. The techniques described in the following section use quantitative measurements to characterize structural properties such as roughness, chain orientation, density distribution, and three-dimensional packing. Surfaces, interfaces, and thin films are probed by scattering methods using these techniques. Some illustrative examples are given and the limitations of these methods are discussed. The complementary set of microscopy techniques is described in Chapter 4.

5.2 Industrial Applications

Adhesion

The structural implication of the interfacial phenomena involving polymers as a substrate (or adherend) or adherate or adhesive is important in a vast range of industries, such as aerospace, automotive, microelectronics, packaging, and textile.[1] In particular, the surface roughness of the adherend, the molecular orientation and chain packing of the polymer in the interfacial layer, and the surface morphology of the polymeric adherend (i.e., texture and percent crystallinity) can profoundly affect the adhesion characterization of the system. Some of the structural aspects (properties) of adhesion are considered below.

Surface roughness is important for enhancing (mechanical) adhesion between a polymer (such as isotactic polypropylene [i-PP], acrylonitrile butadiene styrene [ABS], or high-density polyethylene [HDPE]) and a metal (such as nickel, copper, or aluminum) by interfacial interlocking. The metal could be electrolessly plated on a rough polymer surface using a palladium seed,[2, 3] or the polymer may be deposited by melt (or solution) processing onto the roughened metal.[4, 5] The roughness in semicrystalline polymers such as i-PP, HDPE, and nylons is achieved by relying on the differential rate of etching of crystalline and amorphous regions.[2, 3, 6] The roughness, to enhance adhesion, is usually in the scale of 1–0.1 μm and may be quantified by scanning electron microscopy (SEM) and atomic force microscopy (AFM), which are discussed in Chapter 4. A distribution of root-mean-square (rms) roughness seems to induce better adhesion than a uniform scale roughness. The copper/epoxy peel strength improves from 680 g/cm for 0.3-μm roughness created by dendrites to 2420 g/cm for 0.3-μm dendrites on a 3-μm corrugated surface.[7] The shape of the roughness may also affect the resultant adhesion.

Chain orientation and packing in the interfacial layer is central to the mechanical properties (i.e., toughness) of the interface that contribute to the overall adhesion. In most instances, the interface is highly oriented compared to the bulk. The interface in such a morphology is referred to as a "skin." Poor adhesion can occur because of delamination between the skin and the core[8, 9] or within the highly oriented skin. The ease of crack propagation parallel to the oriented chains is demonstrated in planar-oriented i-PP (Figure 5.1). The oriented sheets of i-PP delaminate like a deck of cards when the sample is cooled to liquid N_2 temperatures and tapped at the edge with a pair of tweezers. The substrate-induced orientation occurs on curing polyimide precursors such as poly(pyromellitic dianhydride-oxydianiline) (PMDA-ODA) and poly(biphenyl dianhydride-*p*-phenylene diamine) (BPDA-PDA). The skin for PMDA-ODA is shown to be ~10 nm as measured by grazing incidence X-ray scattering (GIXS).[10] Surface orientation as a result of the processing of rigid rod polymers such as poly(*p*-phenylene benzobisoxazole) (PBO) reveals a skin of typically <200 nm with higher order and chain orientation parallel to the fiber axis compared to the bulk.[11] Skin formation is also observed in thermotropic liquid crystalline polyesters due to flow fields during molding and extrusion.[12]

Figure 5.1   **SEM micrograph of the freeze fractured edge of a uniaxialiy compressed i-PP sample compressed at 140 °C to a compression ratio of 25×. The sheetlike delamination parallel to the film surface is attributed to the planar orientation of the polymer. In planar texture, the chain axis of the crystal tends to orient parallel to the film surface and the other two axes are randomly oriented with respect to chain axis. Furthermore, the chain-axis is randomly oriented in the plane.**

The formation of skin and surface-induced texturing is also observed in flexible semicrystalline polymers and in the material they interface. The oriented polymer surface of a semicrystalline polymer such as HDPE can texture a vapor-deposited metal such as tin, bismuth, indium, and tellurium in the draw axis.[13, 14] The surface orientation of chains and the textured topography of polymers such as poly(tetrafluoroethylene) (PTFE) caused by rubbing or deformation can orient liquid crystals and other polymers such as HDPE, nylon-6, and poly(caprolactone).[15, 16] Although the size of the skin is not measured for the above-mentioned systems, the effect of surface chain orientation on the structure of the interface and consequently its effect on the practical adhesion is clearly demonstrated.

An interfacial region composed of oriented skin may not have an adverse effect in all cases, for example, in homeotropic orientation due to surface anchoring in poly(benzyl glutamate) (PBLG)[17], polymers with acid end (or side) groups as in L–B films[18], and polymers with silanol groups or coupling agents.[19] Furthermore, in many semicrystalline polymers such as HDPE, i-PP, and nylon-6, if the polymer matrix transcrystallizes on the substrate, inducing row-nucleation[20], the adhesion strength improves significantly even though the chains are oriented parallel to the substrate.[21, 23] Transcrystallinity versus spherulitic morphology at the interface is controlled by the nucleating surface (i.e., the adherend) and processing conditions (i.e., thermal treatment).[24, 25] The improvement in adhesion is attributed to the improvement in the mechanical properties of the transcrystalline layer by about 1½-fold in i-PP and twofold in HDPE.[26]

The adhesion of a polymer surface is related to its surface tension. Thus, in a semicrystalline polymer the change in crystal structure (i.e., polymorphism) and surface density may alter the adhesion strength. For example, in HDPE the surface tension can be altered from 35.7 to 66.8 ergs/cm^2 as the surface crystallinity is altered from 0 to 100%.[25, 27] Recent X-ray reflectivity measurements on polyimide indicate a variation in density at the surface (and hence the surface tension) due to an alteration in curing process.[28] X-ray reflectivity, in principle, may be used to probe thin-skin formation below the 100-nm range. Nylon-6 and metal adhesion for an α-form crystal (where the N–H group is parallel to the surface) is significantly lower compared to the interfacial layer, with predominantly a γ-form crystal (where N–H groups are normal to the surface).[29, 30] Polymorphism and chain orientation at the surface may be probed by GIXS.

Biomedical Applications

In the biomedical and biotechnological industries, interfacial phenomena involving a solid polymer surface, such as an artificial organ, and blood, are very important in the design of a surgical procedure. Biocompatibility with the foreign surface is related to protein adsorption (among other things, such as chemical reaction) on that surface.[31] The preferential rapid adsorption of the plasma (i.e., blood) protein fibrinogen, and the other plasma proteins (mainly allumin, γ-globulin, fibrinogen, and prothrombin[32]), on the artificial organ is believed to be the main reason for blood clotting.[33] Clotting, rather than the organ's malfunction, is often the reason for failure. The clotting process basically involves (in most cases) platelet aggregation at the protein-adsorbed region, resulting in the formation of a fibrin network.[33] Several techniques are used to measure protein adsorption on polymer or metal surfaces, such as ellipsometry,[34, 36] capacitance measurements on oxidized metal (i.e., dielectric) surfaces,[37] and (reversible) protein adsorption process by open cell potential method.[37, 38] The structure of the adsorbed protein (i.e., the chain orientation with respect to the surface), the concentration profile of the adsorbed molecule, the structural rearrangement in some instance due to change in bulk concentration, and the exchange processes involving adsorption and desorption of protein molecules cannot be directly measured by these methods. One possible method would be reflectivity, as discussed above for an analogous problem involving dispersant stabilization. This method can also monitor the clotting process by measuring density changes. Although the technique may be limited for biomedical applications because of beam damage that may occur in the protein molecule, semiquantitative information such as relative change in adsorption may be measured by limiting the exposure time. Furthermore, an understanding of the adsorption on various substrates will be important in designing biosensors[39, 40] and protein purification.[41] The reflectivity technique described in the next section may also have merit in probing antigen–antibody interaction.

Langmuir–Blodgett Films and Other Thin Films

In the past decade, interest in L–B films has intensified because of their potential application in industries such as microelectronics (for microlithography and solid-state devices), biomedical technology (for constructing phospholipid and other bio compatible membranes), and sensor technology (to develop contamination and gas monitors and biosensors). Since the initiation of L–B film research it was obvious that the mechanical and thermal stability of these films would be a major obstacle. The approach taken to overcome these constraints is to synthesize polymeric L–B films. The two basic approaches adopted are (1) depositing an L–B film from polymerizable monomeric amphiphiles and then polymerizing the film by photons or electrons or by condensation reaction[49, 50] and (2) using a preformed polymer to form the L–B film directly.[51, 58] Structurally, the former method produces better order because of better mobility to order prior to the polymerization step. Some of the salient structural features to establish the structure property relationship in polymeric L–B films are packing between the various mesogens (i.e., crystalline or liquidlike), the tilt angle of the mesogens with respect to the normal axis, the conformation, length, and interchain spacing between the spacers, density variations in the film along the thickness direction, and molecular scale roughness of the film. A brief description of some structural aspects of polymeric L–B films is given below.

The photopolymerization moieties for constructing L–B films by the first method (i.e., postpolymerization of precursor L–B films) contain the polymerization functional group in the flexible amphiphiles. The function groups can be one or two double bonds or diacetylenic linkage. The polymerization can be performed either on the air–water interface or after the film is supported on a solid substrate. The general goal is to construct a single- or, preferably, multilayer structure of L–B films with less or no change in the ordered packing of the amphiphiles. Some examples of this class of polymeric L–B films are mentioned below: Cadmium octadecyl fumerate forms an L–B film with hexagonal packing of the amphiphiles.[44] The packing is rearranged to a monoclinic packing on polymerization. Apart from the photopolymerization, successful polymer L–B films are formed by electron,[59] γ-radiation,[60] and condensation reaction[49] from the precursor L–B film. The former has potential electron lithographic application in the microelectronics industry. The polycondensation polymer was demonstrated for amino acid derivatives in biomedical applications.

Diacetylenic amphiphiles are attractive for constructing highly ordered polymeric films that may have long-range π-electron conjugation. Such films will have highly nonlinear optical properties that may have applications in wave-guiding. That diacetylene moieties polymerize in the solid state, maintaining high order,[61] suggests the possibility of constructing well-ordered L–B polymeric films by this route. The solid-state polymerization of L–B films made from monovalent precursors has better order in the polymer than the divalent precursor salts (except for sodium salt).[62] The difference in order is attributed to the topological constraints caused by an electrovalent bond.

Using a preformed polymer that can potentially form L–B films is a more attractive process than the former process. Polypeptides such as poly(γ-methyl L-glutemate) (PMLG) are shown to form L–B films with one to two layers.[63] A general method for making such a polymer is to introduce hydrophilic spacer groups into a water-soluble backbone such as polyethylene oxide (PEO), poly(vinyl alcohol) (PVOH), or poly(vinyl acetate) (PVAc). Polyacrylates such as poly(octadecyl acrylate) (PODA),[64] poly(octadecyl methacrylate) (PODMA),[65] and bidimensional mixtures of poly(isobutyl methacrylate) (PIBM)[52] have been shown to form monolayer and multilayer L–B films. The cross-sectional area of the chain is 21 Å^2, compared to 20 Å^2 for closed packed saturated hydrocarbon chains in an L–B film. The slightly larger area of ~22 Å^2 in PODMA is (probably) attributed to the extra methyl group. The cross-sectional area of PIBM is only 19 Å^2 in spite of the large isobutyl group. A generic idea for forming monolayer and multilayer polymeric L–B films was proposed by Ringsdorf et al.[56]

The lateral packing and ordered structure along the chain in polymeric L–B films have been probed by various techniques such as transmission electron microscopy (TEM),[66, 67] electron diffraction and small-angle X-ray scattering (SAXS),[68] and X-ray diffraction (XRD).[69] Although suitable to obtain the structure of thin L–B films, spectroscopic techniques such as surface-enhanced Raman scattering (SERS)[70] require an appropriate dye. Grazing angle Fourier transform infrared (FTIR)[71, 72] spectroscopy and attenuated total reflection (ATR)[73] spectroscopy are more suitable methods for studying chain orientation in L–B films. Apart from conventional X-ray, neutron, and electron diffraction, helium diffraction has been used to probe low molecular weight L–B films.[74, 75] Due to the low energy of the helium, the technique (easily extendable to polymeric films) is extremely surface-sensitive. It is an appropriate tool to use to obtain the atomic spacing of chain-ends on films. The optical constants and thickness of L–B films have also been measured by spectroscopic phase-modulated ellipsometry (SPME).[76, 77] Although the method is not as direct (i.e., the choice of optical constants has to be made carefully), it is relatively easier to use than the reflectivity technique, which obtains more detailed and direct information on density changes along the thickness of a polymeric L–B film.[78] Another method that has not been exploited for measuring the structure of L–B films and which may reveal information on the three-dimensional structure is GIXS (described later in this chapter).

5.3 Integrated Optics Method

Integrated optics offers several techniques[79] for launching an optical wave in a thin film or an interface to probe its (structure-related) optical properties. In this section some of these methods are discussed with reference to linear and nonlinear optical behavior. Nonlinear optics at the interface is relatively new; consequently, its application as a probe to study the polymer interface is not as fully realized compared to developments in the study of small-molecule monolayers and their interfacial behavior. This section discusses the optical wave-guiding method for studying the linear optical properties of thin films and briefly outlines two methods to probe the surface which utilize the nonlinear nature at the interface.

Wave-Guiding

Optical wave-guiding through dielectric films is a well-established method for signal transmission and a characterization tool for probing thin films. In the wave guide method, a plane wave (optical) is transmitted through a polymer film between two lower refractive index materials. Usually, the substrate is glass and the top medium is air. The wave is launched in the dielectric thin film by coupling an evanescent wave produced in the air gap between the dielectric thin film and a high-refractive-index prism. The signal is generated by two means, depending on the setup: (1) in reflection mode, which is the generally used method for polymer characterization, the signal is the reflected part of the input wave not coupled-in the organic film and (2) in transmission mode, the signal constitutes the wave couple-out of the organic film after having traveled through the film.

An advantage of this technique, in addition to its nondestructive nature, is its relatively simple setup compared to other surface and thin-film characterization tools. The method can be applied to structures similar to the actual application in terms of type of substrate and method for film processing. Since the method is environmentally flexible, it can also be used in studying temperature, humidity, and radiation effects on the optical and dielectric properties of the thin film. Furthermore, the relative short time scale (< 1 s) of the measurement makes this method potentially an on-line technique and allows it to be used as a monitoring scheme.

The method is limited to film thicknesses comparable to the wavelength or the probing light. The films are prepared carefully to avoid losses from bulk scattering due to defects such as air bubbles. The planarity and smoothness of the film are important to reduce surface scattering losses. The scattering losses may be reduced to some extent by using a single-prism coupling method (as discussed below) rather than a two-prism method. The absorption losses are reduced by choosing a probe wavelength that has high transmissivity in the polymer.

Figure 5.2 shows schematics of two methods for measuring the optical constants of an organic dielectric film by the wave-guiding method. In a full-prism coupling geometry (Figure 5.2a), the input beam is coupled to a film of thickness t by producing an evanescent wave in the air gap, d, between the prism and the polymer film. For successful coupling, 80–90% of the energy is transferred into the film. The refractive indices of the four media—substrate, film, air gap, and prism—are such that $n_3 > n_1 > n_0$, n_2. The air gap, d, is a fraction of the wavelength, λ. The thickness of film $t \geq \lambda$. The experiment is performed by changing the input angle θ'_3 (related to θ_3, via refractive index and geometry of the prism) and monitoring the reflected light. At the coupling angle or angles, the reflectivity will go through a minima. The optical constants and thickness of the film are calculated by solving Equation 5.3 for various modes, m, starting from $m = 0$ (corresponding to the lowest coupling angle, θ_3). Half-prism coupling (Figure 5.2b) can be used in reflection or transmission mode. The reflection mode is similar to the full-prism method in Figure 5.2d. In transmission mode, the transmitted signal on the receiving signal is measured. In contrast to the reflection mode, the coupling angle corresponds to a maxima in the measured signal. Since the transmission mode is more sensitive to surface and bulk scattering and adsorption, the reflection mode is more suitable for the analysis of polymer films.

For a successful coupling, the incidence angle, θ_3 (see Figure 5.2), must be such that (1) the wave superimposes constructively in the film (as is detailed in Equation 5.5) and (2) total internal reflections must occur in the prism and the film, satisfying a synchronous condition (Equation 5.2). A high refractive index material for the prism is desirable because it can give a wider range of θ_3. The former condition for constructive superposition, also referred to as the resonance condition, occurs when the total phase change in the light as it travels from x_{n-1} to x_n in the film is $2\pi m$, where the multiple $m = 0, 1, 2, \ldots$ defines the mode of the resonance. It is customary to define the modes by integer m, and the technique is referred to as m-line spectroscopy.

Let the four media—substrate, polymer film, air space, and prism—be designated by indices $i = 0, 1, 2,$ and 3, respectively. The thickness direction and lateral direction are designated as z and x, respectively, as shown in Figure 5.2. The light in the other two media $i = 0$ and 2 (i.e., substrate and air gap) is evanescent wave, with the electric field decaying exponentially because of total internal reflection in media 1 and 3. The exponential constant (i.e., rate of decay) p_i for $i = 0$ and 2 is given by

$$p_i^2 = k_{x,1}^2 - (kn_i)^2 \tag{5.1}$$

For successful coupling of light from the prism to the film, the synchronous condition requires $k_{x,3} = k_{x,1}$, that is,

$$\theta_3 = \sin^{-1}\left(\frac{n_1 \sin \theta_1}{n_3}\right) \tag{5.2}$$

Kinematically, the synchronous condition implies that the spatial phase lag (along the x-direction) in the light falling at the prism surface is identical to the spatial

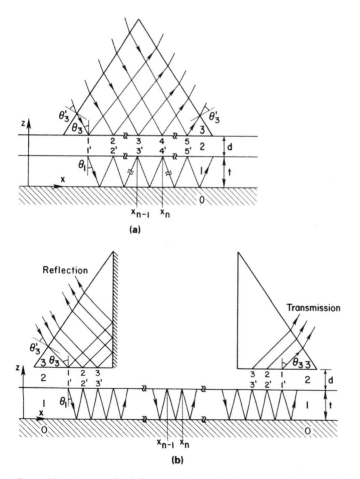

Figure 5.2 Two methods for measuring optical constants of an organic dielectric film by the wave-guiding method.

phase lag (along the x-axis) in the plane wave reflecting off the 1/2 interface in the film. Although the evanescent field is nonradiative—that is, it does not propagate in the z-direction—it couples the energy transmittance from prism to film at ~80% efficiency for uniform air gaps. It is noteworthy that the synchronous condition and Equation 5.1 imply an identical rate of decay for the evanescent wave in the air gap from both the film and prism sides.

The wave mode is excited when the total phase shift in the vertical direction of the wave after one cycle (as depicted in Figure 5.2 by lateral traverse from x_{n-1} to x_n) is $2\pi m$. The resonance condition in the vertical direction implies that

$$2tk_{z,1}(\theta_1) + \phi_{12}^m(\theta_1,\theta_3) + \phi_{10}(\theta_1) = 2\pi m \qquad (5.3)$$

where $k_{z,i} = n_{ik} \cos \theta_i$. The phase shift due to total reflection, ϕ_{ij}, is

$$\phi_{ij} = -2\tan^{-1}\left(\frac{I_{ij}^2 p_j}{k_{z,i}}\right) \qquad \text{where } I_{ij} = \begin{cases} \dfrac{n_i}{n_j} & \text{(TM)} \\ 1 & \text{(TE)} \end{cases} \qquad (5.4)$$

and

$$\phi_{12}^m = \phi_{12} + 2\sin\phi_{12}\cos\phi_{32}\, e^{-2p_2 d} \qquad (5.5)$$

The refractive index, n_1, and thickness, t, of the film can be found by solving Equation 5.3 for two modes. The other modes may be used for consistency. The dependent variables, θ_1, ϕ_{10}, and ϕ_{12}^m are calculated from Equations 5.2, 5.4, and 5.5, respectively, and the independent variable θ_3 is measured.

To envision the experiment it is necessary to see the single-prism coupling operating in the reflection mode. The extension to other coupling methods (i.e., double-prism coupler) and the transmission mode is similar. As shown in Figure 5.2a, the monochromatic light is pumped in at angle θ_3, and the reflected signal at θ_3 on the other face of the prism is monitored. As θ_3 decreases, a minima in the reflected intensity is observed at particular angles corresponding to various coupling modes starting from $m = 0$. Figure 5.3 shows a typical reflected intensity in a single-prism mode from a polyimide film on a quartz substrate. For the input beam in TM or TE mode, refractive index n_i, along the thickness or lateral direction (y-axis in the present case) can be calculated. For the study shown in Figure 5.3, the typical film thickness is ~2.2 µm and the in-plane (TE) and normal (TM) refractive indices were ~1.72 and ~1.64, indicating an extraordinarily high birefringence of ~0.08. In principle the in-plane refractive index in the third direction can be obtained by rotating the sample by 90° around the z-axis. Thus the method can measure in all three directions the refractive index of a thin polymer film.

Nonlinear Optical Method

Two classes of nonlinear optical methods are outlined in this section. The first class involves probing the thin films or interfacial layer by launching a surface wave. The

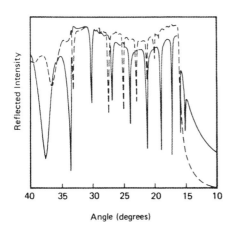

Figure 5.3   Reflectivity curve for a PMDA-ODA polyimide film on quartz using the full-prism method in reflection mode. The refractive index normal and parallel to the film surface is measured by performing the measurement in TE (——) and TM (- - - -) modes, respectively. It is noteworthy that the refractive index in the third direction (parallel to the surface) can be determined by rotating the film 90° and obtaining the reflectivity in TM mode. The angle of incidence, θ, is related to θ_3 as $\theta_3 = \alpha + \sin^{-1}[\cos(\theta + \alpha)/n_3]$, where α is the prism angle.[80] The sharp drop in intensity at ~16° corresponds to the critical angle of the prism. Since the incidence angle is defined with respect to the horizontal plane, the $m = 0$ mode is at the highest angle.

second class of techniques relies on the second harmonic generation (SHG) due to broken symmetry at the interface. The first method, using a surface wave, also probes the linear optical properties of the interface. Both classes are briefly described to give a general idea of these methods in terms of their applications to characterize polymer interfaces and thin films. Although the former method is similar to the wave-guiding technique, it has the distinct advantage of revealing the orientations of various chemical moieties of the organic film. One such method is surface coherent anti-stoke Raman scattering (CARS).[81]

Surface EM-wave techniques   The dispersion relation of a surface wave can be derived by considering a TM (i.e., $E_y = 0$) and a TE ($E_y \neq 0$) electromagnetic (EM) wave at an interface composed of two media, say a and b, such that the wave equation and the boundary conditions are satisfied in the two media and that the evanescent wave field decays exponentially normal to the interface (i.e., z-axis), as decay rates p_a and p_b similar to Equation 5.1. Let the propagation be in the x-direction. The dispersion relation from the boundary condition and wave equation for the TM wave is then given by

$$k_x^2 = \left(\frac{\omega}{c}\right)^2 \frac{\varepsilon_a \varepsilon_b}{\varepsilon_a + \varepsilon_b} \tag{5.6}$$

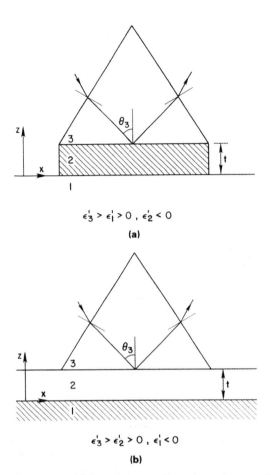

$$\epsilon'_3 > \epsilon'_1 > 0 , \ \epsilon'_2 < 0$$

(a)

$$\epsilon'_3 > \epsilon'_2 > 0 , \ \epsilon'_1 < 0$$

(b)

Figure 5.4   (a) Krestchman configuration, where medium 2 is a metallic film and medium 1 is the polymer (dielectric) sample. The sample may be polymer film deposited on the metallic film or a fluid in contact with the metallic medium. The metal is usually Ag with a typical thickness of 500 Å. (b) In Otto configuration, the sample is a polymer film between the metallic medium 1 and the prism. The experimental setup is similar to the wave-guiding method described earlier.

For positive p_a and p_b, given by Equation 5.1, $k_x^2 > k^2 \epsilon_a$ and $k^2 \epsilon_b$. To satisfy the dispersion relation in Equation 5.6, either the conditions $\epsilon_a < 0$ and $|\epsilon_a| > \epsilon_b$ or vice versa should hold. For metals below the plasma frequency, the dielectric constant is always negative. Thus, to generate surface waves, a metal–polymer interface is ideal. This also makes this method suitable for studying polymer–metal interactions, a characterization crucial to microelectronics industry. Surface EM waves on metal are often referred to as surface plasmon waves. It can further be shown that surface TE waves do not exist.

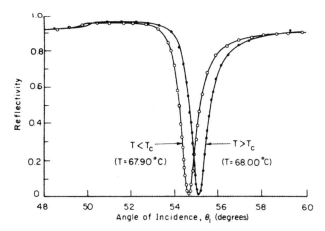

Figure 5.5 Reflectivity of a 1.75 cholesteryl chloride and 1 cholesteryl myristate mixture above and below the cholesteric-to-isotropic phase transition temperature. As described in Figure 5.2, the minima in the reflectivity is related to the refractive indices of the film and is obtained by nonlinear least-squares fitting. The shift in the minima corresponds to first-order phase transition. (Data after Reference 83).

The two geometries often used to launch the surface wave are the Kretschmann and Otto configurations, as shown in Figure 5.4. To demonstrate the principle, we consider the Kretschmann optics. As will be obvious, the equations can easily be modified to the Otto geometry. Figure 5.4a shows the general setup where media 1 and 3 are dielectric (i.e., ε_1, $\varepsilon_3 > 0$) and medium 2 is a metal film (i.e., $\varepsilon_2 < 0$) of thickness t such that the two surface waves at the 1/2 and 2/3 interfaces do not communicate. Furthermore, the surface EM in the TE mode does not exist; therefore only the reflectivity for the TE mode (also referred to as p-polarization) is considered. The reflectivity, $R^{(pp)}$, is given by Fresnel's law[82] as

$$R^{(pp)} = \left| \frac{r_{12}^{(pp)} + r_{23}^{(pp)} e^{i2k_{z,2}D}}{1 + r_{12}^{(pp)} r_{23}^{(pp)} e^{i2k_{z,2}D}} \right|^2 \tag{5.7}$$

For successful coupling, the denominator of the reflectivity in Equation 5.7 should vanish. If κ is ignored then, this condition gives the dispersion relation, which is similar to Equation 5.6. As with the wave-guiding method, a drop in reflectivity is observed at the coupling condition corresponding to the optical constants of the media (see Figure 5.5).

This method seems similar to the wave-guiding technique, but it includes some unique complementary information: (1) the method can study structural changes as a function of environment (i.e., temperature, humidity) at the polymer–metal interface; (2) because it probes the interface by surface EM, the sample is not restricted to thin films—it can even be used for liquid solutions[84]; (3) the coupling is possible even if the interfaces are rough[84]; (4) the nonlinear interaction of the surface EM wave can be used to probe the interface by using spectroscopic techniques such as SERS.

Second-order nonlinearity surface probe In the above-mentioned optical methods, the depth of probe penetration is of the order of λ. (the wavelength of the probing light). However, under special conditions, the nonlinear effects of the polymer ($\chi^{(3)}$) may be utilized to gain sensitivity at less than 100-Å levels. However, the surface-specific technique involving SHG and sum-frequency generation have an advantage over surface CARS in terms of a simpler experimental arrangement and stronger signal. Due to the latter reason and high surface specificity, surface probes based on second-order nonlinearity have sub-monolayer sensitivity.

Figure 5.6 shows the basic model for SHG: ε_1 and ε_2 are the linear dielectric constants for medium 1 and medium 2 (the sample). The SHG arises from the non-linear interfacial layer in the sample characterized by ε_2 and $\varepsilon^{(2)}$ and $\chi_s^{(2)}$ The thickness of the layer t may be much smaller than λ (the wavelength of the probing light). The method has been demonstrated to observe the orientational order parameter for sub-monolayer thicknesses of organic molecules.[86] Thus this technique is more suitable for characterizing thinner films and closer vicinities to the interface compared to the optical methods described earlier.

Briefly, the intensity of the reflected SH wave is calculated as follows[87]: The transmitted linear wave in the sample (medium 2) is given by Fresnel's law as $\underline{e}^t(\omega) = \underline{\underline{F}}\hat{p}e^i(\omega)$, where $\hat{p}$ is the unit polarization vector of the incident beam and $\underline{\underline{F}}$ is the Fresnel factor. The polarization vector, $\hat{p}$, for the TE or s-polarization is $(0 \quad 1 \quad 0)$ (i.e., $e^i = e_{\parallel}^i$) and for the TM or p-polarization is $(-\cos\theta_1 \quad 0 \quad \sin\theta_1)$, implying that $e^i = e_{\perp}^i$. The Fresnel's factor is a diagonal matrix for isotropic media.[82]

The amplitude of the nonlinear polarizability source wave is given by

$$\underline{P}^{NLS}(2\omega) = \underline{\underline{\chi}}_s^{(2)}:\underline{e}^t(\omega)\underline{e}^t(\omega)$$

Then by substituting the source wave as a forcing function in the nonlinear wave equation, and the homogeneous solution of the same equation, one determines the nonlinear transmitted wave in the interfacial layer. The reflected nonlinear wave is simply determined from the transmitted wave by invoking the appropriate boundary conditions at $z = 0$ (in Figure 5.6) to yield[88]

$$I(2\omega) = \frac{32\pi^3\omega^2\sec^2\Theta}{c^3\varepsilon_1(\omega)\,\varepsilon_1^{1/2}(\omega)}\left|\underline{e}^t(2\omega)\cdot\underline{\underline{\chi}}_s^{(2)}:\underline{e}^t(\omega)\underline{e}^t(\omega)\right|^2 I^2(\omega) \tag{5.8}$$

The term $\underline{e}^t(2\omega)$ is the wave in medium 2, given similarly as $\underline{e}^t(\omega)$, except at twice the frequency. Angle Θ is the reflection angle of the second-harmonic wave.

To understand the measurement scheme it is helpful to specify $\underline{\underline{\chi}}_s^{(2)}$, in terms of its components. Examples for the following discussion are polymer thin films, self-assembled structures such as L–B films, or surface-induced-orientated interfaces in semiflexible or rigid rod polymers where the orientation of the layer (or the film) will be cylindrically symmetric around the z-axis and planar with respect to

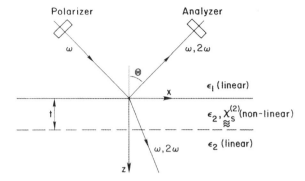

Figure 5.6 Due to the broken (translational) symmetry at the interface, the linear dielectric (polymer) sample is composed of a nonlinear interracial layer of thickness t and linear bulk medium. If polarizer and analyzer are adjusted so that the reflected SHG wave is blocked, this null condition directly provides the segment orientation at the interface given by Equation 5.12.

the x–y plane. Most often, such as in conjugated polymers and liquid crystalline polymers, the hyperpolarizability is dominated by a single axial component, $\alpha^{(2)}_{\xi\xi\xi}$. The total susceptibility of the film is given by

$$\chi^{(2)} = N_s \langle \alpha^{(2)}_{\xi\xi\xi} \rangle \tag{5.9}$$

where the average is taken over all orientations of the molecule in the film and N_s is the surface density.

The average orientation may be determined by a proper choice of the ratios of the susceptibilities as

$$A = \frac{2\chi^{(2)}_{xzx}}{\chi^{(2)}_{zzz} + 2\chi^{(2)}_{zxx}} = \frac{\langle \sin^2\theta \cos\theta \rangle}{\langle \cos\theta \rangle} \tag{5.10}$$

where θ is the angle between the ξ-axis of the molecule and the surface normal (i.e., the z-axis). Planar orientation is assumed.

The measurement of ratio A, which gives an average orientation, can be measured directly by setting the output polarizer so that the second harmonic reflection beam is blocked. If the pump beam is arranged so that $[e^t_\parallel(\omega)/e^t_\perp(\omega)]^2 = 2$, the ratio A is directly measured as

$$A = \frac{-e^t_\perp(\omega)\,e^t_\perp(2\omega)}{e^t_\parallel(\omega)\,e^t_\parallel(2\omega)} \tag{5.11}$$

The same development and measurement scheme can be applied to third-order nonlinearity.[89] This may be an attractive complementary method for two reasons:

(1) it is more prevalent in polymers and (2) it gives different moment averages of the molecular orientation, which may lead to a better understanding of the distribution function.

5.4 Reflectivity

X-ray and neutron reflectivity measure the density or concentration profile of thin films and adsorbed layers. The measurements are absolute if the bulk values are known. X-ray reflectivity depends on electron density changes along the film thickness. Since polymers are low-Z materials, air–polymer and (thin-film) polymer–substrate (high-Z) interfaces are easily accessible by X-ray reflectivity. In neutron reflectivity, contrast arises from the scattering length of the neutron. Since the absorption length for neutrons for substrates such as quartz and silicon is small, both polymer–air and polymer–substrate interfaces can be accessed. Furthermore, by an appropriate labeling of the polymer (i.e., replacing the hydrogen with deuterium), one can extend this method to probe the polymer–polymer interface. However, the requirement of deuteration of the polymer chain makes the method less versatile than X-ray reflectivity. The higher flux and better collimation of X rays from the synchrotron source also makes the resolution of X-ray reflectivity better in some instances where a larger dynamic range (more than 6–7 orders of magnitude) is required.

Sensitivity to surface or interface roughness allows the quantification of roughness at the < 2-nm scale. However, this poses a constraint on the smoothness of the sample and the substrate. Single-crystal silicon, optically polished fused quartz, and sapphire are common substrates employed. The polymer is usually deposited by spin-coating or chemical vapor deposition (CVD). The typical thickness of the deposited thin film or adsorbed layer is $5–10^2$ nm. This makes the method complementary to other profiling techniques such as forward recoil spectrometry (FRS) and secondary ion mass spectroscopy (SIMS), discussed in Chapter 10, with respect to its better spatial resolution but smaller dynamic range.

Next, a theory of reflectivity is outlined to give an appreciation of the quantitative information obtainable by using reflectivity techniques. The refractive index of a material, i, is slightly less than unity for X rays and cold neutrons, and is given by

$$n_i = 1 - \delta_i + i\beta_i \qquad (5.12)$$

where the real part, δ_i, is ~10^{-6} and the imaginary part related to the absorption, β_i, is usually negligible. The real part, δ_i, is related to the wavelength of the probe beam (in a vacuum), atomic weight, Z, and in the case of neutrons, the scattering length of all the elements. The critical angle with respect to a vacuum, $\theta_{c,i}$, is given by $(2\delta_i)^{1/2}$, which is about a milliradian, implying that the incident beam is at glancing angles. The upper-bound angle to measure reflectivity is constrained by two criteria: (1) the scattering from the structure of the sample (i.e., Bragg scattering, amorphous halo) and (2) the intensity of the source, since the intensity drops by 7–9 orders as the incident and reflected angle, θ, increases. The incident beam medium is labeled m and the polymer thin film (or interface) is labeled f. The specular reflectivity (i.e., rocking angle, $\omega = 0$) is given by Fresnel's law as[82]

$$R_F = \left| \frac{\theta - [\theta^2 - \theta_c^2 - i(\beta_f - \beta_m)]^{1/2}}{\theta + [\theta^2 - \theta_c^2 - i(\beta_f - \beta_m)]^{1/2}} \right|^2 \qquad (5.13a)$$

where the critical angle θ_c is given by

$$\theta_c = \sqrt{2(\delta_f - \delta_m)} \qquad (5.13b)$$

Figure 5.7 shows the basic reflectivity setup. Usually, the medium m is a gas or vacuum; therefore, β_m and δ_m can be neglected. From Equation 5.13, total reflectance occurs at $\theta \leq \theta_c$, and for larger angles' reflectivity, R_F drops a θ^4 (consistent with Parod's law). Figure 5.8 shows a typical reflectivity from a polymer thin film and an optically smooth quartz substrate. The thickness of the film measured by the periodic fringes is 468 Å. However, more exact thickness and density measurements are obtained by fitting the theory based on Fresnel's reflectivity (Equation 5.13) to the measured reflectivity. The fit is shown by a solid line for a density of 1.412 g/mm and thickness of 457 Å. The difference in thickness obtained from the fit (which is more accurate) differs from the fringe-periodicity measurement due to possible density variations in the thin film. The latter can be obtained by calculating the Patterson function. We also note that the density corresponding to the critical angle calculated by Equation 5.13b is 1.451 g/cm³. This (significant) apparent increase in density is due to refraction enhancement, discussed quantitatively in the next section (see, for example. Equation 5.23, where the apparent penetration depth decreases). Thus, the density of thin films calculated from θ_c will be higher than that obtained by fitting the full reflectivity curve. Apart from increased accuracy, the latter method also has better resolution.

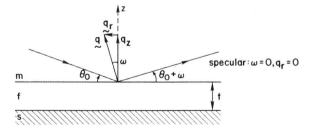

Figure 5.7 The incident and reflected medium, m, is usually a vacuum, gas, or liquid solution for X-ray reflectivity measurements. The medium, m, may also be silicon or quartz for neutron reflectivity experiments. The sample film, f, typically of thickness $t \sim 5\text{--}10^2$ nm, is deposited on a smooth substrate, s, such as single-crystal Si. Under specular conditions, incident and reflected angles are identical, implying in-plane momentum vector $q_r = 0$. Off-specular reflectivity is measured by rocking the sample by ω around the specular condition.

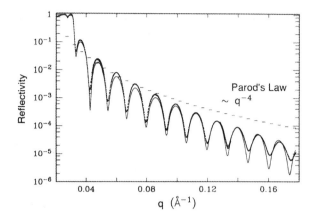

Figure 5.8 X-ray reflectivity of a PMDA-ODA film on optically smooth fused quartz is shown. The thickness t = 457 Å and density = 1.451 g/cm³ are obtained by fitting the theoretical curve (Equation 5.13), shown by a solid line, to the experimentally measured data. The deviation from Parod's law is due to the film roughness modeled by assuming Gaussian roughness with σ ~ 0.6 nm. (After Reference 89.)

As is evident from Figure 5.8, the reflectivity drops at a faster rate than prescribed by Parod's law. This is attributed to the roughness of the reflecting surface and may be quantified as follows: A roughness pair correlation function $g(r)$ may be defined as the mean-square difference in thickness (defined with respect to some arbitrary base plane parallel to the interface) between two locations separated by a lateral distance r (in the x–y plane). Based on the self-affine fractal surface defined by Mandelbrot and simulated by Voss et al., the roughness correlation function is given as[90]

$$g(r) = 2\sigma^2\big[1 - e^{-(r/\xi)^{2h}}\big] \tag{5.14}$$

where σ is rms roughness, ξ, is like a cutoff length below which the surface is fractal-like and for $r \gg \xi$ the surface is smooth. The exponent h, between 0 and 1, defines the fractal dimension, $3 - h$, of the surface. The specular and nonspecular (i.e., rocking with respect to ω) reflectivity for such surfaces is not analytical and is discussed in detail elsewhere.[90] For no long-range correlation (i.e., >100nm), the attenuation of reflectivity due to roughness is given by Gaussian approximation as

$$R = R_F\, e^{-\sigma^2 q_z^2} \tag{5.15}$$

By substituting the Fresnel's reflectivity from Equation 5.13 into Equation 5.15, one can obtain the roughness parameter σ. A reasonable σ for X-ray and neutron reflectivity is around 1 nm. For the data shown in Figure 5.8, $\sigma \approx 0.6$ nm.

Other information that can be calculated from reflectivity is density distribution in thin films or at the interface. Density distribution is an important characterization necessary for understanding the stabilization of colloids and suspensions, structure of L–B films, wetting (i.e., adhesion and composites), and polymer adsorption (from a polymer solution) behavior on various substrates. Under Born approximation (valid for low reflectivities), the reflectivity from a nonuniform (in the z-direction) interface is given by[91]

$$R = R_F \left[\int_{-\infty}^{\infty} dz \left(\frac{-1}{\rho_f} \frac{d\rho(z)}{dz} \right) e^{-iq_z z} \right]^2 \qquad (5.16)$$

where ρ_f is the bulk (electron) density of the film (in case of a thin polymer film on a substrate) or the bulk density of the substrate (when polymer adsorption study is performed). The electron density is related to the bulk density in terms of atomic weight and Avogadro number. A convenient density profile often chosen is error-function-shaped to yield an analytical reflectivity expression identical to the Gaussian function for rough surfaces given in Equation 5.15. In this case σ is the rms average interfacial width. The two effects, due to density and roughness, can be objectively deconvoluted by subtracting the specular reflection due to roughness. The latter is accomplished by extrapolating the scattering from the off-specular diffuse halo due to roughness.

5.5 Grazing Incidence X-Ray Scattering

The refractive index of X rays in a medium is less than unity because their high frequency causes an apparent increase in the phase velocity of light compared to a vacuum, as mentioned in Equation 5.12. Total internal reflection is achieved at the polymer (i.e., media) and air (or vacuum) interface when the angle of incidence is below the critical angle (see Equation 5.13), generating an evanescent wave that will decay exponentially in the medium. The decay rate is similar to Equation 5.1 for optical frequencies with a modification incorporating the absorption of the radiation. As is apparent later in the discussion, the absorption correction, given by the imaginary part of the refractive index in Equation 5.12, modifies the probe distances significantly. In a GIXS experiment, the scattered radiation from the evanescent wave is measured. For surface sensitivity to be achieved, the scattering is also measured at angles below the critical angle. The scattering angle is changed by moving the detector parallel to the film surface, as shown in Figure 5.9.

This method is complementary to the above methods with respect to the length scale of structure measured. Unlike techniques based on refraction, which is rigorously a multiple scattering phenomenon, GIXS measures single scattering related to the atomic structure of the medium. The outcome of GIXS is the crystal structure and (to some extent) its orientation. A detailed analysis also reveals semiquantitative information on the extent of order (i.e., percent crystallinity) and the quality of ordering (i.e., defect density such as paracrystallinity). The latter information,

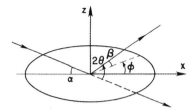

$$2\theta = \cos^{-1}(\cos\beta\cos\alpha\cos\phi - \sin\beta\sin\alpha)$$

Figure 5.9 In GIXS, the grazing angles of the input and exit beams, α and β, are kept constant. The scattering angle, 2θ = cos⁻¹(cos β cos φ cos α − sin β sin α), is changed by varying the azimuthal angle φ.

analogous to that from bulk scattering measurements depends on the order and number of the Bragg reflection. The former information may also (and probably more accurately) be achieved by reflectivity where the total density is measured as a function of depth.

As is reflectivity, GIXS is a nondestructive method that requires smooth surfaces. The requirement of smoothness is more important for GIXS because the roughness will affect the integrated scattered intensity in a manner that may not be uniquely modeled. GIXS is suited for measuring thin films such as L–B films, as is reflectivity, also. GIXS is not suited for measuring the interfaces of two solids, such as polymer–metal, polymer–polymer, and polymer–ceramic (as in oxides). However, it is (indirectly) possible to probe such an interface by having one of the media as a thin film (~10–30 nm). Since polymers are low-Z materials, low scattering intensities require synchrotron radiation sources to perform GIXS experiments. For example, in polyimides the ratio of incident to scattered intensity (for the strongest reflection) is ~10^7. Furthermore, low scattering intensities also make measurements in solution media prohibitively long for a good S/N ratio, unlike reflectivity measurements.

Let the variable without and with a prime designate the input (i.e., incidence) and output (i.e., scattered) beams, respectively. The subscripts 0 and 1 denote the vacuum and polymer media, respectively. The general approach taken to calculate the scattering intensity is as follows: First, the electric field distribution of the incoming X ray is calculated to obtain the intensity distribution "seen" by the sample. Second, the total scattered intensity from the interface exposed to the radiation is computed. Third, the scattered intensity is modified for refraction as the beam traverses out of the sample. The major assumption made in this formulation is the separation of wave distortion due to the refractive index difference at the interface and the scattering (due to the molecular structure) caused by the distorted wave.[92] This assumption—to separate the refraction phenomenon, which is related to multiple scattering, from local density fluctuations (or changes), which are related to single scattering, that give rise to the materials molecular structure factor— simplifies the calculation (and hence the interpretation of GIXS data) significantly. The refraction correction (sometimes referred to as refraction enhancement) is calculated by applying the principal of reciprocity[93] that implies an equivalence between the intensity distribution of the output beam measured in the vacuum (effect) and the source beam due to scattering in the sample (cause).[94]

Let x–z be parallel to the scattering plane. Snell's law for small angles at the vacuum–polymer interface is given by

$$\alpha_1^2 = \alpha_0^2 - \alpha_{1,c}^2 \tag{5.17}$$

where $\alpha_{1,c}$ is given by Equation 5.13b with $\beta_m = \delta_m = 0$. If there is no surface charge at the interface, the boundary condition at the interface entails $k_{0,x} = k_{1,x}$. The boundary condition and the dispersion relation leads to

$$\begin{aligned} k_{1,z} &= \left(k_{0,z}^2 - k_{1,c}^2 - 2ik^2\beta_1\right)^{1/2} \\ &= \Gamma_{1,R} + i\Gamma_{1,I} \quad \text{when } k_{0,z} = k\sin\alpha_0 < k_{0,z} \end{aligned} \tag{5.18}$$

The phase in the (polymer medium) is given by $\underline{k}_1 r (= xk_{1,x} + zk_{1,z})$. Thus, the imaginary part of $k_{1,z}$ will lead to an exponential attenuation of the electric field given by $e^{-\Gamma_{1,I}z}$ in the z-direction. Since the intensity is the square of the electric field, the characteristic depth of penetration corresponding to $1/e$ attenuation is $\Lambda_1 = \frac{1}{2}\Gamma_{1,I}$. The real and imaginary components of $k_{1,z}$ are more explicitly given by

$$\Gamma_{1,R}(k_{0,z}) = \left[\frac{k_{0,z}^2 - k_{1,c}^2 + \sqrt{(k_{0,z}^2 - k_{1,c}^2)^2 + 4k^4\beta_1^2}}{2}\right]^{1/2} \tag{5.19a}$$

and

$$\Gamma_{1,I}(k_{0,z}) = \frac{1}{2\Lambda_1} = \left[\frac{k_{1,c}^2 - k_{0,z}^2 + \sqrt{(k_{0,z}^2 - k_{1,c}^2)^2 + 4k^4\beta_1^2}}{2}\right]^{1/2} \tag{5.19b}$$

The corresponding transmission[82] is given by

$$T_{01} = \frac{E_{1,t}}{E_0} = \frac{2k_{0,z}}{k_{0,z} - k_{1,z}} = \frac{2\alpha_0}{\alpha_0 + \alpha_1} \tag{5.20}$$

If the absorption β_1 is neglected, then for $\alpha_0 \le \alpha_{1,c}$. Equation 5.17 implies that $k_{1,z} = k\sin\alpha_1$ is purely imaginary. The reflectivity implies that $|R_{01}|^2 = 1$, indicating 100% reflection (as would be expected). Furthermore, at the critical angle, Equation 5.20 indicates that $|T_{01}| = 2$. The doubling of the evanescent electric field at the interface may be explained by the constructive superposition of the reflected wave (that corresponds to electric field E_0, since the reflectivity $R_{01} = 1$ at the critical angle) and the incident wave (with the electric field also equal to E_0). Thus GIXS at the critical angle enhances the signal by $|T_{01}|^2 = 4$. The constructive superposition giving rise to a such an enhancement is experimentally confirmed by Dosch.[94]

Let $S'(.)$ be the scattering intensity distribution in sample medium 1 caused by the input beam as a function of input variables (.) to be specified later. The input wave vector $k_{0,z}$ that defines the electric field distribution in the interfacial region (or the thin film sample, depending on the penetration depths, Λ_1) will obviously be one of the variables defining $S'(.)$. The question then is what the (measured) intensity is along output vector k'_0 (in medium 0) due to the source $S'(.)$. If T'_{10} is the transmissivity (analogous to Equation 5.20) from medium 1 to medium 0, then the principle of reciprocity states that the measured intensity is simply $S'(.)|T'_{10}|^2$.

As with the equations describing the input beam, $k'_{0,z} = k \sin \alpha'_0$ and $k'_{0,z} = \Gamma'_{1,R} + i\Gamma'_{1,I}$, where $\Gamma'_{1,R}$ and $\Gamma'_{1,I}$ are given by Equation 5.19 and where $k_{0,z}$ is replaced by $k'_{0,z}$. Thus, similar to Equation 5.20, $|T'_{10}|^2$, the transmissivity for light traveling from medium 1 to 0, is given by

$$T'_{01} = \frac{E'_{0,t}}{E'_1} = \frac{2k'_{1,z}}{k'_{1,z} + k'_{0,z}} = \frac{2\alpha'_1}{\alpha'_1 + \alpha'_1} \tag{5.21}$$

where $k'_{1,z}$, $k'_{0,z}$, α'_1, and α'_0 are all known quantities

The final step is to calculate the structure factor, $S(.)$, arising from the scattering due to the distorted wave. As in the above discussion, $S'(.)$ is the apparent structure factor corrected for the refraction of the input beam, that is, $S'(.) = |T_{01}|^2 S(.)$. Thus the measured intensity at the detector in medium 0 is given by

$$I(.) = |T_{01}|^2 \cdot S(.) \cdot |T'_{10}|^2 \tag{5.22}$$

The corresponding momentum transfer vectors in the vacuum and sample are defined by $q_0 = k'_0 - k_0$ and $q_1 = k'_1 - k_1$, respectively. By applying Snell's law on the z-component of f_0 and $f_{1\bar{1}}$ and for small-angle approximation, the modified penetration depth, Λ, is given by

$$\frac{1}{\Lambda} = \frac{1}{\Lambda_1} + \frac{1}{\Lambda'_1} \tag{5.23}$$

Equation 5.23 implies that the penetration depth can be further lowered when both the input and exit angles are below the critical angle. However, as the angle becomes smaller, the transmissivity decreases according to Equations 5.20 and 5.21, causing a lowering in the signal (see Equation 5.22). The minimum penetration depth, Λ_{min}, is $(4k\alpha_{1,c})^{-1}$ when $\alpha_0, \alpha'_0 \ll \alpha_{1,c}$.

Next, some examples of the structure factor as "seen" by the distorted wave are calculated, that is, $S(.)$. The momentum transfer vector in the sample is given by

$$Q = q_1 = k'_1 - k_1 \tag{5.24}$$

The effect on thickness of the film tends to broaden the Bragg refraction along the z-axis. The broadening is given as

$$\Delta Q_z \approx \frac{5.6}{T} \tag{5.25}$$

Equation 5.25 is also referred to as the Debye–Scherrer formula. Thus, thin films tend to broaden the reflection along the z-scan.

The effects of absorption and evanescent waves (causing an exponential decay of the intensity in the sample) on the structure factor can be calculated by considering the imaginary part of Q_z explicitly. The modified structure factor in the z-direction is given by

$$S(Q) = \frac{\rho_0}{2V} \delta(x) \cdot \delta(y) \frac{1}{1 - e^{-a/s\Lambda} \cos(\Gamma a)} \tag{5.26}$$

The Q_z-broadening is due to the truncation of the crystal plane at $z = 0$ (signified by the $\cos(\Gamma a)$ term) and limited depth of penetration (signified by the $e^{-a/2\Lambda}$ term). For $\Lambda \to \infty$, the exponential term is unity, leading to the ideal crystal truncation rod formula. Scans along the z-axis, referred to as rod-scans, are performed to measure the deviations from the Q_z-broadening (in Equation 5.26) to observe surface reconstruction structure.[95]

The GIXS approximation to observe the structure along the surface can be demonstrated by representing the scattering in terms of the autocorrelation function $\gamma(r)$ as

$$S(Q_x, Q_y, Q_z) = \frac{1}{V} \iint dx\, dy\, e^{i(sQ_x + yQ_y)} \int dz \int dZ H(Z) \cdot H(Z) \cdot \gamma(x,y,z,Z) e^{iQ_z(z - Z)} \tag{5.27}$$

where

$$\gamma(x,y,z,Z) = \iint dX\, dY \cdot \rho(x,y,z) \cdot \rho(x - X, y - Y, Z)$$

The interface or interfaces are located by the shape function $H(z)$, and the exponential decay due to the evanescent wave is included by considering the imaginary part of Q_z. For example, for a film of thickness T, $H(z) = 1$ for $z \in [0,T]$ and $H(z) = 0$ elsewhere. The GIXS approximation to observe just the surface structure is valid when the Q_z dependence vanishes, that is, $e^{iQ_z(z - Z)} \approx 1$. Since $z - Z$ is at most the thickness T of the interface of interest or the thin-film sample thickness, the GIXS approximation where $S(Q_x, Q_y, Q_z) = S(Q_x, Q_y, 0)$ is valid when the grazing angles α_0 and α_0' are such that

$$Q_z \ll 1 \tag{5.28}$$

Furthermore,

$$I \approx S(Q_x, Q_y, Q_z) \int dz |T_{01}|^2 \cdot |T_{10}'|^2 e^{-z/2\Lambda} \tag{5.29}$$

The experiment is performed on a standard four-circle goniometer.[95] For sample with fiber symmetry around the z-axis, ϕ-circle is not required. Figure 5.10 shows the three circle optics, where a symmetric GIXS condition is achieved in a θ–2θ scan. The grazing entrance and exit angles, α_0 and α_0', are given by

$$\sin \alpha_0 = \sin \alpha_0' = -\cos \chi \sin \theta \tag{5.30}$$

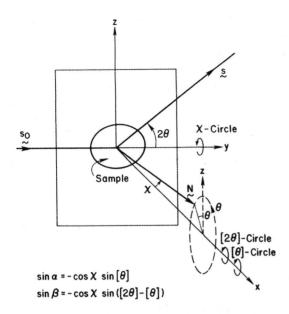

$$\sin \alpha = -\cos X \, \sin [\theta]$$
$$\sin \beta = -\cos X \, \sin ([2\theta]-[\theta])$$

Figure 5.10 The incident and scattered angles with respect to the film surface, α_0 and α_0' or α and β (in Figure 5.9), respectively, are controlled by moving the $[\theta]$, $[2\theta]$, $[\chi]$ circles simultaneously so that $\sin \alpha_0 = -\cos[\chi] \sin[\theta]$ and $\alpha_0' = -\cos[\chi] \sin([\theta] - [2\theta])$. The scattering angle is 2θ and the symmetric GIXD is performed by setting $[\theta] = \frac{1}{2}[2\theta]$.

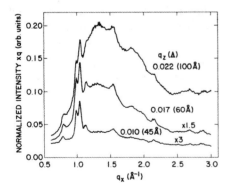

Figure 5.11 GIXS spectra of a PMDA-ODA polyimide formed by CVD under a high vacuum. The penetration depth of the sample probed, as determined by Equation 5.19, corresponds to q_z (shown in the figure), fixed by optics determined by Equation 5.30. The incident and scattering grazing angles, α_0 and α_0' respectively, are equal and constant for a given scattering curve. The 0.6-Å rms-roughness and critical scattering vector, $k_{1,c} = 0.01212$ Å^{-1}, are determined by X-ray reflectivity performed on samples prepared under similar conditions.

The scattering pattern for a polymer thin film may be obtained at various penetration depths determined by the grazing angles α_0 and α_0'. Figure 5.11 shows GIXS spectra for a ~500 Å PMDA-ODA polyimide thin film made by CVD of the monomer in high vacuum.[95] The input and output beam grazing angles α_0 and α_0' are equal and constant for each scan. It is observed that the top ~45 Å is more ordered than the polymer at deeper levels, as is evident from the peaks at $q_x \approx 1.0$ Å^{-1} and the absence of an amorphous halo at $q_x = 1.35$ Å^{-1}. The rms smoothness of the films is ~0.6 Å, as measured by X-ray reflectivity. The line shape of the scattering peak can be modeled by assuming a distribution of the scattering species (such as crystal fraction), $\phi(z)$ in Equation 5.29.[96] Similarly, the line broadening can be quantified by convoluting a "spreading function" on the electric field due to the input and output beams given by $|T_{01}|$ and $|T_{01}'|^2$, respectively. The function is symmetric about $k_{0,z}$ and $k_{0,z}'$ and is related to the optics of the setup. Thus, in principle, quantitative information on changes in crystallinity, chain orientation, and the relative fraction of phases can be obtained from GIXS.

References

1   F. A. Keimel. In *Kirk-Othmer Encyclopedia of Chemical Technology*. Vol. 1, 3rd ed., Wiley, New York, 1979, p. 488.

2   I. A. Abu-Isa. *Polym.-Plast. Technol. Eng.* **2**, 29, 1973.

3   D. R. Fitchmun, S. Newman, and R. Wiggle. *J. Appl. Polym. Sci.* **14**, 2411, 2457, 1970.

4   K. Bright, B. W. Malpass, and D. E. Packham. *Nature.* **223**, 1360, 1969.

5   B. W. Malpass, D. E. Packham, and K. Bright. *J. Appl. Polym. Sci.* **18**, 3249, 1974.

6   P. Blais, D. J. Carlsson, G. W Csullog, and D. M. Wiles. *J. Colloid Interface Sci.* **47**, 636, 1974.

7   D. J. Arrowsmith. *Trans. Inst. Met. Finish.* **48**, 88, 1970.

8   M. Panar et. al. *J. Polym. Sci., Polym. Phys. Ed.* **21**, 1955, 1983.

9   M. Jaffe and H.-N. Yoon. *Polymers For Advanced Technologies.* (M. Lewin, Ed.) VCH Publishers, New York, 1988.

10   B. J. Factor, T. P. Russell, and M. F. Toney. *Phys. Rev. Lett.* **66**, 1181, 1991.

11   R. J. Young, R. J. Day, and M. Zakikhani. *MRS Proceedings.* **134**, 351, 1989.

12   I. C. Sawyer and M. Jaffe. *J. Mat. Sci.* **21**, 1897, 1986.

13   J. Petermann and G. Broza. *J. Mat. Sci.* **22**, 1108, 1987.

14   S. Faghihi, T. Hoffmann, J. Petermann, and J. Martinez-Salazar. *Macromolecules.* **25**, 2509, 1992.

15   J. C. Wittmann and P. Smith. *Nature.* **352**, 414, 1991.

16 T. Takahashi, F. Teraoka, and I. Tsujimoto. *J. Macromol. Sci.-Phys.* **B12**, 303, 1976.

17 J. R. Fernande and D. B. DuPre. *Mol. Cryst. Liq. Cryst., Lett.* **72**, 67, 1981.

18 A. Loschewsy and H. Ringsdorf. *Macromolecules.* **21**, 1936, 1988.

19 E. P. Plueddemann. *J. Adh. Sci. Tech.* **3**, 131, 1989.

20 D. C. Bassett. *Principles of Polymer Morphology.* Cambridge University Press, Oxford, U.K., 1981.

21 H. Schonhorn, F. W. Ryan, and R. H. Hansen. *J. Adhes.* **2**, 93, 1970.

22 H. Schonhorn and F. W. Ryan. *J. Polym. Sci., A-2.* **7**, 105, 1969.

23 H. Schonhorn and F. W. Ryan. *Adv. Chem. Ser.* **87**, 140, 1968.

24 D. R. Fitchmun and S. Newman. *J. Polym. Sci., A-2.* **8**, 1545, 1970.

25 H. Schonhorn. *Macromolecules.* **1**, 145, 1968.

26 T. K. Kwei, H. Schonhorn, and H. L. Frisch. *J. Appl. Phys.* **38**, 2512, 1967.

27 H. Schonhorn and F. W. Ryan. *J. Phys. Chem.* **70**, 3811, 1966.

28 R. F. Saraf, C. Thompson, S. Anderson, and G. B. Stephenson. In press.

29 S. Ueda and T. Kimura. *Kobunshi Kagaku.* **15**, 243, 1958.

30 I. A. Abu-Isa. *J. Polym. Sci., A-1.* **9**, 199, 1971.

31 N. F. Rodman and R. G. Mason. *Thromb. Diath. Haemorrh., Suppl. No. 40.* 145, 1970.

32 S. W. Kim, R. G. Lee, H. Oster, D. Coleman, D. J. Andrade, D. J. Lentz, and D. Olsen. *Trans. Am. Soc. Artif. Intern. Organs.* **20**, 449, 1974.

33 S. W. Kim and R. G. Lee. *Advances in Chemistry.* Series No. 145, Applied Chemistry at Protine Interface, Washington, DC, 1975.

34 J. A. DeFeijter, J. Benjamins, and F. A. Veer. *Biopolymers.* **17**, 1759, 1978.

35 P. A. Cuyper, J. W. Corsel, M. P. Jansen, J. M. M. Kop, W. T. Hermens, and H. C. Hemker. *J. Biol. Chem.* **258**, 2426, 1983.

36 U. Jonsson, M. Malmquist, and I. Ronnberg. *J. Colloids Interface Sci.* **103**, 360, 1985.

37 B. Ivarsson, P.-O. Hegg, I. Lundstrom, and U. Jonsson. *Colloids Surfaces.* **13**, 169, 1985.

38 T. Arnebrant, B. Ivarsson, K. Larsson, I. Lundstrom, and T. Nylander. *Progr. Colloid Polymer Sci.* **70**, 62, 1985.

39 T. Moiizumi. *Thin Solid Films.* **160**, 413, 1988.

40 M. Sriyucdthsak, H. Yamagishi, and T. Moiiziumi. *Thin Solid Films.* **160**, 493, 1988.

41 J.-C. Janson and P. Hedman. In *Advances in Biochemical Engineering.* Vol. 25, (A. Fiechter, Ed.) Springer Verlag, Heidelberg, 1982.

42 R. Ackerman, D. Naegele, H. Ringsdorf. *Makromol. Chem.* **175**, 699, 1974.

43 J. B. Lando and T. Fort. In *Polymerization of Organized System.* Midland Macromolecular Monograph, Vol. 3, (H. G. Elias, Ed.) Gordon and Breach, New York, 1977.

44 D. Naegele, J. B. Lando, and H. Ringsdorf. *Macromolecules.* **10**, 1339, 1977.

45 Y. Tanaka, K. Nakatama, S. Iijima, and Y. Maitani. *Thin Solid Films.* **133**, 165, 1985.

46 A. Laschewsy and H. Ringsdorf. *Macromolecules.* **21**, 1936, 1988.

47 B. Tieke and G. Lieser. *J. Colloid. Sci.* **88**, 471, 1982.

48 M. Ozaki, Y. Ikeda, and I. Nagoya. *Synth. Met.* **28**, C801, 1989.

49 K. Fakuda, Y. Shibasaki, and H. Nakahara. *J. Macromol. Chem.* **A15**, 999, 1981.

50 S. J. Valenty. *Macromolecules.* **11**, 1221, 1978.

51 M. Puggelli and G. Gabrielli. *J. Colloid Interface Sci.* **61**, 420, 1977.

52 K. Naito. *J. Colloid Interface Sci.* **131**, 218, 1989.

53 R. H. Tredgold, R. A. Allen, P. Hodge, and E. Khoshdle. *J. Phys. D: Appl. Phys.* **20**, 1385, 1987.

54 R. H. Tredgold. *J. de Chim. Phys.* **85**, 1079, 1988.

55 H. Ringsdorf, G. Schmidt, and J. Schneider. *Thin Solid Films.* **152**, 207, 1987.

56 R. Elbert, A. Laschewsky, and H. Ringsdorf. *J. Am. Chem. Soc.* **107**, 4134, 1985.

57 M. B. Biddle, J. B. Lando, H. Ringsdorf, G. Schmidt, and J. Schneider. *J. Colloid Polym. Sci.* **266**, 806, 1988.

58 J. Schnieder, H. Ringsdorf, and J. F. Rabolt. *Macromolecules.* **22**, 205, 1989.

59 G. Fariss, J. B. Lando, and S. Rickert. *Thin Solid Films.* **99**, 305, 1983.

60 A. Banerjie and J. B. Lando. *Thin Solid Films.* **68**, 67, 1980.

61 *Polydiacetylenes—Synthesis, Structure, and Electronic Properties.* (D. Bloor and R. R. Chance, Eds.) Martinus Nijhoff Publishers, Dordrecht, The Netherlands, 1985.

62 F. Kajzar and J. Messier. *Thin Solid Films.* **99**, 109, 1983.

63 C. S. Winter and R. H. Tredgold. *Thin Solid Films.* **123**, L1, 1985.

64 S. J. Mumby, J. F. Rabolt, and J. D. Swalen. *Thin Solid Films.* **133**, 161, 1985.

65 S. J. Mumby, J. F. Rabolt, and J. D. Swalen. *Macromolecules.* **19**, 1054, 1986.

66 A. K. M. Rahman, L. Samuelson, D. Minehan, S. Clough, S. Tripathy, T. Inagaki, X. Q. Yang, T. A. Skotheim, and Y. Okamoto. *Synth. Metal.* **28**, C237, 1989.

67 W. M. Heckel, M. Losche, H. D. Mattes, and H. Mohwald. *Thin Solid Films.* **133**, 83, 1985.

68 G. Lieser, B. Tieke, and G. Wagner. *Thin Solid Films.* **68**, 77, 1980.

69 F. Keneko, M. S. Dresselhaus, M. F. Rubner, M. Shibata, and S. Kobayashi. *Thin Solid Films.* **160**, 327, 1988.

70 Y. J. Chen, G. M. Carter, and S. K. Tripathy. *Solid State Commun.* **54**, 19, 1985.

71 S. J. Mumby, J. F. Rabolt, and J. D. Swalen. *Thin Solid Films.* **133**, 161, 1985.

72 S. J. Mumby, J. F. Rabolt, and J. D. Swalen. *Macromolecules.* **19**, 1054, 1986.

73 A. Camel, T. Fort, and J. B. Undo. *J. Polym. Sci., A-1.* **10**, 2061, 1972.

74 C. E. D. Chidsey, G.-Y. Liu, P. Rowntree, and G. Scoles. *J. Chem. Phys.* **91**, 4421, 1989.

75 C. E. D. Chidsey, D. N. Loiacono, T. Sleator, and S. Nakahara. *Surf. Sci.* **200**, 45, 1988.

76 R. Benferhat, B. Drivillon, and P. Robin. *Thin Solid Films.* **156**, 295, 1988.

77 Y. Ishino and H. Ishida. *Langmuir.* **4**, 1341, 1988.

78 W. Jark, T. P. Russell, G. Comelli, and J. Stohr. *Thin Solid Films.* **199**, 161, 1991.

79 R. G. Hunsperger. *Integrated Optics: Theory and Technology.* 2nd ed., Springer Verlag, Berlin, 1984.

80 T. P. Russell, H. Gugger, and J. D. Swalene. *J. Polym. Sci., Polym. Phys. Ed.* **21**, 1745, 1983.

81 Y. R. Shen and F. Demartini. In *Surface Polaritons.* (W. M. Agranovich and D. L. Mills, Eds.) North-Holland, Amsterdam, 1982, p. 629.

82 M. Born and E. Wolfe. *Principles of Optics.* 6th ed., Pergamon, New York, 1980.

83 N.-M. Chao, K. C. Chu, and Y. R. Shen. *Mol. Crys. Liq. Cryst.* **67**, 201, 1981.

84 D. M. Kolb. In *Surface Polaritons.* (W M. Agranovich and D. L. Mills, Eds.) North-Holland, Amsterdam, 1982, p. 299.

85 H. Raether. In *Surface Polaritons.* (W. M. Agranovich and D. L. Mills, Eds.) North-Holland, Amsterdam, 1982, p. 331.

86 T. F. Heinz, H. W. K. Tom, and Y. R. Shen. *Phys. Rev. A.* **28**, 1883, 1983.

87 N. Bloembergen and P. S. Preshan. *Phys. Rev.* **128**, 606, 1962.

88 T. Rasing, Y. R. Shen, M. W. Kim, and S. Grubb. *Phys. Rev. Lett.* **55**, 2903, 1985.

89 G. Berkovic and Y. R. Shen. In *Nonlinear Optical and Electroactive Polymers.* (P. N. Prasad and D. R. Ulrich, Eds.) Plenum, New York, 1988, p. 157.

90 S. K. Sinha, E. B. Sirota, S. Garoff, H. B. Stanley. *Phys. Rev. B.* **38**, 2297, 1988.

91 I. M. Tidswall, B. M. Ocko, P. S. Pershan, S. R. Wassermann, G. M. Whitesides, and J. D. Axe. *Phys. Rev. B.* **41**, 1111, 1990.

92 P. Selenyi. *Compt. Rend.* Paris, **157**, 1408, 1913.

93 R. S. Becker, J. A. Golevechenko, and J. R. Patel. *Phys. Rev. Lett.* **50**, 153, 1983.

94 H. Dosch. *Phys. Rev. B.* **35**, 2137, 1987.

95 R. F. Saraf, C. Dimitrakopoulos, S. P. Kowalczyk, and M. F. Toney. Unpublished results.

96 M. F. Toney, T. C. Huang, S. Brennan, and Z. Rek. *J. Mater. Res.* **3**, 351, 1988.

6

Surface Thermodynamics

THOMAS B. LLOYD and DAVID W. DWIGHT

Contents

6.1 Introduction

The thrust of this chapter is on quantifying surface energetics and using those results to predict wetting, spreading, and adhesion on polymer surfaces. Such measurements can be an efficient way to infer surface composition (contamination in particular). Potential correlations with other properties such as the durability of adhesive bonds are addressed.

The primary practical objective in this application of surface thermodynamics is to predict the thermodynamic work of adhesion from constants that characterize each independent material. We assume that the work of adhesion is primarily the sum of two contributions, one arising from London–Lifshitz "dispersion" attractive energies (computed from the root mean square of the individual surface energies) and the other from Lewis acid–base (or electron donor–acceptor) interaction energies. In order for one to predict the effects of interphase modifications, often the acid–base terms are of greatest importance because the dispersion force component effectively cancels out in the difference.

The properties of laminated and composite materials depend upon the interfacial adhesion of the matrix to the particle or fiber surface, as we and others have demonstrated for inorganic fillers such as silica, alumina, calcium carbonate and carbon black, and reinforcing fibers of glass and Teflon®. Usually, chemical modification of the surface is necessary to provide the strong acid–base interactions with the matrix required for tough composites. This approach can be extended to films, coatings, and adhesives.

Analysis of the chemistry and bonding of the interfaces, both before and after failure is important for two reasons: (1) to assign the thermodynamic characteristics to specific atomic and molecular structures, thus opening the door for analysts to understand the scaling laws from atomic to microscopic to macroscopic levels and (2) to develop a molecular-level understanding of the locus and mechanism of failure, which is essential to determine whether it is reasonable to expect the thermodynamic work of (formation of) adhesive bonds to correlate with failure (e.g., if failure is actually cohesive, other factors must be taken into account).

Structure–property relationships derived on this basis will facilitate molecular engineering of interphases and optimize systems properties. (See other chapters in this book for details of spectroscopic and microscopic techniques.)

The surfaces of polymers are of relatively low energy[1] compared to metals and many inorganic materials. Often polymer surfaces are modified to increase their interaction with inks, adhesives, and coatings. Thus it becomes necessary to characterize the surface chemistry quantitatively in order to improve products. This chapter introduces the concepts of polymer surface thermodynamics and provides references to guide the reader toward a deeper understanding. The phenomena of polymer adsorption onto solids from solution are not dealt with in any depth.

In order to quantify the surface energetics of solid polymers, one must measure phenomena wherein materials of known physical characteristics are interacting with the solid polymer. This gives the information needed to predict, at least qualitatively and often quantitatively, the interaction of materials of interest with the polymer surface, for example, the strength of adhesive bonds.

6.2 Theory of Wetting, Spreading, and Adhesion

The attractive forces available in polymers for both cohesion and adhesion are essentially of two types: London–Van der Waals forces and Lewis acid–base or electron acceptor-donor forces. There are other attractive forces (see Equation 6.1), but if the molecules have no net charge, these two dominate. Ross and Morrison[2] give a good description and discussion of attractive and repulsive forces among molecules and condensed phases.

London–Van der Waals (dispersion) forces are rather strongly attractive within homogeneous condensed phases and across interfaces in the two-phase systems considered here. They are the only intermolecular forces in fully saturated hydrocarbons such as polyethylene. For instance, the dispersion force contribution to surface tension of water (γ^d) is 30%, with H-bonding accounting for 70%. For glycerol, γ^d is 58% of the total and for formamide 68%.

Most polymers are Lewis acids or bases or both. For instance, polyesters contain basic oxygens in the ester linkage as well as acidic carboxylic acid and alcohol end groups. These enter into reactions of self-association and interactions with plasticizers, solvents, and fillers. Acid–base interactions at surfaces with adhesives and other polymers are of great importance in terms of the strength of individual bonds as well as their number per unit area.

The thermodynamic work of adhesion, W_{SL}, at a solid(S)-liquid(L) interface is the reversible Helmholtz free energy change per unit area of interface. It is the sum of various types of forces per unit area:

$$W_{SL} = W_{SL}^{\text{dispersion}} + W_{SL}^{\text{acid base}} + W_{SL}^{\text{dipole dipole}} + W_{SL}^{\text{charge dipole}} + \ldots \qquad (6.1)$$

Hydrogen bonding is a subset of W^{a-b}. Similarly, the work of cohesion within each condensed phase can be separated into additive free energies or surface tensions, γ_L^d, $\gamma_L^{a\,b}$, γ_L^{dipole}, etc.[3] (e.g., $\gamma_L = \gamma_L^d + \gamma_L^{ab}$) Usually the dispersion and acid–base components are dominant in microelectronic systems. Electrostatic terms are important in fine particle adhesion and electret films.

The Young equation is the most common basis for interpreting the contact angle, θ:

$$\gamma_{LV} \cos \theta = \gamma_{SV} - \gamma_{SL} \qquad (6.2)$$

Buff[4] added a term to the Young equation to include the reduction of free energy of the solid by the adsorption of the vapor of the liquid. Thus the Young equation becomes $\gamma_{LV} \cos \theta = \gamma_{SV} - \gamma_{SL} - \pi_e$. The variable π_e is called the equilibrium spreading pressure. The value of π_e on polymer surfaces is small and is usually neglected.

Combining Equations 6.1 and 6.2 leads to

$$W_{SL}^{\text{total}} = \gamma_L(1 + \cos\theta) = W_{SL}^d + W_{SL}^{ab} \qquad (6.3)$$

Fowkes showed that the dispersion term γ_S^d can be calculated from γ_L^d and by a geometric mean expression[5]:

$$W_{SL}^d = 2(\gamma_S^d \gamma_L^d)^{1/2} \qquad (6.4)$$

In order for γ_S^d to be determined for a polymer surface, "neutral" liquids are used which have only dispersive interactions ($W_{SL}^{ab} = 0$) and known γ_L^d. For the solid polymer, Equations 6.3 and 6.4 become

$$\gamma_S^d = \frac{1}{4}\gamma_L^d(1 + \cos\theta)^2 \qquad (6.5)$$

It is important to predict values of W_{SL}^{ab} at solid–liquid or solid–solid interfaces, since they relate to adhesion. Some workers have erroneously used an expression $W_{SL}^{ab} = 2(\gamma_S^{ab} \gamma_L^{ab})^{1/2}$. This geometric mean expression has been shown to be invalid for the acid–base contribution.[6, 7] Two valid, but distinctly different methods have been published: the method of Fowkes and Mostafa[8] and the method of van Oss et al.[7, 9] The former method makes use of the molar heats of acid–base interaction between the two materials in contact, $-\Delta H_{SL}^{ab}$ (see Adsorption and

Calorimetry in Section 6.3), and the interfacial concentration, n_{ab}, of acid–base bonds (in mol/m^2):

$$W_{SL}^{ab} = -fn_{ab}\Delta H^{ab} \tag{6.6}$$

where f is a constant (close to unity in some cases) for converting enthalpy of interfacial acid–base interaction into free energies of interfacial acid–base interaction. This method was tested with the benzene–water interface using Drago's E_A and C_A (see Adsorption and Calorimetry in Section 6.3) constants for water [5.01 and 0.67 (kJ/mol)$^{1/2}$, respectively] and the E_B and C_B constants for benzene [0.75 and 1.8 (kJ/mol)$^{1/2}$, respectively]; these yield a ΔH^{ab} of −5.0 kJ/mol. If each interfacial benzene molecule lies flat at the interface, occupying 0.50 nm^2 each, $n_{ab} = 3.3$ μmol/m^2, and $W_{12}^{ab}/f = 16.5$ mJ/m^2, which is very close to the value measured at 20 °C; that is,

$$W_{12}^{ab} = \gamma_1 + \gamma_2 - \gamma_{12} - 2(\gamma_1^d \gamma_2^d)^{1/2}$$
$$= 72.8 + 28.9 - 35.0 - 2(22 \cdot 28.9)^{1/2} = 16.3 \text{ mJ/m}^2 \tag{6.7}$$

The method of van Oss et al. also recognizes that interfacial acid–base interaction requires that acidic sites of one phase interact with basic sites of the other, so that if either phase is neutral or if both phases have only basic or only acidic sites, there can be no acid–base interaction. In this approach the form of the geometric mean equation remains, but this is not a geometric mean of the "polar" (acid–base) contribution to the surface free energy:

$$W_{12}^{ab} = 2(\gamma_1^\ominus \gamma_2^\oplus)^{1/2} + 2(\gamma_1^\oplus \gamma_2^\ominus)^{1/2} \tag{6.8}$$

where $\gamma_1^\ominus$ is a measure of the surface basicity of phase 1, $\gamma_1^\oplus$ is a measure of the surface acidity of phase 1, and $\gamma_2^\ominus$ and $\gamma_2^\oplus$ are measures of the surface basicity and acidity, respectively, of phase 2. The units of these functions are the same as those for the surface energy (mJ/m^2), and the values of these functions can be determined from measured values of W_{12}^{ab} on the assumption that for water $\gamma_1^\ominus = \gamma_1^\oplus = \gamma_1^{ab}/2 = 25.4$ mJ/m^2 at 20 °C. Reference 9 lists $\gamma^\oplus$ and $\gamma^\ominus$ values for a number of polymeric solids, and the attractive free energy of polymers in various solvents is calculated. Good,[10] in a review of contact angle, adhesion, and wetting, shows an increase of both $\gamma^\ominus$ and $\gamma^\oplus$ by the corona treatment of polypropylene.

Although this approach avoids the difficulties of using the geometric mean of the actual polar components of the surface free energy, values of $\gamma^\ominus$ and $\gamma^\oplus$ for the test liquids are still under investigation. Furthermore, the use of a single parameter to characterize acidity or basicity ignores the hard and soft character (Pearson) of acids and bases.[8]

The driving force of adsorption from solution is the loss of Helmholtz free energy, ΔU, for the following change:

$$\text{solute}_{\text{solution}} + \text{solvent}_{\text{adsorbed}} = \text{solute}_{\text{adsorbed}} + \text{solvent}_{\text{solution}}$$

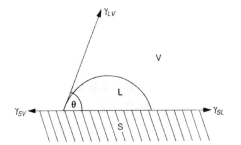

Figure 6.1 The contact angle θ resulting from surface and interfacial forces in equilibrium.

If adsorption occurs spontaneously, ΔU must be negative and the change of enthalpy, ΔH, is also negative (exothermic). This is so because the entropy change, ΔS, is negative due to the increase in order of the solute molecules at the surface, making the last term positive in

$$\Delta U_{ads} = \Delta H_{ads} - T\Delta S_{ads} \tag{6.9}$$

When the solute is a polymer, there is a high entropic cost of bringing the long chains into the adsorbed layer. The picture is complex, with some polymer segments bound to the surface ("trains") separated by other segments ("loops") extending away and more mobile "tails" which are bound to the surface at only one end. There is also the complication of varied solvation of the chains in the bulk solution versus the interface.

Napper[11] deals with polymer adsorption from a statistical thermodynamic viewpoint. Also, deGennes[12, 13] studies adsorbed polymer conformation and thermodynamics. As stated previously, we do not deal at any length with adsorbed polymers here. One more note on entropy: If ΔH_{ads} is measured, ΔU can, at least, be approximated by

$$\Delta U = -RT \ln K \tag{6.10}$$

where K = (solute concentration, surface)/(solute concentration, solution), R is the gas constant, T the absolute temperature, and ΔS is calculated using Equation 6.9.

6.3 Experimental Methods

Contact Angles

Flat solids and films Commonly, the surface energetics of polymers have been characterized through the measurement of contact angles of drops of test liquids using a goniometer (see Figure 6.1). Lower contact angles indicate higher surface energies. The Young equation (Equation 6.2) states the relationship between the various surface free energies and the contact angle θ.

Zisman[14] found, for a wide range of polymer surfaces, that γ_{LV} of a homologous series of liquids plotted against cos θ was linear. The linear equation

$$\cos\theta = 1 + b(\gamma_{LV} - \gamma_c) \tag{6.11}$$

Polymer	γ_c at 20 °C, mJ/m^2
Polymethacrylic ester of ϕ'-octanol	10.6
Polyhexafluoropropylene	16.2
Polytetrafluoroethylene	18.5
Polytrifluoroethylene	22
Poly(vinylidene fluoride)	25
Poly(vinyl fluoride)	28
Polyethylene	31
Polytrifluorochloroethylene	31
Polystyrene	33
Poly(vinyl alcohol)	37
Poly(methyl methacrylate)	39
Poly(vinyl chloride)	39
Poly(vinylidene chloride)	40
Poly(ethylene terephthalate)	43
Poly(hexamethylene adipamide)	46

Table 6.1   Critical surface tensions of various polymeric solids.[14]

introduces the term γ_c, or critical surface tension for wetting. When the plot is extrapolated to cos θ = 1 (θ = 0), $\gamma_c = \gamma_{LV}$, that is, the surface tension required for the spreading of the liquid on the polymer. Thus the procedure allows for a relative assessment of the surface tension of the solid polymer. However, the expression holds for situations where only dispersion forces are involved, that is, where the polymer or the liquid, or both, are saturated hydrocarbons or otherwise have no Lewis acidity or basicity. Table 6.1 gives typical values of γ_c for various polymers.

Good[15] suggests that cos θ be plotted versus $\gamma_{LV}^{-1/2}$ for more precise extrapolation of the data. He further suggests plotting $\gamma_{LV}(1 + \cos\theta)^2/4$ versus γ_{LV} and fitting the best horizontal straight line. Here the intercept with the ordinate defines γ_c.

Using the geometric mean (Equation 6.4), Fowkes[16] derived the following equation:

$$\cos\theta = -1 + 2\sqrt{\gamma_S^d}\left(\frac{\sqrt{\gamma_L^d}}{\gamma_L}\right) \qquad (6.12)$$

Plotting cos θ versus $\sqrt{\gamma_L^d}/\gamma_L$ will give an estimate of γ_S^d, which is essentially equal to Zisman's γ_c when test liquids having only dispersion forces are used. The slope $2\sqrt{\gamma_S^d}$ is the London dispersion force component of the surface free energy of the solid but not the total surface free energy. Table 6.2 lists some γ_S^d values for low-energy surfaces determined by this method.

Polymer	γ_S^d, mJ/m^2
Polyhexafluoropropylene (PHFP)	18.0
Polytetrafluoroethylene (PTFE)	19.5
Parafin wax	25.5
Polytrifluoromonochloroethylene (Kel-F)	30.8
Polyethylene	35.0
Polystyrene	44.0

Table 6.2   **Dispersion component of the surface free energy of polymers from contact angles.**

Several other methods for estimating the surface tension of polymer solids using contact angles have been reviewed and compared by Hata et al.[17] They recommend plotting $\gamma_{LV}\cos\theta$ versus γ_L for a series of pure liquids, including various Lewis acids and bases as well as neutral ones, and equating γ_s with the maximum in the curve.

Contact angle hysteresis   Most polymeric surfaces are chemically heterogeneous to some degree and exhibit some roughness. As a result, the contact angle measured after the drop volume is increased (advancing) differs from the angle measured after the volume is decreased (receding). Hysteresis is the difference $\theta_a - \theta_r$; it is a function of the heights of energy barriers between the metastable states of the liquid drop. It is generally agreed that the advancing angle represents the low-energy component of the surface, whereas the receding angle characterizes the high-energy component of heterogeneous surfaces. Good's paper[15] is recommended for its discussion of contact angles as well as surface free energy of solids. He suggests that both θ_a and θ_r be measured for the added information one can gain on the nature of the surface.

The influence of roughness on contact angle hysteresis is dealt with by Dettre and Johnson[18] by comparing predictions based on computer-generated model surfaces with laboratory experiments. Polyethylene, polypropylene, and fluorocarbon polymer surfaces were physically configured to various geometries and measured wettabilities were found to be consistent with computed values from the energy-barrier model.

Particulate materials and fibers   On flat surfaces (films), the contact angle is usually measured for single sessile drops using a telescope with adjustable cross hairs. This procedure is often used on powders that have been pressed to give a flat surface. However, this method should be used with caution since the angle is influenced by the roughness of porous, compacted surfaces.

A better way to approach the measurement is to use a Bartell cell, which uses a compacted plug of panicles or fibers. The liquid fills the pores and the pressure required to remove it is related to the contact angle by the Laplace equation,

$$\Delta p = \frac{2\gamma\cos\theta}{r} \qquad (6.13)$$

where r is the average cylindrical pore radius in the plug. The value of r is determined by using a liquid of low surface tension that has cos θ = 1. Liquid flows are measured versus pressure and both advancing and receding angles can be determined.

An alternative, related procedure was introduced by Washburn[20] based upon the equations of Poiseuille (for the flow of liquids in capillaries) and Laplace (which introduces the pressure due to the curvature of the liquid surface). In the Washburn equation

$$\frac{dl}{dt} = \frac{\gamma r \cos \theta}{4 \eta l}$$ (6.14)

where l is the capillary length, r is the radius, γ is the surface tension of the liquid, η is its viscosity, and t is time. As in the Bartell procedure, r is measured on a plug of packed particles or fibers by use of a liquid having zero contact angle. However, the rate of liquid penetration in the Washburn procedure is measured without a pressure drop.

Another dynamic procedure is based on the measurement of surface tension by the Wilhelmy plate method in conjunction with the Young equation and has been automated by the Cahn Instrument Co. The contact angle is determined using the equation

$$\cos \theta = \frac{F}{p \gamma_L}$$ (6.15)

where F (in dynes) is the force of the meniscus of the liquid at a fiber or film interface, p (in centimeters) is the length of the interface and γ_L (in dyne centimeters) is the surface tension of the liquid. The polymer can be fiber, coated wire, or film, but p must be determined accurately. The force is measured with a micro-balance as the liquid in contact with the specimen is raised or lowered, giving advancing and receding contact angles over chosen areas of the specimen. Bystry and Penn demonstrated the differences in surface energetics of various brittle thermosets using this procedure on resins cast onto wires and plates.

Adsorption and Calorimetry

A quantitative understanding of the forces at polymer surfaces requires the measurement of thermodynamic quantities. The driving force of adsorption is $\Delta U_{ads} = \Delta H_{ads} - T \Delta S_{ads}$. ΔH_{ads} can be measured directly by calorimetry and indirectly by shifts in IR spectral peaks as shown below. Of particular interest are the heats of interaction for adsorption from solution of (1) probes of known acid–base character to establish the number and strength of the sites on a polymer surface and (2) soluble polymers onto chosen surfaces to determine the strength of interaction. Certainly $-\Delta H_{ads}$ is proportional to the free energy of adsorption expressed as ΔU_{ads} or as the work of adhesion W_{SL}. In adsorption from solution the dispersion force contribution to the energy effectively cancels, assuming that solute–solvent and solute–surface dispersion interactions are identical. Thus, $\Delta H_{ads} = \Delta H_{ads}^{ab}$. Fowkes et al.[22] introduced the expression

$$f = \frac{W_{SL}^{ab}}{n_{ab}(-\Delta H_{ads}^{ab})} \tag{6.16}$$

The constant f is the factor for converting heats of interfacial acid–base interaction to free energies of the event; n_{ab} is the interfacial concentration of acid–base bonds in the two dimensional and solution (mol/m^2). The value of f is near unity in some cases, but Vrbanac and Berg[23] concluded from their studies of acidic, basic, bifunctional and neutral polymers interacting with a series of pure liquids, again of varying functionality, that f is significantly less than unity.

ΔH_{ads}^{ab} is a useful measure of the strength of adsorption. The Drago concept of the enthalpy of 1:1 complex formation (Lewis acid–base) is useful here.[24] Drago et al. developed the four-constant expression $-\Delta H^{ab} = C_A E_A + C_B E_B$ starting with arbitrary values of $C_A = E_A = 1.0$ for iodine and determining E and C values for some 40 acids and 40 bases by measuring heats of complex formation of selected pairs. Calorimetric determination of ΔH_{ads}^{ab} of materials with others of known Drago constants[24] allows determination of the Drago C and E constants of the surface sites of a polymer of interest. For instance, the values of $-\Delta H_{ads}^{ab}$ (in kJ/mole) for an acidic polymer reacting with three bases of known C_B and E_B results in three linear Drago equations:

$$E_A = \frac{-\Delta H_{ads}^{ab}}{E_B} - C_A \frac{C_B}{E_B} \tag{6.17}$$

where the intercepts ($-\Delta H_{ads}^{ab}/E_B$) and slopes (C_B/E_B) are known. Each pair of equations allows the calculation of the two unknowns, C_A and E_A, of the surface sites. Thus three values for E_A and C_A result and can be averaged.[25] In this paper, the E_A and C_A values for α-Fe$_2$O$_3$ were determined and the adsorption of poly(vinylpyridine) from benzene studied on the same surface. Adsorption took 6 min and stopped, but the heat continued for 19 min more as the polymer rearranged to bring 52% of its basic sites into bonds with the FeOH sites.

Calorimetric procedures with solids require that the polymer be a powder of >1 m^2/g for a reasonable signal intensity. Commercial flow microcalorimeters are quite sensitive and, coupled with a downstream detector of solute concentration, allow both the ΔH_{ads}^{ab} and the amount adsorbed to be calculated versus elapsed time. A common mode of operation is to equilibrate the bed of powder with flowing solvent and then pump through a solution where the probe, small molecule or polymer, is the solute in the same solvent. Equilibrium is established wherein the surface adsorbs the maximum possible at that concentration. Varying the concentration with several runs produces an adsorption isotherm and allows an assessment of site energy distribution, presuming the higher energy sites are occupied first.

Care must be taken to avoid diffusion of the probe molecules into the polymer; sterically hindered probes are effective.[26]

Often the polymer of interest is in solution and the adsorbing surface can be polymeric or not, such as an oxide pigment. Since polymers of commercial interest are insoluble in neutral hydrocarbons (solvents with only dispersion forces), they must be dissolved in a solvent having some acidity or basicity. This introduces another interaction into the measurement: the endothermic heat of dissociation of the solute–solvent complex. Some quantitative accounts have been reported, but comparative work has been more valuable. An example is that of Fowkes and Mostafa[8] illustrating the competition of basic calcium carbonate and solvents (acidic, neutral, and basic) for interaction with basic polymethylmethacrylate and acidic chlorinated poly(vinyl chloride). The same solutions were studied using acidic silica as the adsorbent. Vrbanac and Berg[23] define the approach to acquire precise thermodynamic data using equations

$$W_{SL}^{ab} = f n_{ab} (-\Delta H_{ads}^{ab}) \tag{6.16}$$

and

$$f = \left(1 - \frac{d \ln W_{SL}^{ab}}{d \ln T}\right)^{-1} \tag{6.18}$$

on the pure liquid–polymer systems described above. They suggest that more data are needed to confirm quantitative agreement between W_{SL}^{ab} from wetting versus independent measurements of n_{ab} and $-\Delta H_{ads}^{ab}$.

Partyka et al.[27] describe a differential, titrating calorimetric procedure to determine ΔH_{ads} of molecules from solution onto solids. They illustrate this with the adsorption of poly(ethylene glycol) octaphenyl ether (Triton X-100) from water onto silica and carbon black.

Adsorption isotherms Isotherms of gases adsorbing on particulate polymer surfaces can be used to determine surface area and average equivalent spherical diameter. This is the Brunauer–Emmett–Teller (BET) analysis, wherein the gas is physically adsorbed (dispersion force) and the number of molecules in a close-packed monolayer is calculated. Knowing the area covered per molecule yields the total surface area. A good discussion of gas adsorption to yield surface area and porosity is given by Gregg and Sing[28] and by Hiemenz[29].

The spreading pressure, π, (see contact angle theory, in the discussion between Equations 6.2 and 6.3) can be determined from the adsorption isotherm of the vapor onto the polymeric solid surface. According to the Gibbs adsorption equation

$$\pi = RT \int_0^{p_0} \Gamma \, d \ln p \tag{6.19}$$

the spreading pressure is related to the Gibbs surface excess, Γ (moles per unit area), and the pressure of the vapor, p_0, at a constant temperature. R is the gas constant. Therefore, measuring Γ at various pressures allows a determination of π.

The Langmuir equation for the adsorption isotherm is applicable to gas adsorption as well as to solute adsorption from solution. The theory postulates adsorption–desorption equilibrium from a dilute solution in which the adsorbed two-dimensional solution is ideal. The linear form is

$$\frac{C_{eq}}{\Gamma} = \frac{1}{K_{eq}\Gamma_m} + \frac{1}{\Gamma_m} C_{eq} \tag{6.20}$$

Plotting C_{eq}/Γ, where C_{eq} is the bulk molar concentration of the adsorbate at equilibrium, versus C_{eq}, where γ is moles adsorbed per m^2, will yield a straight line if the Langmuir model holds. The slope of the line is $1/\Gamma_m$ where Γ_m is the number of moles adsorbed at monolayer coverage. Knowing Γ_m allows us to calculate the equilibrium constant, K_{eq}, from the intercept, $1/K_{eq}\Gamma_m$. Using data from isotherms at two temperatures, we can calculate the enthalpy of adsorption using the van't Hoff equation:

$$\Delta H_{ads} = \left(\frac{RT_1 T_2}{T_2 - T_1} \right) \ln \frac{K_{eq1}}{K_{eq2}} \tag{6.21}$$

It would seem that the surface area of a solid polymer adsorbent could be determined if the area were large enough to establish Γ_m and the cross-sectional area of the adsorbed molecule were known. However, because of the complexity of intermolecular forces between solute, solvent, and solid, such a procedure is fraught with difficulties. Therefore, the surface area of adsorbents is best found using gas adsorption.

Small molecules adsorbing from a dilute solution onto polymer surfaces may be expected to follow the Langmuir model. However, one must be wary about the adsorption of polymers from a solution onto solids. An example will be given where experimental results agreed with the Langmuir expression.[30] Here a mixed phosphoric acid ester with a nonylphenoxy heptaethyleneoxy adduct of 1120 mol wt (Gafac RE-610, GAF Corp.) adsorbed from a dilute mixed solvent solution onto iron oxide. Concentrations ranged up to 15 µM and equilibrium times were 5 days. The plot was linear and revealed that, at monolayer coverage, each molecule occupied 2.2 nm^2 as calculated from the slope $1/\Gamma_m$. The concentration of RE-610 was then increased to the millimolar range (2–20 mM) and a new linear Langmuir plot resulted due to the adsorption of a second layer. Here the packing was tighter and the area per molecule was 0.72 nm^2. The experimental isotherm for the two stages agreed very well with the calculated values. The spread-out conformation of the first layer was a result of the 12 basic ether sites reacting with the predominantly acidic FeOH sites. Then at higher concentration the second layer adsorbed through the acidic P–OH group onto the free ether sites of the first layer. Here there was an on-end configuration with the oxyethylene chains up. In this experiment the analysis of the amount adsorbed was done by equilibrating the iron oxide/polymer solution under gentle agitation, centrifuging the iron oxide with its polymer layer and analyzing the bulk solution for a decrease in RE-610 via atomic emission spectrometry for phosphorus.

This example and others of low molecular weight polymers, as reviewed by Kipling[31], follow the Langmuir model, and thermodynamic data can be derived. High molecular weight polymers, however, present complexities due to molecular weight distribution, solubility, and chain flexibility. Kipling reviews attempts to develop model mathematical expressions for some complex adsorption situations. On the other hand, Koral et al.[32] found the adsorption of poly(vinyl) acetate (1×10^5 to 9×10^5 mol wt) from benzene onto iron powders to be Langmurian at low and at high concentration. At high concentration, where the polymer segments interfere with one another, the Sinha–Frisch–Eirich isotherm has the same shape as the Langmurian[33].

There is considerable interest in establishing adsorption isotherms where the particles are too small or their density too close to that of the medium for convenient separation of the bulk solution by sedimentation. Emulsion polymers represent one such case. Such systems can be studied by means of the serum replacement technique, a procedure which also avoids any conformational changes that might occur upon centrifugation. Here a series of solutions varying in adsorbate concentration are equilibrated serially in a stirred cell having an inlet and an outlet with a semipermeable membrane filter. The cell is pressurized to recover a small sample of bulk solution for analysis. Alternately, a desorption isotherm can be established by pumping in solvent, at a slow (10 mL/h) rate, and continuously monitoring the effluent concentration downstream, say with a calibrated refractive index detector. The rate is slow enough that equilibrium is always maintained. A mass balance at several points yields the adsorbed amounts and the isotherm[34].

New techniques for adsorption isotherms on flat solids and films Polymer film surfaces can be characterized thermodynamically by a new procedure using contact angles. Fowkes et al.[22] report the use as a solvent of methylene iodide (CH_2I_2), a liquid of high surface tension which has dispersion–force interactions only and forms finite contact angles with most polymers. Solutions of various concentrations (0.01–10 M) of Lewis acid and base probe molecules in CH_2I_2 are evaluated for surface tension and contact angle. Plotting the adhesion tension ($\gamma_L \cos \theta$) versus the natural log of concentration ($\ln c$), then determining the slopes at several points, provides a means to obtain the Gibbs surface excess (Γ) at any concentration in solution using Equation 6.22. When the Young equation (Equation 6.2) with γ_{SV} = constant is substituted into the Gibbs equation,

$$\Gamma = -\frac{1}{RT} \frac{d\gamma_{SL}}{d\ln c} = \frac{1}{RT} \frac{d(\gamma_{LV} \cos \theta)}{d\ln c} \qquad (6.22)$$

Now we can plot c/Γ versus c and use Equation 6.20 to calculate the Langmuir parameters, Γ_m and K_{eq} Determining the isotherm at two or more temperatures allows the calculation of ΔH_{ads}^{ab} through the van't Hoff equation (Equation 6.21). Fowkes et al.[22] used phenol (acidic) in CH_2I_2 to study the surface basicity of polymethylmethacrylate. The surface concentration of ester groups was only 0.55 µmole/m^2. This indicates that the ester groups were largely buried below the surface. The heat of phenol adsorption was –20 kJ/mole, in good agreement with the literature.[35, 36]

Ellipsometry as a means of determining the thickness and refractive index of adsorbed films is a valuable tool. The surface must be extremely smooth, which indeed limits the choice of substrate. However, the measurements are fast and adsorption can be followed with time in situ. References 37 and 38 are good for both experiment and theory. The papers reported here deal largely with metal surfaces, and one[37] reports the adsorption of polystyrene onto chromium. Bäckström et al.[39] report studies of the removal of triglycerides from polyvinylchloride surfaces.

Work reported to date has dealt with the amount and conformation of polymers adsorbing onto inorganic solids and smaller molecules adsorbing onto polymers. However, no one has used the technique to establish isotherms. Casper[40] suggests this approach.

Inverse Gas Chromatography

A technique for studying the thermodynamic and interactive nature of polymer surfaces is inverse gas chromatography (IGC). Here the polymer is placed in a gas chromatographic column as the stationary solid phase. It may be as a film on the wall, coated on inert particles, or as particulate or fibrous polymer. Probe gases of a known chemical nature are introduced into the inert carrier gas stream and the retention time is measured. Retention time is related to polymer probe interaction energy and the equilibria existing between adsorbed and gas phase molecules. The technique is simple and rapid, temperature can be varied readily, and the apparatus is not expensive. It yields thermodynamic information and can be used to establish adsorption isotherms. This section gives a brief view of IGC, and the reader is referred to ACS Symposium Series 391[41] for methodology, surface and interface characterization, and other IGC topics, well referenced.

The primary measured quantity in IGC is the specific retention volume, V_g, the volume (corrected to STP) of carrier gas needed to elute a probe from one gram of stationary phase.

$$V_g = \frac{273.16}{T} \cdot \frac{760}{P_o} \cdot \frac{V_N}{w} \tag{6.23}$$

where T is the absolute temperature of the column, P_o is the outlet pressure, w is the weight of polymer (grams), and V_N is the net volume of carrier gas, that is, $V_N = $ flow rate$_{(STP)}$ × (time$_{probe}$ – t$_{marker}$). The inert marker (methane or air) establishes the dead volume. Typically, a plot of log V_g versus $1/T$ is linear with changes in slope signaling phase transitions. Based upon the assumption that the distribution of site energies can be represented by a single mean value, Gray and coworkers[42, 43] developed a procedure for determining the dispersive component of surface thermodynamic properties from IGC data. Given a probe (vapor) at infinite-dilution in the gas stream the free energy of desorption is $\Delta G^\circ_{des} = -\Delta G^\circ_{ads}$ for transition from a reference adsorption state (π°) to a reference gas phase (P_o). Then

$$\Delta G^\circ_{des} = RT \ln V_N \times \text{constant} \tag{6.24}$$

where the constant equals

$$\frac{P_o}{sg\pi_o} = \frac{\text{partial pressure probe}}{(\text{specific area} \times \text{wt})_{solid} \times \text{spreading pressure probe}}$$

One of two reference states are chosen: either that of Kemball and Rideal[44] (where $P_o = 1.013 \times 10^5$ Pa and $\pi_o = 6.08 \times 10^{-5}$ Nm^{-1}) or that of De Boer and Kruyer[45] (where $P_o = 1.013 \times 10^5$ Pa and $\pi_o = 3.38 \times 10^{-4}$ Nm^{-1}). Thus,

$$\Delta G^\circ_{des} = RT \ln V_N + K \tag{6.25}$$

where K includes the constant of Equation 6.24 and ΔG°_{des} approximates the work of adhesion of the probe onto the surface per unit area, or

$$\Delta G^\circ_{des} \approx NA W_a \tag{6.26}$$

where N is Avogadro's number and A is the area per probe molecule. Unfortunately, the values for A are not easy to determine unambiguously. The dispersive component of the work of adhesion of the gaseous probe is

$$W_a^d = 2(\gamma_S^d \gamma_L^d)^{1/2}$$

and combining this expression with Equations 6.29 and 6.30 gives

$$RT \ln V_N = 2N(\gamma_S^d)^{1/2} A (\gamma_L^d)^{1/2} + \text{constant} \tag{6.27}$$

Gray found $\gamma_S^d = 40$ mN/m for polyethylene terephthalate film in good agreement with five workers using contact angle techniques. He found that In V_N varied linearly with the number of carbon atoms in a series of n-alkanes.

Schultz et al.[46] and Lara and Schreiber[47] plotted $RT \ln V_N$ versus $A(\gamma_L^d)^{1/2}$ and obtained a straight line for a series of n-alkanes, as shown in Figure 6.2.

The slope of the alkane line, $2N(\gamma_S^d)^{1/2}$, gives the dispersive contribution to the free energy of the surface. The acid–base (or specific) contribution is calculated from the distance above the alkane line, or $\Delta G^\circ_{ab} = RT \ln (V_N/V_{N\text{alkane}})$. In Figure 6.2, the surface has both acidic and basic sites. A plot of ΔG° versus log P_o can be treated in an analogous fashion to Figure 6.2.[48] If IGC experiments are carried out at two temperatures, the enthalpy can be determined by

$$\Delta H^\circ = \frac{RT_1 T_2}{T_2 - T_1} \ln \frac{V_{N_1}}{V_{N_2}} \tag{6.28}$$

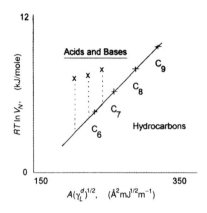

Figure 6.2 Schematic plot of $RT \ln V_N$ versus $A(\gamma_L{}^d)^{1/2}$ for various probes on a polar surface.

which is equivalent to Equation 6.21. The entropy then is available by means of Equation 6.9.

Saint-Flour and Papirer[49] developed a procedure for the acid–base reactions of probes with polymer surfaces using Gutman acceptor and donor numbers AN_p and DN_p for the probe *(p)* in the linear expression

$$\frac{\Delta H_{ab}}{AN_p} = K_A \frac{DN_p}{AN_p} + K_D \tag{6.29}$$

where K_A and K_D are acid and base constants for the solid polymer. Knowing the Gutman numbers for a series of probes interacting with a polymer surface to give $-\Delta H_{ab}$ yields K_A (slope) and K_D (intercept).

Schultz et al. extended this approach to calculate a specific interaction parameter, A, for acid–base interactions of polymeric matrices (m) and carbon fibers (f):

$$A = K_{A_{(f)}} K_{D_{(m)}} + K_{A_{(m)}} K_{D_{(f)}} \tag{6.30}$$

Parameter A was found to correlate well with shear strength for the carbon fiber/epoxy composites.

Infrared (FTIR) Spectroscopy

Acid–base interactions result in shifts (Δv) of the infrared spectral peaks of both functional groups. Let us say an alcohol (acidic) interacts with an electron-rich carbonyl (basic) on the surface of a polymer, and we see shifts in both characteristic peaks. The magnitudes of the shifts are related to the energy of the new hydrogen bond. Often the Δv is linearly related to the enthalpy of the interaction,[50] but the relationship must be calibrated.

Drago et al. found the Δv_{C-OH} for phenol forming 1:1 complexes with various bases in CCl_4 solution was linearly related to the $-\Delta H$ determined calorimetrically,[51] $-\Delta H^{ab}_{Kcal/mol} = 0.0103\Delta v_{OH} \ (cm^{-1}) + 3.08$. Assuming the dispersion force of the CCl_4 was the same on the phenol as on the complexes, the measured $-\Delta H^{ab}$ and Δv_{OH} were due to acid–base interactions. Fowkes et al.[52] established an equivalent expression in the same way for the carbonyl shift of ethyl acetate interacting with acids:[51] $-\Delta H^{ab}_{Kcal/mol} = 0.236\Delta v^{ab}$ (cm^{-1}). With the same acids on poly(methylmethacrylate), the Δv^{ab}_{CO} was found to be 63% of that for ethyl acetate.

Limited studies have been reported for Fourier transform infrared (FTIR) spectroscopy on polymer surfaces, but the four-volume *FTIR—Application to Chemical Systems*[53] contains an article by Jakobsen[54] detailing internal and external FTIR reflectance procedures for polymer surface studies. He reports qualitative conformational findings on several polymer systems, but the procedures could be used to measure Δv. Internal fracture surfaces were studied by D'Esposito and Koenig[55] using transmission FTIR to identify the products of chain scission reactions during the elongation of several polymers. Infrared microscopy allows the sampling of areas as small as 5 µm in diameter. One instrument (the IRµs/SIRM, manufactured by Spectra-Tech, Inc.) is offered that can be used in either the reflectance or grazing incidence mode to obtain spectra of polymer layers as thin as 20 Å on metal.

The use of FTIR to measure Δv at polymer surfaces is a promising method for gathering thermodynamic and conformational information. Little's book on infrared spectra of adsorbed species[56] is a good source of basic material. Yarwood[57] reviews studies on ultra-thin organic films (single monolayer) using attenuated total reflectance and reflection–absorption techniques. His examples deal with characterizing potential electronic device structures but give no thermodynamic information. A good bibliography is included.

References

1   W. A. Zisman. *Adhesion and Cohesion*. (P. W. Weiss, Ed.) Elsevier, Amsterdam, 1962, pp. 176–208.

2   S. Ross and I. D. Morrison. *Colloidal Systems and Interfaces*. Wiley, New York, 1988, pp. 205–246.

3   F. M. Fowkes. *J. Adhesion*. 4, 155–159, 1972.

4   F. P. Buff. *Encyc. of Phys.* (S. Flügg, Ed.) Springer Verlag, Berlin, 1960, pp. 281-304.

5   F. M. Fowkes. *Ind. Eng. Chem.* 56 (12), 40–52, 1964.

6   F. M. Fowkes. *J. Adhesion Sci. Tech.* 4 (8), 669–691, 1990.

7   C. J. van Oss, R. J. Good, and M. K. Chaudhury. *Langmuir.* 4, 884-891, **1988**.

8   F. M. Fowkes and M. A. Mostafa. *IEC Prod. Res. Dev.* 17, 3, 1978.

9   C. J. van Oss and R. J. Good. *J. Macromol. Sci.-Chem.* A26 (8), 1183-1203, 1989.

10 R. J. Good. *J. Adhes. Sci. Tech.* **6**, 1992.

11 D. H. Napper. *Polymeric Stabilization of Colbidal Dispersions.* Academic Press, New York, 1983.

12 P. G. deGennes. *Macromolecules.* **14**, 1637, 1981.

13 P. G. deGennes. *Adv. in Coll and Int. Sci.* **27**, 189, 1987.

14 W. A. Zisman. *Contact Angle, Wettability and Adhesion.* Adv. in Chem. Ser., Vol. 43, American Chemical Society, Washington, DC, 1964, pp. 1-51.

15 R. J. Good. *Surface and Col. Sci.* Vol. 2. (R. J. Good and R. R. Stromberg, Eds.) Plenum, New York, 1979.

16 F. M. Fowkes. *Chem. and Physics of Interfaces.* American Chemical Society, Washington, DC, 1965, chapt. 1.

17 T. Hata, Y. Kitazaki, and T. Saito. *J. Adhesion.* **21**, 177-194, 1987.

18 R. H. Dettre and R. E. Johnson. Monograph No. 25. Soc. Chem. Ind., London, 1967, chapt. 6.

19 F. E. Bartell and H. J. Osterhoff. *Colloid Sym. Monog.* 5th Nat. Symp. 1927 **5**, 113-134, 1928.

20 E. W. Washburn. *Phys. Rev.* **17** (2), 273-283, 1921.

21 F. A. Bystry and L. S. Penn. *Surf. Interfaces Analysis.* **5**, 98, 1983.

22 F. M. Fowkes, M. B. Kaczinski, and D. W. Dwight. *Langmuir.* **7**, 2464, 1991.

23 M. D. Vrbanac and J. C. Berg. *J. Adhesion Sci. Tech.* **4**, 255-266, 1990.

24 R. S. Drago, G. C. Vogel, and T. E. Needham. *J. Am. Chem. Soc.* **93**, 6014, 1971.

25 S. T. Joslin and F. M. Fowkes. *I&EC Prod. Res. Dev.* **24**, 369, 1985.

26 F. M. Fowkes, K. L. Jones, E. Li, and T. B. Lloyd. *Energy and Fuels.* **3**, 97, 1989.

27 S. Partyka, M. Lindenheimer, S. Zaini, E. Keh, and B. Brun. *Langmuir.* **2**, 101–105, 1986.

28 S. J. Gregg and K. S. W Sing. *Adsorption, Surface Area and Porosity.* Academic Press, New York, 1967.

29 P. C. Hiemenz. *Principles of Colloid and Surface Chemistry.* Marcel Dekker, New York, 1977.

30 Y. C. Huang, F. M. Fowkes, and T. B. Lloyd. *J. Adhesion Sci. Tech.* **5**, 39-56, 1991.

31 L. L. Kipling. *Adsorption from Solution of Non-electrolytes.* Academic Press, London, 1965, pp. 154-157.

32 J. Koral, R. Ullman, and F. R. Eirich. *J. Phys. Chem.* **62**, 541, 1959.

33 R. Sinha, H. L. Frisch, and F. R. Eirich. *J. Phys. Chem.* **57**, 584, 1953.

34 S. M. Ahmed, M. S. El Aasser, F. J. Micale, G. W. Poehlein, and J. W. Vanderhoff. *Polymer Colloids II.* (R. M. Fitch, Ed.) Plenum, New York, 1980, p. 265.

35 T. D. Epley and R. S. Drago. *J. Am. Chem. Soc.* **89**, 5770, 1967.

36 T. K. Kwei, E. M. Pearce, F. Ren, and J. P. Chen. *J. Polym. Sci.—Polym. Phys. Ed.* **24**, 1597, 1986.

37 *Ellipsometry in the Measurement of Surfaces and Thin Films.* (E. Passaglia, R. R. Stromberg, and J. Kruger, Eds.) NBS Misc. Pub. 256, Washington, DC, 1964.

38 *Recent Developments in Ellipsometry.* (N. M. Bashara, A. B. Buckman, and A. C. Hall, Eds.) North Holland, Amsterdam, 1969.

39 K. Bäckström, B. Lindman, and S, Engström. *Langmuir.* **4**, 872, 1988.

40 L. A. Casper. Ph.D. dissertation. Lehigh University, Bethlehem, PA, 1985.

41 *Inverse Gas Chromatography.* ACS Symposium Series 391. (D. R. Lloyd, T. C. Ward, and H. P. Schreiber, Eds.) American Chemical Society, Washington, DC, 1989.

42 G. M. Dorris and D. G. Gray. *J. Coll. Interface Sci.* **77**, 353, 1980.

43 J. Anhang and D. G. Gray. *J. Appl. Polym. Sci.* **27**, 71, 1982.

44 C. Kemball and E. K. Rideal. *Proc. Roy. Soc.* **A187**, 53, 1946.

45 J. H. DeBoer and S. Kruyer. *Proc. K. Ned. Akad. Wet.* **B55**, 451, 1952.

46 J. Schultz, L. Lavielle, and C. Martin. *J. Coatings Tech.* **63**, 81, 1991.

47 J. Lara and H. P. Schreiber. *J. Coatings Tech.* **63**, 81, 1991.

48 E. Papirer and H. Balard. *J. Adhesion Sci. Tech.* **4**, 357, 1990.

49 C. Saint-Flour and E. Papirer. *Ind. Eng. Chem., Prod R&D.* **21**, 666, 1982.

50 L. J. Bellamy. *Advances in Infrared Group Frequencies.* Methuen, London, 1968, chapt. 8.

51 R. S. Drago, G. C. Vogel, and T. E. Needham. *J. Am. Chem. Soc.* **93**, 3203, 1977.

52 F. M. Fowkes, D. O. Tischler, J. A. Wolfe, L. A. Lannigan, C. M. Ademu-John, and M. J. Halliwell. *J. Poly. Sci.* **22**, 547, 1984.

53 *FTIR-Application to Chemical Systems.* (J. R. Ferraro and L. J. Basile, Eds.) Academic Press, New York, 1978-85.

54 R. J. Jakobsen. *Inverse Gas Chromatography.* Vol. 2, ACS Symposium Scries 391. (D. R. Lloyd, T. C. Ward, and H. P. Schreiber, Eds.) American Chemical Society, Washington, DC, 1989, chapt. 5.

55 L. D'Esposito and J. L. Koenig. *Inverse Gas Chromatography.* Vol. 1, ACS Symposium Series 391. (D. R. Lloyd, T. C. Ward, and H. P. Schreiber, Eds.) American Chemical Society, Washington, DC, 1989, chapt. 2.

56 L. H. Little. *Infrared Spectra of Adsorbed Species.* Academic Press, New York, 1966.

57 J. Yarwood. *Spectroscopy.* **5**, 34, 1990.

7

Surface Modification of Polymers

NED J. CHOU and CHIN-AN CHANG

Contents

7.1 Introduction

The polymer and plastics industry can be dated either from 1870 with the initial production of celluloid or from 1907 with the announcement of the phenolic resins.[1] Its development was spurred by the acute shortage of corrosive-resistant metals in the mid 1930s during the Second World War and further nurtured by the war efforts in the 1940s. The industry has made such big strides in the following four decades that polymers and plastics now find pervasive use in practically every branch of engineering. Surface treatment or modification has been an indispensable step in the fabrication of polymer and plastic products from films, fibers, and laminates to intricate parts and structures. Even when plastic welding technology was in its embryonic stage, it was realized that the weld strength could be greatly improved by proper surface preparation, which involved, among other things, mechanical roughening of the surfaces to be joined.[2]

Most polymeric surfaces are inert and hydrophobic. The purpose of surface treatment is to modify the surface layer of the polymer in order to enhance its wettability, weldability, and printability, its resistance to crazing, or its adhesion to metals or polymers, while retaining the desirable properties of the polymer in the bulk. Surface modification can be achieved in many ways, from wet-chemical means to dry treatment methods. Historically, many of these schemes of surface modification have been developed empirically in production practices. As polymers find increased applications in the "hi-tech" fields of microelectronics and biomedicine, the schemes of surface modification and engineering have evolved from simple surface roughening and corona discharge treatment to more advanced—and sometimes costly—techniques using ion beams, photografting, plasmas, sputtering, and γ-ray or electron irradiation.

Parallel to the evolution of surface modification technology has been the rapid development of a number of sophisticated surface-sensitive and surface-oriented spectroscopic and imaging techniques, which are particularly suited for the characterization of polymer surfaces and interfaces. Perhaps most significant of all is the wide-ranging development of X-ray photoelectron spectroscopy (XPS) by Siegbahn and his co-workers during the period 1955–1970.[3] This work was responsible for transforming surface modification and engineering from empiricism to scientific studies. The first documented use of XPS in surface treatment studies was by von Brecht et al.,[4] who in 1973 studied the effect of chemical pretreatment to enhance adhesive bonding of polytetrafluoroethylene (PTFE). It is virtually impossible to find a surface modification study in the years that followed that does not use XPS to monitor the changes in chemical composition and structure of the polymer surfaces. For researchers to relate the observed changes in macroscopic property (wetting, adhesion, etc.) to the changes observable on a microscopic and molecular scale, more information is needed than what XPS alone can provide. The last few years have witnessed a dramatic increase in the employment of the so-called multi-technique approach to surface modification studies. In a multitechnique investigation, a number of sophisticated, mutually complementary characterization techniques are used either simultaneously or in tandem to obtain a more complete picture regarding the relationship between the changes in microscopic characteristics and the changes in surface properties brought about by modification. Despite the vast knowledge we have so far amassed on polymer surfaces, it is still very difficult to elucidate the nature of such relationships. To do this successfully, we must be able to address three key issues:

1 A reproducible method of modification can be developed for polymer surfaces whose structures are well understood at the molecular level.

2 A detailed study can be made of the structure-reactivity relationship exhibited by the materials under study.

3 The molecular basis of gross surface properties can be elaborated through the use of specific, well-defined changes in the microscopic structure of the materials.

In this chapter we focus on the first issue, that is, on the methods developed for the surface modification of polymers. Whenever appropriate we also discuss the molecular basis of the macroscopic property changes in the systems which have been surface modified.

Surface modification can be achieved either by wet-chemical means or by "dry" treatment methods. The schemes of treatment are many and can range from simple surface roughening by mechanical means to sophisticated irradiation techniques.

We therefore cover the modification techniques under two headings: (1) Wet-Chemical Modification of Polymer Surfaces and (2) Dry Modification Techniques. We deal with the techniques widely accepted in industrial practices as well as those still in an experimental stage of development. We discuss each of these techniques in terms of their fundamental understanding, scope of applicability, and experimental implementation.

7.2 Wet-Chemical Modification of Polymer Surfaces

The wet-chemical treatment of polymer surfaces has been a part of the manufacturing process since the inception of polymer technology. It has been used to improve the wettability, printability, bondability, and antistatic properties of polymeric materials. Although different chemicals and treatment procedures have been adopted to achieve the same objective for various groups of polymers, there is a considerable degree of commonality to the treatment:

- chemical reagents used in solution are highly reactive for the polymers to be treated

- the treatment usually results in the oxidation of surfaces

- a modified surface layer different in composition, structure, or functionality from the bulk is formed in the processes following the oxidation.

We therefore examine wet-chemical techniques according to the objective of the treatment and its application in the production practice, instead of by category of polymer.

Surface Roughening and Seeding in Plating Operations

Surface roughening by chemical methods was an important processing step in the production of metallized plastics long before the study of surface modification began to attract the attention of chemists and surface scientists. In the electroless or electrolytic plating processes, wet-chemical etching can shorten the roughening operation from tens of hours to a few minutes. It should be noted, however, that a more rigorous process control is required for chemical etching than for mechanical roughening because the etchant may cause severe surface pitting and chain scission, with the formation of a weak boundary layer on the surface.

Plastics of the phenol and urea-formaldehyde type can be treated in an acid etch composed of

Sulfuric acid (66° Baume)	256 parts per volume
Nitric acid (40° Baume)	128 parts per volume
HCl (1.2 sp. gr.)	1 part per volume
H_2O	32 parts per volume

The treated parts should be rinsed immediately to remove the excess acid and neutralized in a 10% Na_2CO_3 solution.

For thermoplastic polymers, the most popular etchant is an aqueous chromic acid solution. One of the time-tested recipes is a 2-min soak in the solution

H_2SO_4	100 cm^3
$K_2Cr_2O_7$	15 g
H_2O	50 cm^3

An efficient treatment for thermosetting-type plastics consists of a 3-min immersion in the solution

Hydroquinone	400 cm^3
Pyrocatechin	100 cm^3
Acetone	4000 cm^3

which will produce the desirable etching of the surfaces.

In plating technology, "seeding" is often accomplished by "sensitizing" and "activation" in two separate steps. It consisted in treating the roughened and thoroughly cleaned polymer surface as follows:

1 immersion for 1–2 min in a "sensitizing" solution, followed by a thorough cleaning in running deionized water; and

2 activating the sensitized surface by immersion for a fraction to a couple of minutes in a seeding solution, again followed by rinsing in running water.

The "sensitizing" solution usually is either a chromic acid solution or a dilute acidic solution of stannous chloride, whereas the seeding solutions are Pd, Au, or Pt salts in dilute acids. The seeding procedure clearly indicates that the "seeded" layer is comparable in thickness to an adsorbed layer.

It had been suggested as early as the 1940s that the function of the sensitizing treatment is to furnish Sn(III) ions which, in the activation step, reduce the Pd ions to form an "invisible" film of metallic Pd on the substrate, acting as a catalyst for electroless plating.[5–9] The contemporary view, however, no longer regards seeding as a surface modification procedure, but holds that sequential immersion in SnCl$_2$ and PdCl$_2$ solutions constitutes a two-step catalyst and that the polymer surface is equivalent to the catalyst support in conventional catalysis. Among the Pt metal catalysts, Pd-based two-step systems are the most popular reagents used in the plating industry.

It has been shown that chemically modified polymer surfaces can affect the activity of electroless catalysts in different ways.[10–12] Photochemically oxidized polystyrene (PS) surfaces, for instance, have a poisoning effect on the electroless catalyst, which can be reversed by soaking the photooxidized PS surfaces in an alkaline solution before catalysis.[11]

Extensive studies have been made in the field of conventional supported catalysis in the last two and half decades. Various surface science techniques including XPS, Auger electron spectroscopy (AES), electron energy-loss spectroscopy (EELS), low-energy electron diffraction (LEED) techniques, thermal desorption spectroscopy (TDS) and electron microscopies (scanning electron microscopy [SEM] and transmission electron microscopy [TEM]) have been employed to study simulated and applied catalyst systems. Much progress has been made in the study of the chemical change (particularly in oxidation state) of the dopants during the setup and use of catalysts and in the study of the interaction between the dopant and the support. (For an excellent survey on the successful application of electron spectroscopy to conventional supported catalysts see Reference 13.) However, the XPS study of Pd-based electroless catalysts on PS has been less than successful.[14] The Pd and Sn spectra of the original and variously treated surfaces arc indistinguishable. The XPS data also indicate that the catalysts on the platable (activated) and unplatable (deactivated) surfaces are structurally similar and are adsorbed to the surfaces in similar concentrations. The mechanism of deactivation by various treatments therefore cannot be clearly defined. However, the lack of mechanistic understanding does not prevent researchers and engineers from trying to manipulate the activation–deactivation effect of the surface treatment to achieve a selective (patterned) plating of plastic surfaces.[10–12, 14]

Surface Treatment for Bondability

One of the primary objectives of surface treatment is to enhance the adhesion of a polymer to a metal or ceramic or to another polymer. The chemical reagents used for the enhancement treatment vary with the type and chemical structure of the polymers to be treated.

Polytetrafluoroethylene (PTFE) The polymer $(C_2F_4)_n$ is conventionally classified as a thermoplastic even though at processing temperatures the material is actually a semisolid gel rather than a melt. It is fabricated by the techniques of powder metallurgy or ceramic processing. The chemical and thermal stability of PTFE is well-known. It can be used continuously at 260 °C without measurable degradation.[14] Even at elevated temperatures for long periods of time, sulfuric acid, nitric acid, aqua regia, hydrofluoric acid, alkaline solutions, and hydrogen peroxide have no effect on the polymer. It is attacked only at temperatures above 150 °C by fluorine, chlorine trifluoride, and molten alkaline metals. Although several hundred solvents have been tested none can dissolve PTFE below 300 °C. These remarkable properties of PTFE also mean that the material is difficult to bond to other material or to fabricate by heat-sealing, hot-gas welding, etc. A "cementable" surface, that is, one with adhesive bonding, on this material can be obtained by a pretreatment that involves treating the material for a few seconds in (1) a 1% solution of metallic sodium in liquid anhydrous ammonia or (2) a sodium-naphthalene-tetrahydrofuran solution.[1, 15] After washing and drying, the treated surface can be bonded satisfactorily with conventional adhesives.

XPS studies of modified PTFE indicate that the surface treatment gives rise to surface defluorination with the production of an unsaturated carbon layer which reacts with O_2/H_2O upon subsequent exposure to air or in the cleaning procedure that follows. Similar results of defluorination and oxygen and nitrogen incorporation can be achieved by exposing PTFE to a glow discharge in ammonia.[1] The increased wettability and adhesion of the treated PTFE can be reversed by a nitric-perchloric acid treatment, which XPS shows to be due to the removal of the modified layer.

Polyolefins   In this class of polymers are low- and high-density polyethylene (LDPE, HDPE), polypropylene (PP), and copolymers involving ethylene, propylene, and higher α-olefins, which dominate the polymer industry in terms of the volume of annual production and the scope of end products. Surface modification of these polymers has been successfully applied in many of their fabrication processes.

Several commercial modification techniques can be used to improve the adhesion characteristics of polyolefins. Printing on untreated PE surfaces is generally unsatisfactory, but excellent ink adhesions can be achieved if the surfaces are modified. For films and sheets, the most effective method is to pass an electric discharge over the PE surface. The next common method is to subject the surface to an open gas flame while the material runs over a water-cooled roller to prevent overheating. For intricate or irregular shapes, chromic acid etching is preferred. Chromic acid etching can cause severe roughening of polyolefin surfaces.[16] A systematic study by Briggs et al.[17] shows that the water contact angle on LDPE and HDPE surfaces decreases with etching, whereas the contact angle on the PP surface exhibits a minimum, which is reached after a brief exposure, followed by an increase to a plateau value close to that of an untreated surface. Reflection IR measurements detect chemical modifications (oxidation and sulfonation) on the treated LDPE but not on the HDPE or PP surface. Adhesion strength data on LDPE, HDPE, and PP joints using an epoxy adhesive indicate that adhesion improvement is correlated with the degree of oxidation but not sulfonation. With the aid of XPS analysis these observations can be explained as follows: (1) the thickness of the oxidized layer on PP is much less than that on LDPE; (2) the degree of oxidation of the PP surface reaches a steady state value with a continuous loss of the material into solution, whereas LDPE continues to oxidize in depth with time. The mechanism of adhesion enhancement in this case is believed to be due primarily to surface oxidation rather than the interlocking effect of surface roughening, although controversy exists as to which of the two has a greater effect on adhesion.

Polyimides (PI)   PI, a new class of thermosets, are characterized by excellent thermal, chemical, and mechanical stability. Once fully cured, PI are resistant to all acids, weak bases, and common organic solvents and can withstand temperatures as high as 400 °C[18]. Thin PI films have a breakdown strength in excess of 5 MV · cm^{-1} and a dielectric constant comparable to that of SiO_2, with a value between 3.02 and 4.0 depending on the film thickness.[19, 20] The use of PI as insulating layers in microelectronic multilevel interconnect structures dictates that PI layers must adhere well to substrates (Si or SiO_2), to metal lines, and to another PI layer. Patent literature contains a number of citations for chemical treatments to improve metal adhesion to PI.[21] Adhesion promoters are used to enhance PI adhesion to Si and SiO_2.[22–25]

A new PI layer can be put on a previously cured layer by spin-casing polyamic acid resin dissolved in an *N*-methyl pyrrolidone (NMP) medium. Two factors affect the adhesion in this case: the cleanliness of the cured PI surface, and the degree of attack on the old surface by the solvent of the newly deposited resin.

A chemical treatment that can affect the adhesion between two cured PI films has been reported in several investigations.[26-28] It consists in modifying the surface by base hydrolysis (e.g., 2–5 h in an aqueous 0.25 M NaOH solution) followed by protonolysis (e.g., 0.1 M aqueous solution of acetic acid) to form a thin film of polyamic acid on the fully cured PI. When two modified films are pressed together and heated, a strong bond between them will be formed following imidization of the modified layers. Both hydrolysis and protonolysis can be carried out at room temperature, and modified films can be air- or vacuum-dried and should be kept free of contamination. The peel strength of the polymer–polymer joint usually exceeds the cohesive strength of the PI, that is, greater than 125 g/mm.[29] Although the treatment is effective for self-adhesion enhancement, it is not practiced commercially because the production demand does not justify its complicated engineering implementation. Furthermore, in multilevel interconnect technology, the chemical treatment to enhance self-adhesion is supplanted by the dry method of water vapor plasma treatment.[†]

Surface Treatment Using Adhesion Promoters

Silane coupling agents are known to be very effective in bonding organic polymers to minerals and ceramics. The earliest use of chemical adhesion promoters in the microelectronic industry involves the application of hexamethyldisilazane (HMDS), which is known to be monofunctionally reactive with OH radicals. The material is selected as the adhesion promoter for resist polymers on SiO_2, which is known to be covered with SiOH called "silanols."[30] The promoter is usually applied in a dilute 0.3–7 wt. % solution (in acetone or xylene) by the conventional spinning technique. It has been shown that liquid phase HMDS application, effective for thermal oxides, is not adequate for spin-on glass, P-doped low-pressure chemical vapor-deposited oxides or plasma-deposited oxides.[31] For the latter three types of oxide substrate, a two-step double promoter process is necessary, involving the successive application of a liquid solution of vinyltrichlorosilane (VTS) and 3-chloropropyltrimethoxysilane, followed by a successive cure cycle at 90 °C in N_2 before a photoresist application. Because of its process complexity, the double promoter treatment has been replaced by the vapor phase HMDS application, which proves to be effective for all four types of oxide substrates.

Vapor phase HDMS treatment can now be conducted in a commercial machine, where the oxide substrates are dehydrated in situ at 150 °C in dry N_2 and HMDS vapor primed at the same temperature. XPS studies suggest that vapor phase HMDS

---

† R. D. Goldblatt. Private communication.

Organofunctional Group	Chemical Structure	Trade Name
Epoxy	$CH_2-CHCH_2OCH_2CH_2CH_2Si(OCH_3)_3$ (epoxide O bridge on CH_2-CH)	DC® Z-6040
Methacrylate	$CH_2=\overset{CH_3}{\underset{|}{C}}-CO-O-CH_2CH_2CH_2Si(OCH_3)_3$	DC® Z-6030
Diamine	$H_2NCH_2CH_2NHCH_2CH_2CH_2Si(OCH_3)_3$	DC® Z-6020
Vinylbenzyl cationic	$V.B.\overset{HCl}{NH}CH_2CH_2NHCH_2CH_2CH_2Si(OCH_3)_3$	DC® Z-6032

Table 7.1   Typical organofunctional silanes for organic-to-organic bonding.

treatment yields superior adhesion results because the process removes carbon contamination very efficiently and covers the treated surface with covalently bonded methyl groups, which are responsible for enhanced adhesion to polar resin of the photoresist.

Although organofunctional silanes are usually known as "coupling agents" for bonding organic polymers to mineral or oxide substrates, certain silane primers which are effective in bonding dissimilar polymers to glass are also effective as polymer–polymer adhesion promoters for these dissimilar polymers. Typical organofunctional silanes for polymer–polymer bonding are shown in Table 7.1.[32] In practice, these promoters or primers are generally applied not in a simple monolayer but in a multilayer of oligomeric siloxanes. The oligomers are initially soluble and fusible and may bond with polymers by simple copolymerization or by one of two possible mechanisms:

1 If a liquid polymer or polymer solution contacts the promoter/primer while it is soluble and fusible, the promoter may interdiffuse into the polymer to an extent determined by the solubility of the two phases.

2 If the fusible silane films include organofunctional groups that can react chemically with the polymer, the interdiffused layer may become an interpenetrating polymer network, which provides even better adhesion across the interface.

A silane primer film which interpenetrates different polymers should interpenetrate two such polymers brought into intimate contact and bond them to each other. Several formulations of modified silane primers have been developed to exploit the interpenetration effect.[33] Among them is the formulation of partially prehydrolyzed silane in a solvent, which is effective in bonding various coatings and thermoplastics in general. A typical recipe for such a formulation is

Silane	50 parts
Methanol	50 parts
Water	5 parts
Dicumyl peroxide (with unsaturated silanes)	1 part

The mixture should be allowed to stand at room temperature for a few hours to equilibrate the oligomeric structure. The concentrated primer may then be diluted to 10% solids in any convenient solvent (e.g., isopropanol).

Surface Modification by Photochemical Methods

The use of organofunctional silanes as adhesion promoters is not the only method to provide polymer surfaces with desirable functionalities. Photografting is a photochemical technique whereby functional can be chemically attached (grafted) to the polymer without an intervening phase. In principle, grafting polymer surfaces creates "macroradicals" from which a chain reaction can be started by adding unsaturated monomers. Macroradicals can be formed either by high-energy irradiation such as γ rays[34–36] or UV irradiation together with a photoinitiator.[37] In the latter case only the initiator absorbs the UV radiation, leaving the polymer bulk unaffected, and grafting takes place only at the interface between the polymer and surrounding medium from which the initiator and the grafting monomer are supplied.

Benzophenone and its derivatives are the most commonly used photoinitiators. When irradiated with UV light, benzophenone is excited to its first singlet state and through intersystem crossing rapidly transforms to the triplet state. Benzophenone in its triplet state can abstract hydrogen from a donor, namely, the polymer substrate, leaving a macroradical on the surface which can attack the unsaturated monomer and start grafting.[38] The benzhydryl radicals produced are too bulky to polymerize by themselves. They will participate in terminating reactions by radical combination. The grafting process is depicted schematically in Figure 7.1.

A critical issue in photografting is the selection of an appropriate solvent. To be used in liquid phase grafting, the solvent should be inert to the initiator, and the solution should wet the polymer substrate. Furthermore, it should also be a solvent towards the graft chains. Its penetration into the polymer will affect the depth of grafting. Acetone has been found to be a good solvent for grafting since it directs grafting to the top surface layers.[39] Ethanol also works as a grafting solvent, but the grafting goes deeper into the substrate; that is, it is not as surface-specific as acetone.

Photografting also can be performed in the vapor phase. Proponents of this approach contend that the constraint imposed by the use of solvents in liquid phase grafting can be removed[40] and there will be a greater choice of initiators and monomers. Photografting in liquid or vapor phase has been successfully performed on a number of polymer substrates, including those of LDPE, LDPE, PS, and PET (polyethlene terephthalate).[41] A wide variety of monomers were used in these experiments, from simple acrylic acid, acrylamide, and acrylonitrile to glycidal acrylate (GA) and glycidal methacrylate (GMA), which contain epoxy groups.

The effect of grafting can be evaluated by a number of surface-sensitive analytical techniques: contact angle measurement, XPS, attenuated total reflection-Fourier transform infrared (ATR-FTIR), and static secondary ion mass spectrometry (SIMS)

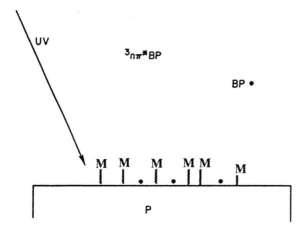

UV

$^3_{n\pi^*}$BP

BP •

M M M MM M

P

M = VARIOUS FUNCTIONAL GROUPS

P' + M ⟶ PM' ⟶ PM$_n$M'

Figure 7.1 Schematic representation of photografting with benzophenone (BP) as an initiator: X is the monomer for graft polymerization; $^3n\pi^*$BP is BP in its triplet state; and *BP* is the benzhydryl radical.

techniques. Successful grafting of GA and GMA to LDPE substrates, for instance, has been confirmed by ATR-IR analysis, which showed that grafting of GA occurred at a much higher rate than for GMA.[42] In Figure 7.2 the ATR-IR spectra between 1300 and 1900 cm^{-1} for GA- and GMA-grafted LDPE are shown with that of ungrafted LDPE. The carbonyl band at 1735 cm^{-1} is a clear sign of a successful grafting. It should be noted that during grafting homopolymers also form on the surface. It is important to remove these homopolymers and adsorbed chemicals thoroughly before the analysis is undertaken.

Photografting is not yet a commercial process, although attempts have been made to demonstrate that the process can be carried out continuously as a thin liquid film containing the initiator and the monomer is applied to the polymer surface and irradiated with UV light.[43]

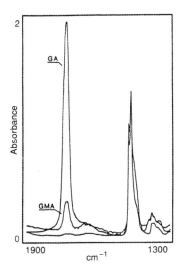

Figure 7.2   ATR-IR spectra near carbonyl absorption near 1735 cm^{-1} for GA-, GMA- and ungrafted LDPE. Note that the faint carbonyl signal for the ungrafted LDPE is due to surface oxidation.[44]

7.3 Dry Modification Techniques

The dry modification treatment (as opposed to wet-chemical methods), usually carried out in a vapor or gas phase or in a rarified ambient, may be classified according to the methodology employed:

- modification by high- or low-energy ion beams

- corona discharge and plasma treatment

- application of adhesion promoters by vacuum or plasma-enhanced deposition.

Modification by High- and Low-Energy Ion Beams

In an ion beam treatment the sample or workpiece is irradiated or bombarded with energetic ions. The kinetic energy of the bombarding ions may range from thousands to millions of electron volts depending on the nature of the treatment, which dictates the appropriate penetration depth and the choice of ion species and fluence. In this energy range, which is far above the chemical binding energies, the forces between an impinging ion and target atoms are strongly repulsive and collisions can occur in cascades. In such collisions a projectile ion is deflected or scattered through some angle while the struck atom gains some energy. This is a process called "nuclear stopping," in which the projectiles lose their kinetic energy via elastic collisions. In addition, inelastic processes associated with electronic excitation or ionization of host atoms also cause the projectiles to lose their energy by "electronic stopping." In general, inelastic collisions are much more frequent when the projectiles have high energy, and elastic collisions become more important after the ions have slowed. Thus, the projectiles will continue to travel through the sample until they come to a stop after they have lost their excess energy via electronic and nuclear stopping. The study of ion–solid interaction has advanced to such a stage that stopping (rate of energy loss) and range (penetration depth) calculations can be made for any ion species in a given solid with an accuracy of a few percent compared with experimental data.[45]

At low (keV) energies, the ion–target interactions are confined to the first few surface layers of the treated material, whereas at high (MeV) energies the range of ions may be several micrometers deep into the substrate. It is primarily this difference in interaction depth that determines the nature and technique of ion beam treatment.

Based on the current mechanistic understanding of adhesion phenomena, several potential adhesion enhancement mechanisms can be expected of an ion beam treatment: (1) surface cleaning by sputtering away extraneous contaminants and the creation of unsaturated surface functionalities and dangling bonds; (2) development of surface morphology, which promotes mechanical interlocking between the films and the substrates; (3) interface mixing and graded hardening of polymeric substrates via ion-induced carbonization; and (4) tailoring of interface chemistry by appropriate ion implants. It is obvious that the latter two effects can be produced only by ion beams of intermediate (20–400 keV) and high (MeV) energies.

At present, ion beam treatment as a means to enhance adhesion between metal films and polymeric substrates is still in an exploratory stage of development. Except for broad beam ion sputtering, it is far from being competitive in terms of cost performance when compared with other dry processes such as plasma modification and vacuum deposition of promoter layers.

In this section we give an overview of ion irradiation studies and point out promising aspects of the technique.

Equipment and technique The most important piece of equipment in ion beam technology is the accelerator, which generates the beam and directs it onto a sample or workpiece. Accelerators of all energies are now commercially available. At low ion energies modular accelerators known as "ion guns" often form part of a larger processing equipment; at the high end of the energy spectrum a linear accelerator with its beam lines can occupy several large rooms (Figure 7.3*a*). In between are machines of 20–400 keV ion energy, widely used in the semiconductor industry as ion implanters. The most popular ion guns used in low energy ion beam treatment are broad-beam Kaufman guns,[44, 46] which have now superseded the early versions of Penning or high-voltage discharge guns.[47] In Kaufman guns, an externally applied magnetic field is used to increase the ionization probability via electron cyclotron heating and a screen grid accelerator system is used to extract a broad beam from the discharge chamber. As the electron cyclotron resonance (ECR) technique, originally developed in the fusion and propulsion technologies,

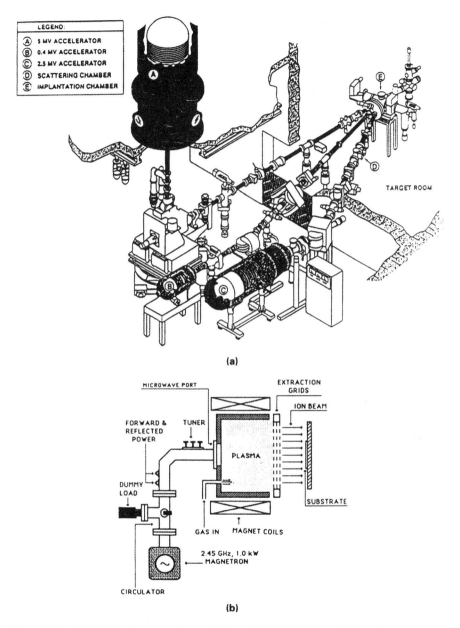

Figure 7.3 A triple ion linear accelerator facility at Oak Ridge National Laboratory.

is extended to low-pressure processing applications, a microwave version of the Kaufman gun has evolved where the EC heated dc discharge is replaced by a micro-wave ECR discharge (Figure 7.3*b*).[48]

High-energy ion beam treatment Early studies on the high-energy ion beam treatment have been conducted using old nuclear physics research machines from the 1950s. Many of these studies involve ill-defined experiments dealing with a number of metals on a wide variety of insulating substrates,[49] which constitute nothing more than a "quick-and-dirty" probing at possible adhesion enhancement in various film/substrate combinations. In these experiments metal films were first deposited on the substrates, and the samples were then irradiated with ion beams of various species at various energies and doses. Inert gas ions such as He, Ne, Ar, and Kr and reactive ions such as N, O, Cl, and F were used, and ion energies were chosen to place the projectiles near or far beyond the interface. Since neither the sample preparation nor the ion beam treatment was carried out under UHV conditions, undocumented contamination from hydrocarbons and other impurities could have easily obscured the identification of the basic processes responsible for ion-induced enhancement. It is not surprising, therefore, that after evaluating all the pertinent data available through 1986, Baglin[49] concluded that the mechanisms of ion-induced adhesion enhancement have been "identified" only by inference from secondary experimental evidence.

More recently, however, modern accelerators have been built with configurations reflecting the needs of materials science groups. Ion beam treatment can be performed in machines with a greatly improved vacuum so that hydrocarbon cracking and the contamination of carbon and other impurities no longer present a problem. As metal/polymer adhesion becomes increasingly important in a variety of modern technologies, better designed and carefully monitored experiments have been performed to investigate and identify the basic processes of ion-induced enhancement in various metal/polymer combinations. In a series of investigations, for example, Galuska[50] examined the adhesion enhancement produced by Si and Kr ion beams in Ni/polyester, Ni/polyimide, and Ni/Si/polyimide structures. The samples were prepared by depositing 30-nm Ni films on ultrasonically cleaned polymer substrates or on PI substrates with a 10-nm layer of predeposited Si. Although the sample preparation and ion beam treatment were not carried out under UHV conditions, surface science techniques such as AES and XPS were used to monitor surface and interface contaminants. During ion irradiation the specimens were maintained at a substrate temperature between 90 and 100 °C, but the annealing effect of substrate heating was not investigated. The ion energy was chosen to place the majority of the ions at or beyond the interface of interest. The maximum ion dose was determined by the onset of extensive microscopic void or island formation in the ion-processed films. Ion-induced mixing and morphology changes were examined by Auger profiling and cross-sectional transmission electron microscopy (XTEM), and interface chemistry by XPS depth profiling. The results of Galuska's studies highlighted the complexify of ion-induced adhesion enhancement and emphasized the fact that beam-induced effects are system specific, that is, equivalent ion beam treatments can produce different effects in different film/substrate combinations, and that similar beam effects may enhance adhesion in one system but not in the other. A case in point is the observation that Kr[+]-induced interface mixing can enhance adhesion in Ni/Polyimide structures but is completely ineffective in Ni/polyester specimens.[50] Furthermore, Kr[+]-induced interface mixing does not produce a graded Ni/polyimide interface, whereas Si[+] implant does. Another interesting observation is that Kr[+] irradiation in a Ni/Si/PI structure not only produces the ion mixing effect but also brings forth the formation of a SiO_2 layer at the Si–PI interface as a result of ion-induced liberation of oxygen from the PI substrate. This is in contrast to the result of Si[+] implant in the Ni/PI specimen, where maximum adhesion is obtained due to the formation of a graded interface with new Ni–Si, Si–O, and C–O chemical bonds.

Low-energy ion bombardment Surface modification by low-energy ion beams first gained popularity when chemists and material scientists began to use electron spectroscopies (AES and XPS) as surface-sensitive analytical tools in the late 1960s. In the early days of surface science studies the sample surfaces, in most cases, had to be cleaned in situ by ion sputtering before useful spectra could be obtained.[51] UHV-compatible ion guns developed for this purpose also served well in studies of ion-induced adhesion enhancement in metal/polymer structures. Since these guns can be installed in a large UHV system, metal films can be deposited in situ onto the sputtered substrates without exposing the latter to possible ambient contamination.

Experimentally, the generally subscribed procedure consists in the following:

1 preparation of polymer substrates by proven methods of synthesis followed by, if necessary, ultrasonic degreasing and cleaning of the substrate surfaces,

2 characterization by XPS and SEM of the substrate surface chemistry and morphology before and after ion sputtering at various energies and doses,

3 in situ metallization and, if possible, characterization of interface chemistry by XPS, and

4 determination of adhesion enhancement by comparing the results of peel, scratch, or tensile tests on the untreated and ion-processed metal/polymer samples.

Most of the low-energy ion bombardment studies have focused on fluorocarbon polymer substrates presumably because polymers such as PTFE and fluorinated ethylene-propylene (FEP) (known generically as teflon) were chosen for their chemically inert surfaces and ion-induced adhesion enhancement was first discovered in the Au/teflon specimens.[52–55] XPS monitoring indicates that low-energy Ar[+] bombardment removes F ions preferentially and the unstable radicals thus created give rise to branching and cross-linking of polymer chains.[56] SEM examination reveals that a 500-eV Ar ion beam can produce a textured morphology on the surface of teflon substrates, which becomes discernible after 15 s of sputtering (Figure 7.4).[57] A similar treatment of PI usually results in little or no change in surface morphology.[58] However, changes in molecular structure of the surface do occur in the same range of low ion energies and doses: the PI surface loses its oxygen in the carbonyl groups, leaving behind anions, which may be neutralized upon exposure to ambients containing moisture.[59] Adhesion enhancement in Cu/teflon

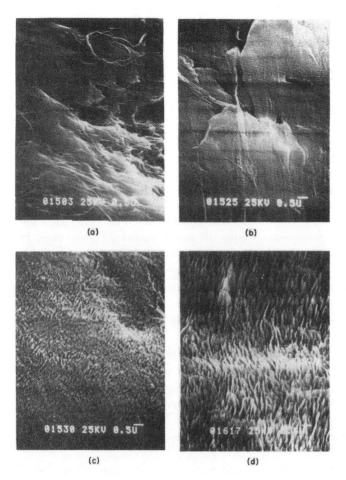

Figure 7.4 Development of surface morphology of a teflon film subject to 500-eV Ar⁺ ion sputtering: (*a*) untreated surface and with different sputter times: (*b*) 15 s, (*c*) 1 min, and (*d*) 5 min.

samples measured in terms of peel strength is found to increase with sputter time from 1 g/mm to a plateau of 50 g/mm after 80 s of sputtering (Figure 7.5).[57] The enhancement can be attributed to two factors: (1) the surface morphology change which increases metal/polymer contact areas and mechanical interlocking and, to a lesser extent, (2) the creation of CF_3 sites, which are responsible for the increased bonding with metal films.[60] In support of the latter postulate, Chang et al.[61] have presented the results of an experiment in which five different metals— Au, Cu, Al, Cr, and Ti—were deposited on PTFE, FEP, and PFA (FEP with per-fluoroalkoxy side chains) substrates, and the peel strengths for the metal/FEP and metal/PFA samples were found to be consistently higher than for the metal/PTFE

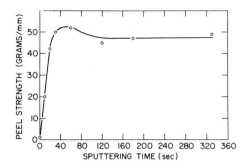

Figure 7.5   Peel strength of an ion-irradiated Cu/teflon specimen as a function of sputter time.

counterparts. They suggested that the presence of CF_3 side-chain units in FEP and PFA is responsible for the increased peel strength in the metal/FEP and metal/PFA samples.

Effect of electron and γ-ray irradiation   The ionizing effect of high-energy ion beams on adhesion can be separately evaluated by irradiating metal/polymer samples with energetic electrons and γ rays. Low-energy (< 1500 eV) electron bombardment of PTFE surfaces results in defluorination and cross-linking, which improves adhesion of PTFE to thin metal films deposited in situ. Variable takeoff angle XPS measurement indicates that the depth of surface modification by low-energy electron irradiation is less than 3 nm.[62] A similar bond modification has been observed in the interface of the Au/teflon structures exposed to high-energy γ-ray fluxes[63] and in the PTFE surfaces irradiated with soft X rays.[64] A β- and γ-ray treatment is used commercially to modify photoreflectance (PR) blood plasma packages. Cross-linking imparts surface rigidity and raises the softening temperature, both of which are beneficial in terms of package handling.

If bond rearrangement is the primary mechanism contributing to adhesion enhancement, then electron irradiation may supplant ion processing as a means of improvement since electrons penetrate deeper than ions of the same energy and cause no displacement damage in the substrates. However, the problem of heat dissipation in insulating substrates during high-dose electron irradiation should not be overlooked.

Adhesion enhancement and ion beam treatment   *Mechanisms of ion-induced enhancement*   Physical processes accompanying the passage of charged particles through matter have been extensively covered in many reports.[65] In the energy range of ion beam treatment, the primary processes consist in the displacement of electrons (ionization), the displacement of atoms from lattice sites, and the excitation of both atoms and electrons without displacement. Nuclear reactions can be, in principle, produced at these ion energies (up to 10 MeV), but can occur to an appreciable extent only when materials of high cross section are irradiated (e.g., the

(d,p)-reaction of C-12 irradiated by deuterium at 0.8 to 1.2 MeV).[66] The secondary effects of ion-solid interactions consist of further excitation and disruption of the structure by the struck electrons and atoms.

The aggregate of cascaded primary and secondary effects gives rise to permanent changes in the material:

- changes in surface chemistry and morphology due to sputtering

- branching and scission of polymer chains following bond breakage and cross-linking

- void or island formation due to coalescence of vacancies or preferential sputtering

- structural disorder due to displacement damage

- changes in interface chemistry due to ion implant

- changes in interface morphology and boundary delineation due to an unusual atomic transport effect known as "ion beam mixing."[67]

Identification of these effects with the adhesion enhancement mechanism in a metal/polymer combination is difficult and often becomes a subject of controversy. This is not only because both adhesion and radiation effects are very complex physical phenomena, but because many of the radiation effects cannot be directly observed or unequivocally characterized. Furthermore, effects accompanying ion irradiation that are observed in some systems do not occur in others. Several mechanisms of ion-induced enhancement have been, however, identified with a reasonable degree of confidence:

1 Enhancement via mechanical interlocking and increased contact area as a result of surface morphology development due to ion sputtering.[59, 60]

2 Formation of a graded interface with new chemical bonding by "ion beam mixing" either with high-energy inert gas ions or with reactive ion implant.[50]

3 Creation of new chemical bondings across the interface by modifying the polymer substrate surfaces with ion sputtering or with electron and photon irradiation.[63, 64, 68]

4 A combination of several or all of the above mechanisms.

Ion beam treatment has been shown to improve the hardness and wear resistance of polymer surfaces.[69] Such improvements usually result in the increased peel strength, scratch resistance, or pull strength of a metal/polymer system. It is, however, debatable whether the improvement is an enhancement mechanism since the increase is merely a substrate effect which can occur in the absence of any increase in adhesion strength.[70, 71]

Prospects of ion beam treatment   For high-energy ion beam treatment to become a viable processing technique for adhesion enhancement, one must take advantage of the effects that cannot be achieved by other processing techniques. Heavy ion irradiation has the practical limitations of requiring high-energy accelerators and rather thin metal films. It also poses the problem of possible radiation damage and contamination in the underlying substrate. Thus, the most promising approach to adhesion enhancement by high-energy ion treatment appears to be interface tailoring by implanting appropriate reactive ion species in the metal/polymer structures. Optimized, this will create a graded interface with new chemical bonds, which will increase adhesion very substantially, as is demonstrated in the case of Si implant in the Ni/Pi system.[50] In addition, the treatment can be carried out at intermediate ion energies in an ion implanter, which has now become standard production equipment in the semiconductor industry. Since the implantation is performed after the metal/polymer structure has been fabricated, interface contamination can be minimized. However, one should bear in mind that other basic properties of the metal/substrate system must not be adversely affected by the treatment.

The development of the broad-beam Kaufman source and its ECR microwave version has put low-energy ion sputtering on a new footing to compete with other plasma-based techniques. Since ion sputtering can remove those metals not susceptible to plasma or reactive ion etching, it is likely to find a niche in the processing of multilayer metal films on polymer substrates.

As pointed out earlier, the effect of ion-induced adhesion enhancement observed in one metal/polymer system may not be operative in another, even if the ion beam treatments are equivalent in the two systems. We conclude this section with the comment that a great deal of research effort involving well-defined experiments in uncontaminated or well-characterized metal/polymer systems is needed for the ion beam treatment to find practical engineering applications in the metallization of polymers.

Plasma Modification of Polymer Surfaces

Extensive studies have been performed to characterize and understand the plasma processes which are finding an ever-expanding variety of applications from reactive ion etching in microelectronic chip processing to polymerization and plasma-assisted chemical vapor deposition in xerography and solar cell fabrication. There is a substantial body of literature regarding the mechanism and kinetics of these processes. Since these topics have been summarized in several reviews[72–75] we focus only on those studies which pertain to the modification of polymer surfaces, a subject which has attracted increasing attention since the recent deployment of polymers in advanced microelectronic packaging modules and at the chip level.[76, 77] Qualitative as it may be, sufficient knowledge has been accumulated from these studies to help optimize a plasma-based surface modification experiment.

Equipment selection A group of Canadian researchers has compared the etching and surface modification performance of a large-area microwave reactor with or without radio frequency (rf) bias and an rf parallel plate reactor under virtually identical conditions.[78, 79] The results of their studies reaffirmed what have been recognized or inferred in previous investigations performed under diverse conditions (different plasma parameters and system configurations):

1 A microwave discharge produces a higher concentration of useful chemical precursors than an rf discharge under identical conditions.

2 The concentration of the precursors varies with the composition of the plasma gas and exhibits a maximum at a certain optimum composition.

3 The concentration of the precursors increases with increasing gas pressure and reaches a maximum when increased recombination and collisional quenching begin to affect the generation of precursors.

4 Energetic ion bombardment due to the self-bias in an rf discharge or due to the rf bias in a microwave discharge has the consequence of resputtering and other impact-related processes such as dissociation and desorption of adsorbates, decomposition and preferential desorption of one component of the surface material, formation of defects, short-range diffusion, etc. For polymer substrates it often results in the cross-linking and scission of polymer chains. Whether the impact-related effects are beneficial for surface modification must be assessed on a case-by-case basis.

It appears that unless uniformity is an overriding issue, microwave discharge is more advantageous than rf discharge for surface treatment studies. Although parallel plate reactors have been standard equipment in the semiconductor industry, they may be, to a large extent, replaced by microwave ECR reactors with rf bias, which can operate efficiently at pressures below 1 mTorr and yet can have the benefit of ion bombardment when necessary. During plasma processing, ion and electron bombardment of reactor walls and powered electrodes causes the desorption of adsorbed impurities, which are the dominant source of plasma pollution. In a new reactor, the processing chamber can be separated from the discharge chamber, and the plasma stream and active chemical species reach the sample by diffusing downstream. The effect of pollutants may be minimized.

Choice of gas species Hydrogen, oxygen, nitrogen, H_2O, NH_3, CF_4, and their binary mixtures, or their mixtures with inert gas carriers such as He, Ar, and Ne are the gas species most frequently used for the plasma processing of polymers. Until recently, modification studies have focused mostly on linear homopolymers such as PE, PTFE, and PS.[80, 81] With the increasing use of PI in microelectronics, investigations on PI have recently multiplied.

Nuzzo and Smolinsky[82] investigated the surface modification of PE using rf plasmas of oxygen, water, and hydrogen and concluded that various oxidation functionalities from alcohols to carboxylic acids were incorporated in the films exposed to an oxygen or water plasma, whereas alkyl radicals were found on the polymer surface exposed to a hydrogen plasma. Evans et al.[83] studied the H_2O and hydrogen rf-plasma modification of PS surfaces. Various oxidation functionalities incorporated in the surface region following H_2O treatment were found to be chemically reduced by subsequent hydrogen plasma treatment. Klemberg-Sapieha et al.[79] used microwave plasmas of N_2 and NH_3 with a variable rf bias to modify PE and PI and found that C–N and C=N bonds were incorporated in the treated films. Darchicourt[84] investigated the microwave oxygen plasma modification of PE and PS and observed an increased oxygen coverage attributable to C–O bonds. Chou et al.[85] and Leu and Jensen[86] treated PI in microwave oxygen plasmas and attributed the increased surface oxygen to ketones, peroxy, and polycarbonate acid or ester groups. Celerier and Machet[68] reported improved copper–PTFE adhesion by treating the substrate in self-biased rf plasmas of oxygen and argon and suggested the formation of Cu–O–C bonds at the interface. Nakamae et al.[87] described the H_2O plasma treatment of PEP, which improved film wettability and the adherence of Co films deposited on PEP substrates. They concluded from XPS data that the surface oxidation products were alcohols, ketones, peroxides, and carbonate functional groups. Goldblatt et al.[30] reported a tenfold increase in self-adhesion and fivefold increase in Cu/PI adhesion when the polymer films were treated with an rf H_2O plasma. An increase in oxygen coverage of the treated PI films was identified as ketone and hydroxyl functionalities. Common to these findings is the observation that the modified layers of the treated films are no thicker than ~20–30 nm and that there is great similarity among the oxidized functionals incorporated on the surfaces exposed to the plasmas of different gas species.

The treatment of PI in rf and microwave oxygen plasmas produces an increase in surface oxygen, which XPS interprets as the formation of peroxy (C–O–O–C) and carbonate-type groups on the surface.[84, 85] PI films exposed to an H_2O or O_2 plasma exhibit a similar degree of surface oxidation[30, 80, 81] because the water plasma has a high density of oxygen atoms comparable to that of an oxygen plasma. An NH_2 plasma causes consistently higher nitrogen incorporation on PE and PI surfaces than an NH_3 plasma: nitrogen is bonded predominantly in imine (C=N) groups in the former case, but in the amine (C–N) groups in the latter. The difference is attributable to the plasma chemistry. In an N_2 plasma the predominant species is activated N_2 with 10% atomic N, whereas in an ammonia plasma there exist mainly fragments of NH_n, ($n = 0,1,2$).

At present the selection of gas species for surface modification is made more or less intuitively on the basis of empirical knowledge. In many of the plasma processing studies the chemistry of the plasma gas has not been well-characterized. Although optical emission spectroscopy and mass spectrometry can provide information regarding active species in the plasma environment, the emission intensities and mass peaks are not directly relatable to the concentration of the detected species. As such, it is difficult to determine which of the many active species plays a dominant role in the reaction. To compound the issue, the plasmas in the modification studies are frequently polluted with impurities desorbed from the reactor walls or electrodes. The impurity concentration can reach several mol % in a typical parallel plate reactor and usually consists of oxygen and water.[88] The effects of these impurities on plasma chemistry and plasma-surface interaction often were overlooked, but unintentional contamination of treated surfaces has been observed in many studies.[78, 84] The cleanliness of a plasma reactor system and the plasma purity are therefore prerequisites to fundamental studies of surface modification.

Whereas with plasma etching, etch rate is one of the primary concerns, the choice of gas has only to fulfill the objective of obtaining a modified layer with a uniform distribution of appropriate functionals. The use of binary mixtures such as CF_4/O_2 is usually discouraged because the fluorine contamination of reactor walls and electrodes is difficult to remove. Admixtures with inert or other gases, which will either increase the density of useful precursors via Penning effect[89] or lower the power to sustain a discharge, may be optimized to improve the effectiveness and uniformity of treatment without adversely contaminating the treated surfaces. Since modified polymer surfaces are unstable and will change with time or upon exposure to ambient, metallization of the treated polymer should be done immediately following the modification and in situ, if possible.

Characterization techniques and modified surface layers Not only do the chemical structure and composition of the modified layer depend on the chemical precursors in the plasma gas, but they depend upon the properties and chemical structure of the substrate, as well. Driven by a demand for quantification in the engineering applications of plasma processing, particularly in the area of plasma-assisted chemical vapor deposition, optical emission techniques[79, 90] such as actinometry and laser-induced fluorescence (LIF) have been used to characterize and quantify the concentration and spatial distribution of active species in the plasmas. In spite of the progress made in understanding the precursors in clean, well-defined systems, quantitative data for surface modification in conventional reactors are still missing.

Modified polymers are generally examined ex situ after plasma treatment. XPS is the primary tool used to determine the changes in surface composition and bonding. FTIR, ion scattering spectroscopy (ISS), SIMS, and sometimes labeled chemical derivation[82] and other unconventional techniques[91] have been used as complementary techniques to help resolve the ambiguity of XPS results. Hook et al.,[92] for instance, have used multiple techniques (XPS, ISS, SIMS, and FTIR) to study the rf $H_2O/$ Ar plasma treatment of polymethylmethacrylate (PMMA) of different tacticity. They found that the top layer of the treated PMMA was dominated by oxygen-rich functional and that a greater degree of surface modification occurred in a structurally more constrained PMMA (isotactic and syndiotactic). They also identified the presence of OH groups on the surfaces by FTIR, but suggested that these functional groups could be either C–O–H or adsorbed H_2O molecules. Modifying polypropylene in a microwave plasma of Ar/O_2 mixtures results in an increased oxygen coverage, which FTIR and XPS attribute to C–O bonds.[84] Radioisotopic labeling chemicals have been used to derivatize polymer surfaces treated with oxygen and water plasmas, and the results of derivatization provide information regarding the concentration and identification of active sites on the treated surfaces.[82] The isotope approach has also been used to study microwave water plasma-treated PI surfaces.[92] The hydroxyl groups incorporated in the modified layer during the treatment were unequivocally identified and quantified by a combination of multiple internal reflection (MIR)-FTIR and deuterium forward scattering techniques.

There is ample evidence that plasma-treated surfaces are further modified upon exposure to reactor or room ambient: the moisture uptake from the reactor or room ambient,[90, 92] unintentional oxygen incorporation,[78, 84] and the well-known aging effect,[94] just to name a few. The literature data on molecular structure and functionalities of the modified surfaces should therefore be viewed in that light. Cognizant of this uncertainty, many investigators have resorted to in situ multiple monitoring techniques. Chou et al.[85] used in situ XPS and ex situ FTIR transmission and MIR methods to investigate the etching mechanism of PI in CF_4/O_2 downstream microwave plasmas. Leu and Jensen,[86] on the other hand, used the in situ FTIR reflection-absorption technique and ex situ XPS to study PI and PMMA surfaces during etching in the downstream of microwave $NF_3/CF_4/O_2$ plasmas. The advantages of in situ measurement and deployment of mutually complementary techniques were clearly demonstrated in these two investigations. Direct evidence was obtained, for instance, which showed that formation of polyfluorinated benzene compounds and aliphatic CF_n type bonding was responsible for etching inhibition at high CF_4 and NF_3 concentrations.

Other Dry Treatment Techniques

Apart from abrasion, polishing, and other mechanical means of surface treatment, which were usually developed empirically, there are two interesting dry modification schemes we have not covered so far. One appears to be simple and is likely to be ignored; the other is promising in concept, but requires a great deal of exploratory work before its implementation in engineering applications.

Application of promoter layers   One of the simple ways to improve metal–polymer adhesion is to deposit, in a vacuum, an adhesion promoter on the substrate before the deposition of the primary metal. The use of a ~100-Å Ti or Cr layer in a Cu/Cr/PI or Cu/Ti/PI structure, for example, is widely practiced in the semiconductor industry. Thin vacuum deposits (10–30 Å) of metals such as Pd and W can serve as a prenucleation (or seeding) layer for the selective deposition of metals from organometallic substances in the so-called nanofabrication technology which has attracted attention in the last few years.[94] Several plasma polymerized films can be formed on a variety of substrates.[95] A polysiloxane is prepared by the polymerization of hexamethyldisiloxane or tetramethylsilane in an Ar/O_2 plasma. A PS is formed in a plasma of styrene monomers. Although these films were initially developed for use as scratch-resistant coatings and passivation layers, their potential application as adhesion promoters should not be ignored.

Vapor phase photografting as a surface modification technique can introduce a variety of functionalities onto the polymer surfaces. Grafting is generally performed in a liquid phase. A polymeric substrate is placed in contact with a solution containing the photoinitiator and the monomer. UV irradiation excites the initiator and causes it to extract hydrogen from the substrate, leaving active sites on the surface, which now behaves as a macroradical attacking the unsaturated monomer from the surrounding media and starts grafting. The use of a photoinitiator limits grafting at the interface between the substrate and the media because only the initiator absorbs the UV radiation, leaving the bulk of the substrate unaffected. For the technique to be adapted to dry treatment, both the initiator and the monomer should be supplied from the vapor phase. It has been shown that surfaces of high- and low-density PE and PS can be successfully grafted using benzophenone (photoinitiator) and acrylic acid (monomer) in the vapor phase.[96, 97] UV irradiation of the substrate through the vapor phase results in a 90% poly(acrylic acid) coverage for PS, 63% for LDPE, and 56% for HDPE, the coverage being hydrophilic and stable at room temperature. Other monomers such as GA can be similarly grafted to PE to form a 72% polyGA coverage. Such grafted surfaces contain epoxy groups which can be utilized to introduce other functionals such as amines to the surface.[98, 99]

7.4 Summary

The successful fabrication of metallized plastics, from decorative consumer items to advanced microelectronic packaging structures, requires good interfacial adhesion between thin metal films and polymer substrates. Most polymeric materials that possess desirable bulk properties have inert and hydrophobic surfaces. In order to tailor the surfaces to specific needs without altering the bulk properties, one may employ a variety of surface modification techniques. The most common approach is to oxidize the surface, thus raising its energy. Surface oxidation can be achieved by treatment in an oxidizing solution, and by corona discharge, by blending the polymer with a reactive surfactant which will segregate to the interface. In recent years elaborate methods of modification have been developed: surface grafting, plasma etching, polymerization, etc. In this chapter we have focused our discussion on techniques in which the use of wet chemicals can be avoided. As our discussion indicates, the technology of dry surface modification is still in its infancy. We do not have the knowledge to provide us with the predictive power necessary to design a treatment process. With all the sophisticated surface-sensitive analytical tools and characterization techniques at our disposal, we are still far from having a complete picture of what is taking place on the modified surface at a molecular level. However, with an extensive research effort motivated by the promise of dry treatment methods and considering the advances that have been made in the last few years, we will witness a rapid growth of the technology in the near future.

In closing this chapter we note that cost performance is the factor that determines whether a technological process will find an engineering application. Most of the dry treatment methods covered in this chapter are costly and unproven in industrial practices. In the vast field of plastics, only empirically proven dry processes such as corona and low pressure glow discharge are applied in production practices. Only a few of the "dry" techniques have been adopted for use in the semiconductor industry, where the performance of a process may outweigh its cost in the larger scheme of things.

References

1 G. F. Kinney. *Engineering Properties and Applications of Plastics.* Wiley, New York, 1957, pp. 7, 66–69.

2 J. A. Newmann and F. J. Bockhoff. *Welding of Plastics.* Reinhold, New York, 1959, p. 12.

3 K. Siegbahn, C. N. Nordling, A. Fahlman, R. Nordberg, K. Hamrin, J. Hedman, G. Johansson, T. Bermark, S. E. Karlsson, I. Lindgren, and B. Lindberg. *ESCA: Atomic, Molecular and Solid State Structure Studied by Means of Electron Spectroscope.* Almquist and Wiksells, Uppsala, Sweden, 1967.

4 H. von Brecht, F. Meyer, and H. Binder. *Makromol. Chem.* **33**, 89, 1973.

5 J. R. I. Hepburn. *J. Electrodepositers' Tech. Soc.* **17**, 1, 1941.

6 J. R. Rasmussen, E. R. Stedronsky, and G. M. Whitesides. *J. Am. Chem. Soc.* **99**, 4736, 1977.

7 J. R. Rasmussen, D. E. Bergbreiter, and G. M. Whitesides. *J. Am. Chem. Soc.* **99**, 4746, 1977.

8 E. C. Marboe and W. A. Weyl. *Glass Industry.* **26**, 3, 1945.

9 H. Narcus. *Metallizing of Plastics.* Reinhold, New York, 1960, p. 28.

10 A. M. Mance and R. A. Waldo. *J. Electrochem. Soc.* **135**, 2729, 1988.

11 A. Aviram, V. I. Mayne-Banton, and R. Srinivasan. U.S. Patent 4,440,801, 1984.

12 I. Nakamichi, Y. Ishikawa, and Y. Hirose. *Jpn. J. Appl. Phys. Part 2.* **26**, L228, 1987.

13 T. K. Barr. In *Practical Surface Analysis by Zuger and X-Ray Photoelectron Spectroscopy.* (D. Briggs and M. P. Seah, Eds.) Wiley, New York, 1983, pp. 283–358.

14 A. M. Mance, R. A. Waldo, and A. A. Dow. *J. Electrochem. Soc.* **136**, 1667, 1989.

15 D. W. Dwight and W. M. Riggs. *J. Coll. Interface Sci.* **47**, 650, 1974.

16 G. C. S. Collins, A. C. Lowe, and D. Nicholas. *Euro. Polym. J.* **9**, 1173, 1973.

17 D. Briggs. In *Surface Analysis and Pretreatment of Plastics and Metals.* (D. M. Brewis, Ed.) Applied Science, London, 1982, Chapt. 9 and references cited therein.

18 D. Briggs, V. J. I. Zichy, D. M. Brewis, J. Comyn, R. H. Dahm, M. A. Green, and M. B. Koiezko. *Surf. Interface Anal.* **2**, 107, 1980.

19 A. M. Wilson. *Thin Solid Films.* **83**, 1345, 1981.

20 L B. Rothman. *J. Electrochem. Soc.* **127**, 2216, 1980.

21 G. Sammuelson. *ACS Org. Coat. Plast. Chem. Div., Extended Abstracts.* **43**, 446, 1980.

22 W. B. Linsey. U.S. Patent 3,791,589, 2 Jan. 1968.

23 J. Hermer. U.S. Patent 3,770,528, 6 Nov. 1973.

24 H. Knorre and E. Meyer-Simon. U.S. Patent 3,702,285, 7 Nov. 1972.

25 M. A. DeAngelo. U.S. Patent 3,791,848, 12 Feb. 1974.

26 A. Saiki, K. Sato, S. Harada, T. Tsunoda, and Y. Oba. U.S. Patent 4,001,870, 4 Jan. 1977.

27 A. Saiki, S, Harada, and Y. Oba. U.S. Patent 4,040,083, 2 Aug. 1977.

28 K.-W. Lee and S. P. Kowalczyk. In *Metallization of Polymers.* (E. Sacher, J.-J. Pireaux, and S. P. Kowalczyk, Eds.) ACS Symposium Series 440, American Chemical Society, Washington, DC, 1990, p. 179.

29 M. M. Plechaty and R. R. Thomas. *J. Electrochem. Soc.* **39**, 810, 1992.

30 R. D. Goldblatt, L M. Ferreiro, S. L. Nunes, L P. Buchwalter, N. J. Chou, J. E. Heidenreich, R. R. Thomas, and S. E. Molis. IBM Internal Report RC15152, 1989.

31 B. E. Wagner, J. N. Helbert, E. H. Poindexter, and R. D. Bates. *Surface Sci.* **67**, 251, 1977 and the references cited therein.

32 J. N. Helbert, F. Y. Robb, B. R. Svechovsky, and N. C. Saha. In *Surface and Colloid Science in Computer Technology.* (K. L. Mittal, Ed.) Plenum, New York, 1985, p. 133.

33 E. D. Plueddemann. In *Surface and Colloid Science in Computer Technology.* (K. L Mittal, Ed.) Plenum, New York, 1985, pp. 143–153.

34 B. P. Ratner. *J. Appl. Polym. Sci.* **22**, 643, 1978.

35 K. Kaji. *J. Appl. Polym. Sci.* **32**, 4405, 1986.

36 G. Oster. *J. Polym. Sci. Polym. Symp.* **34**, 671, 1959.

37 S. Tazuke and H. Himura. *J. Polym. Sci. Polym. Lett. Ed.* **16**, 497, 1978.

38 J. F. Rabek. *Mechanisms of Photophysical and Photochemical Reactions in Polymers.* Wiley, New York, 1987, p. 272.

39 K. Allmer, A. Hult, and B. Ranby. *J. Polym. Sci. Polym. Chem. Div.* **26**, 2099. 1988.

40 N. J. Chou, R. D. Goldblatt, J. R. Heidenreich, and D. Tortorrella. *IBM Technical Disclosure Bulletin.* **33**, 360, 1991.

41 R. D. Goldblatt, J. M. Park, R. C. White, L. J. Matienzo, S. J. Huang, and J. P. Johnson. *J. Appl. Polym. Sci.* **37**, 335, 1989.

42 K. Kilmer, A. Hult, and B. Ranby. *J. Polym. Sci.: Part A. Polym. Chem.* **27**, 1641, 1989.

43 B. Ranby, Z. M. Gao, A. Hult, and P. Y. Zhang. ACS Symp. Ser. Chemical Reactions on Polymers. In press.

44 W. C. Lathem. *J. Spacecraft and Rockets.* **6**, 1237, 1969.

45 A. D. Marwick. *Met. Trans.* **20A**, 2627, 1989.

46 H. R. Kaufman. *Advances in Electronics and Electron Physics.* (L. Marton, Ed.) Academic Press, New York, 1974, pp. 265–373.

47 Sidenius. "Low Energy Ion Beams." *Inst. Phys. Conf. Series*, No. 38. (K. G. Stephens, J. H. Wilson, and J. L. Moriozzi, Eds.) Institute of Physics, Bristol and London, 1978.

48 E. Ghanbari, I. Trigor, and T. Nguyen. *J. Vac. Sci. Technol.* **A7** (3), 918, 1989.

49 J. E. E. Baglin. In *Surface and Colloid Science in Computer Industry.* (K. Mittal, Ed.) Plenum, New York, 1987, pp. 211–233.

50 A. A. Galuska. *J. Vac. Sci. Technol.* **B8** (3), 470, 1990.

51 R. E. Weber and W. T. Peria. *J. Appl. Phys.* **38**, 4355, 1967.

52 M. Rost, J. J. Erler, H. Gienpack, P. Fiedler, and C. Weissmantel. *Thin Solid Films.* **20**, 515, 1974.

53 M. J. Mirtich and J. S. Sovey. *J. Vac. Sci. Technol.* **16**, 809, 1979.

54 W. K. Fisher and J. C. Corelli. *J. Polymer Sci.* **19**, 2465, 1981.

55 R. Michael and D. Stulik. *Radiat. Eff. Lett.* **87**, 9, 1985.

56 D. R. Wheeler and S. V. Pepper. NASA Technical Memo No. 83413, 1983.

57 C.-A. Chang, J. E. E. Baglin, A. G. Schrott, and K. C. Lin. *Appl. Phys. Lett.* **51**, 103, 1987.

58 B. J. Blachman and M. J. Vasile. *J. Vac. Sci. Technol.* **A7**, 2709, 1989.

59 K. S. Sengupta and H. K. Birbaum. *J. Vac. Sci. Technol.* **A9** (6), 2928, 1991.

60 C.-A. Chang. *Appl Phys. Lett.* **51**, 1236, 1987.

61 C.-A. Chang, Y.-K. Kim, and A. G. Schrott. *J. Vac. Sci. Technol.* **A8**, 3304, 1990.

62 J. A. Kelber. J. W Rogers, Jr., and S. J. Ward. *J. Mater. Res.* **1** (5), 717, 1986.

63 C. J. Sofield, J. S. Woods, C. Wild, J. C. Riviere, and L. S. Welch. *Proc. Mat. Res. Soc.* **25**, 197, 1984.

64 D. R. Wheeler and S. V. Pepper. *J. Vac. Sci. Technol.* **A8**, 4046. 1990.

65 G. J. Dienes and G. H. Vineyard. *Radiation Effects in Solids.* Interscience, New York, 1957.

66 G. Amsel, J. P. Nadai, E. D'Artenmare, D. David, E. Girad, and J. Moulin. *Nucl. Instr. and Meth.* **92**, 482, 1971.

67 R. S. Averback. *Nucl. Instrum. Methods B.* **15**, 675, 1986.

68 A. Celerier and J. Machet. *Thin Solid Films.* **148**, 323, 1987.

69 E. H. Lee, M. B. Lewis, P. J. Blau, and L. K. Mansur. *J. Mater. Res.* **6**, 610, 1991.

70 K. S. Kim. Report No. UILU-ENG 85-6003. University of Illinois, March 1985.

71 M. H. Bernier, J. E. Klemberg-Sapieha, L. Martinu, and M. R. Wertheimer. In *Metallization of Polymers.* (E. Sacher, J. J. Pireaux, and S. Kowalczyk, Eds.) American Chemical Society, Washington, DC, 1990, p. 147.

72 F. Jensen. In *Plasma Deposited Thin Films.* (J. Mort and F. Jansen, Eds.) CRC Press, Boca Raton, FL, 1986, pp. 2–17.

73 A. R. Reinberg. *J. Electronic Mater.* **8** (3), 345, 1979.

74 E. Kay, J. Coburn, and A. Dilks. In *Topics in Current Chemistry.* (E. Kay, J. Coburn, and A. Dilks, Eds.) Springer-Verlag, Berlin, 1980, pp. 1–40.

75 S. Veprek. *Plasma Chemistry and Plasma Processing.* **9** suppl., 29S, 1989.

76 K. Niwa, N. Kamechara, K. Yokouchi, and Y. Imanada. *Adv. Cer. Mater.* **2**, 832, 1097.

77 S. Lebow. In *Proceedings.* 30th Electronic Components Conf., 1980, pp. 307–309.

78 A. M. Wrobel, B. Lamontagne, and M. R. Wertheimer. *Plasma Chemistry and Plasma Processing.* **8** (3), 315, 1988.

79 J. E. Klemberg-Sapieha, O. M. Kuttel, L. Martinu, and M. R. Wertheimer. *J. Vac. Sci. Technol.* **A9** (6), 2975, 1991.

80 D. T. Clark and A. Dilks. *J. Polym. Sci. Polym. Chem. Ed.* **17**, 957, 1979.

81 D. T. Clark and A. Dilks. *J. Polym. Sci. Polym Chem. Ed.* **15**, 2321, 1977.

82 R. G. Nuzzo and G. Smolinsky. *Macromolecules.* **17**, 1013, 1984.

83 J. F. Evans, J. H. Gibson, J. F. Moulder, J. S. Hammond, and H. Goretzki. *Z. Anal. Chem.* **319**, 841, 1984.

84 D. Darchicourt, E. Bloyet, P. Leprince, and J. Marec. *J. Vide Couches Mines* (France). **237** suppl., 53, 1987.

85 N. J. Chou, J. Paraszczak, E. Babich, J. Heidenreich, Y. S. Chaug, and R. Goldblatt. *Microelectron Eng.* **5**, 375, 1986.

86 J. Leu and K. F. Jensen. *J. Vac. Sci. Technol.* **A6**, 2948, 1991.

87 K. Nakamae, K. Yamaguchi, S. Tanigawa, K. Sumiya, and T. Matsumoto. *Japan J. Chem. Soc.*, **11**, 1995, 1987.

88 H. Winters. In *Proc. Intern. Summer School on Plasma Chem.* (S. Veprek, Ed.) Atami, Japan, Aug. 1987.

89 E. Nasser. In *Fundamentals Gaseous Ionization and Plasma Electronics.* Wiley-Interscience, New York, 1971, p. 214.

90 R. W. Dreyfus, J. M. Jasinski, R. E. Walkup, and G. S. Selwyn. *Pure Appl. Chem.* **57**, 1265, 1985.

91 N. J. Chou, A. D. Marwick, R. D. Goldblatt, L. Li, J. E. Heidenreich, and J. R. Paraszczak. *J. Vac. Sci. Technol.* **A10** (1), 248, 1992.

92 T. J. Hook, J. A. Gardella, Jr., and L. Salvati. *J. Mater. Res.* **2** (1), 117, 1987.

93 F. D. Eggito, F. Emmi, R. S. Horwath, and V. Vukanovic. *J. Vac. Sci. Technol.*, **B3**, 893, 1985.

94 D. R. Gagnon and T. J. McCarthy. *J. Appl. Polym. Sci.* **29**, 4335, 1984.

95 B. Lecohier, J. M. Phillipoz, and H. van den Bergh. *J. Vac. Sci. Technol.* **B10**, 262, 1990.

96 Y. Segui and B. Ai. *J. Appl. Polym. Sci.* **50**, 6567, 1976.

97 A. M. Wrobel, M. Kryszewski, and M. Gazicki. *Polymer.* **17**, 673, 1976.

98 K. Allmer, A. Hult, and B. Ranby. *J. Polymer Sci. Part A. Polymer Chem.* **26**, 2099, 1988.

99 K. Allmer, A. Hult, and B. Ranby. *J. Polymer Sci. Part A. Polymer Chem.* **27**, 1641, 1989.

Adhesion

STEVEN P. KOWALCZYK and JUNG-IHL KIM

Contents

8.1 Introduction

Adhesion of polymers, whether to metals, ceramics, or other polymers, or of a polymer to itself (self adhesion), has become an increasingly important topic as polymers find increasing applications in such diverse industries as aerospace, automotive, domestic household appliance, and microelectronics. The manufacturing of adhesives has become a multibillion dollar industry.[1] The scale of adhesion ranges from the immobilization or anchoring of a protein to barnacles sticking to ship hulls, to protective coatings of cables, to aircraft structures. In the microelectronics industry, increasingly complex and shrinking multilevel structures create challenges in the fabrication of multilevel structures approaching the nanometer scale, which contain a plethora of interfaces that must be cohesive and must withstand a variety of hostile operating ambients.

Although the role of adhesion in a product is usually a secondary consideration, it is often the key enabling property, in the sense that the polymer is usually chosen for a certain property, such as low dielectric constant, when used as an insulating layer in microelectronic structure. However, successful technological implementation requires that the structure "holds" together, that is, has good *adhesion*. Thus in this case, the adhesional property of the polymer can produce a compromise in the selection of the polymer in that a polymer with a higher dielectric constant may be selected over one with a lower dielectric constant if the lower dielectric constant polymer has unsatisfactory adhesive properties. An example is Teflon®, which has a superior dielectric constant to many polymers currently implemented in microelectronics. Because of Teflon's poorer adhesive properties, however, it is not yet widely employed as a dieletric layer material. Such compromises are common for many polymer applications, thus the desire to measure adhesion and understand the factors controlling adhesion, so that it becomes possible to engineer improved adhesion intelligently. Some examples of engineered adhesion are given in Chapter 7.

The importance of polymer adhesion to technological implementation is ubiquitous—from the above-mentioned multitude of layers in multilevel electronic devices to the aluminum lug joined to a composite tube in lightweight racing bicycles, to decorative paint on automobile bumpers, to protective layers for neural prosthetic devices. It should also be noted at the outset that adhesion is not a one-way street: strong adhesion is desired, but in some cases weak or no adhesion can also be desirable—thus one desires the Teflon to adhere strongly to the frying pan but does not want the omelet to adhere to the Teflon at all. Another example of where poor adhesion is desirable is of a part to a mold. Even intermediate strength adhesion can be desirable, as for example with the now-ubiquitous self-stick removable notepaper, Post-it® notes.

Both experimental measurement and theoretical treatments of adhesion have become increasingly sophisticated. In this chapter, we briefly discuss some theoretical aspects of adhesion, describe the experimental methods for the determination of adhesion, and emphasize the difference between intrinsic interfacial adhesion and engineering peel strength. The role of locus of failure analysis is described. Some methods for the modification of adhesion are enumerated, a number of which are which are discussed in detail in other chapters in this book. Another important adhesion concern only briefly discussed here is the degradation of adhesion upon environmental aging.

As previously mentioned, adhesion is really a secondary property, allowing the many advantageous properties of a particular polymer to be utilized; thus, many aspects of adhesion are touched upon in the other chapters of this book: surface thermodynamics in Chapter 6, structure and morphology in Chapters 4 and 5, surface modification in Chapter 7, and fracture mechanics in Chapter 9. Also, there already exist a number of excellent reviews on adhesion. Thus, this chapter emphasizes and highlights the salient points and guides the reader to the appropriate literature. A brief section listing additional reading material is given near the end of this chapter, including a reference to the more extensive earlier bibliography of Mittal, which also includes some pertinent ASTM standards.[2]

8.2 Physical Meaning of Adhesion

Adhesion is the union or sticking together of two surfaces. Adhesion energy is a thermodynamic quantity, that is, the work needed to separate two surfaces. The actual interfacial adhesion is on the microscopic level of molecular interactions, depending on chemical bonding and other intermolecular attractive forces.[3] We call this intrinsic or interfacial adhesion. In reality, the work in separating two materials from each other includes mechanical contributions in addition to molecular-level interactions. Thus one must consider continuum mechanics to understand the viscoelastic contribution to the work of separation. We call this engineering adhesion.

Other important contributions must be considered, such as surface roughness, where there can be mechanical interlocking from dendritic structures (the "Velcro" effect) or an increase of contacted surface area. Interdiffusion, for polymer–polymer interfaces, in particular for self-adhesion, may lead to a "healing" of the interface, that is, a disappearance of the interface. This leads us to a simple question with a complex answer: "Can adhesion—interfacial adhesion—be measured?" This has been discussed fully by many authors, in particular by Mittal, in numerous reviews. The answer to that question depends on how carefully one defines one's terms and how the measurement is performed. What one desires is often a measure of the effectiveness of the strength of the interface, a measure of how changes in preparation (such as surface modification treatments) affect adhesion, and a measure of how well the interface can withstand the rigors of aging and stress before catastrophic fatigue occurs. Such information can be obtained only with an understanding of what the adhesion test is measuring. A rigorous determination of interfacial adhesion is far from straightforward, however.

8.3 Theoretical Analysis of Adhesion

As previously indicated, one must distinguish between intrinsic microscopic interfacial adhesion and what is measured in a practical adhesion test, such as the widely used peel test. Most tests, like the peel test, depend not only on the interfacial molecular interactions between the two adhering materials but on the degree of

Cr Thickness, μm	Peel Strength, g/mm
0.01	61.3
0.05	52.0
0.1	45.7
0.5	36.2

Table 8.1   Peel strength of Cr/Cu film as a function of Cr thickness (total film thickness is constant at 15 μm).

stress in the film and the mechanical properties of the film and substrates. This can be visually observed in the curling of the film after it has become unloaded, clearly demonstrating the plastic deformation occurring during the testing. It is also evidenced by the variability of adhesion measurement as a function of sample preparation and measurement techniques. Thus the analysis of an adhesion test must include the force going into plastic deformation during the test.

A theoretical analysis of the mechanical effects in the peel strength of thin films has been performed by Kim and co-workers.[4-7] They partition peel strength into its various components. This work clearly shows that the properties of the adherend (elastic modulus, thickness, yield strength, etc.) and also the elastic properties (compliance) of the substrate are important to the measured peel strength. They need to be known to extract the intrinsic adhesion. Table 8.1 shows peel strength behavior as Cr thickness increases in a Cr/Cu film while constant total thickness is maintained.[5] The peel strength decreases as the Cr thickness increases, demonstrating the effect of plastic deformation, which changes because of the difference in ductility between Cu and Cr (Cu is more ductile). Table 8.2 compares peel strength as a function of total thickness.[5] Peel strength increases with decreasing film thickness. Clearly, in this case the intrinsic adhesion is not changing because the interfaces are identical. This behavior just reflects the plastic deformation in copper. These effects are also illustrated in Figure 8.1. Figure 8.2 shows the experimental results compared with the theoretical analysis of Kim and Kim.[4] A good fit is obtained taking into account the film properties, suggesting that the intrinsic adhesion can be estimated by extrapolation to infinite thickness.[5] In this work, they derived a universal peel diagram (Figure 8.3) to extract the intrinsic adhesion from the peel force and film thickness.[4] They showed that the plastic work can be two orders of magnitude higher than the interfacial fracture energy. Thus the intrinsic adhesion is only a small contribution to the measured peel strength. These factors, peel rate, peel geometry, and film thickness, need to be controlled so that reliable information can be extracted from the peel test.

Recently there has also been much promising theoretical effort to use molecular quantum chemical calculations to understand the interfacial interactions at metal–

Total Cr/Cu Thickness, μm	Peel Strength, g/mm
15	52.0
21	45.7
40	39.4
52	35.8
77	31.2

Table 8.2   Peel strength of Cr/Cu film as a function of thickness (Cr thickness is constant at 0.05 μm).

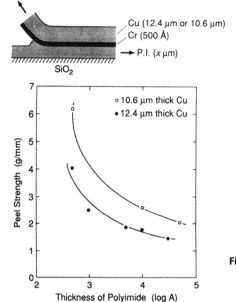

Figure 8.1   **Peel strength as a function of polyimide thickness for two different copper film thicknesses. (After Kim et al.[5])**

polymer interfaces.[8-17] It is hoped that developing an understanding of chemical bonding interactions can lead to the identification of key reactive moieties, which then can intelligently be modified to tailor the desired adhesion. These calculations help one understand interfacial mechanisms and interpret surface spectroscopic results from techniques such as X-ray photoelectron spectroscopy (XPS) and high-resolution electron energy-loss spectroscopy (HREELS). These are large-scale calculations which deal primarily with metal atoms interacting with the polymer. One approach to simplify them is to use model compounds, small molecules with the same reactive sites as polymers.[17, 18]

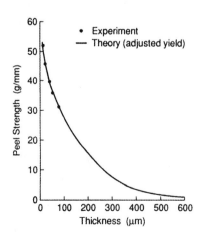

Figure 8.2   **Peel strength as a function of film thickness by experiment and theory. (After Kim et al.[5])**

ADHESION   Chapter 8

8.4 Experimental Measurement of Adhesion

The basis of an adhesion test is that a load is placed on the sample until failure is induced. A number of techniques have been used in the attempt to measure adhesion—Mittal has enumerated over 250. Several of the more common techniques are

- Scotch tape

- scratch

- indentation

- pull

- peel

- lap shear

- bend

- blister

- frequency domain dielectric spectroscopy.

These tests can be quite simple, such as the simple qualitative—and infamous but ubiquitous—Scotch tape test.[19] Another simple test is the scratch test,[20] which has evolved from the use of a fingernail to a cotton-tipped swab to a stylus to an atomic force microscope (AFM). More quantitative tests include the pull test,[19] peel test,[4–7, 21] lap shear test,[20] flex test,[22] and indentation test.[19, 23] The peel test, which is the most used in the microelectronics industry (see Figure 8.4 for a

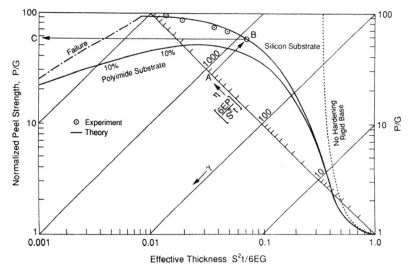

Figure 8.3   Universal peel diagram (After Kim and Kim.[4])

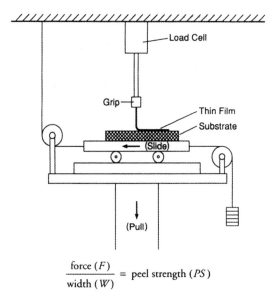

$$\frac{\text{force }(F)}{\text{width }(W)} = \text{peel strength }(PS)$$

Figure 8.4   Schematic of 90° peel test.

schematic of the 90° peel test), uses a flexible strip bonded to an underlying substrate. The strip is pulled from the substrate, with the force depending on parameters such as peel angle, peel strength, thickness of film, mechanical properties of the film, and the quantity that one most often wants to determine—the interfacial adhesion. Our discussion above concerning the peel test shows the caution one must use when attempting to extract quantitative information about intrinsic adhesion from the peel test, since that analysis shows it does not measure interfacial adhesion but a peel strength (*PS*) or engineering adhesion

$$PS = IA + MW$$

where *IA* is intrinsic adhesion and *MW* mechanical work. One wants to maintain *MW* constant in a series of tests to be able to make direct comparisons. With the caution that a comparison can only be made to similar films under similar conditions, one can obtain limited but quite useful information. It is also important to simulate the loading of the structure that is of interest, because the failure mode induced in a test structure may not be identical to that in an actual part. Sophisticated techniques have been developed that take advantage of advanced microelectronic processing techniques, as in the case of the blister test.[24–26] Recently, a promising nondestructive technique based on frequency domain dielectric spectroscopy was developed by Narducci and co-workers.[27] This technique correlates dielectric response to adhesion failure and can be useful in analyzing corrosion-induced adhesion failure.

8.5 Standardization of Test Procedure

As demonstrated above for the peel test, the interpretation of an actual measurement in terms of intrinsic adhesion is not straightforward, but the use of continuum mechanics analysis can help extract intrinsic adhesion. However, one would prefer to extract useful information directly. This can be done by careful standardization of the test employed. Such standardization allows the assessment of effects such as surface pretreatments or resistance to aging and environmental insults. Again, using the peel test for our case example, a group of IBM researchers has assessed what the key factors were to standardize the 90° peel test.[28] Among the factors to be strictly controlled were film thickness and peel rate. Such standardization is required for all mechanical adhesion tests; it is only with standardization that meaningful comparisons of data can be made.

8.6 Locus of Failure

For a complete description of adhesion, one needs to know not only the adhesion strength but also where the failure occurs, that is, the locus of failure. The locus of failure is necessary to understand the mechanism of failure. The failure of adhesion between two materials A and B can be cohesive, that is, the failure is in the weaker of the two materials. Failure can be interfacial, that is, at the interface between A and B. The failure can be in the region near the interface, where, due to bonding interactions, pretreatments, promoters, etc., a modified region is produced that is weaker than A or B—this is what we usually mean by adhesional failure. Often an intermediate layer C, an adhesive layer, is added between A and B, and this layer can be the weak link.

The autopsy tools used to determine the locus of failure are either surface-sensitive techniques or microscopic techniques. The surface-sensitive techniques, discussed in Chapter 3, are XPS, Auger spectroscopy, secondary ion mass spectrometry (SIMS), and ion scattering spectrometry (ISS); these autopsies are performed on both sides of the peel. By using these techniques, one can determine if the failure is cohesive, adhesive, or interfacial. The current generation of instrumentation for these techniques has a much improved spatial resolution as well as imaging capabilities. Microscopic techniques, which are discussed in Chapters 4 and 5, are also employed for locus of failure analysis[29]: transmission electron microscopy (TEM) and secondary electron microscopy. Another useful technique is scanning acoustic microscopy, which can be employed in situ while a structure is under bias.[30, 32] Nuclear magnetic resonance (NMR) imaging is another promising technique,[33] and dye penetration is another commonly employed tool. To summarize, the techniques used for locus of failure analysis are

- XPS

- scanning Auger microscopy

- SIMS

- Rutherford backscattering spectroscopy (RBS)

- ISS

- scanning acoustic microscopy

- NMR imaging

- scanning electron microscopy (SEM)

- transmission microscopy

- optical microscopy.

Figure 8.5 shows an example of XPS being applied to determine the locus of failure, in this case from a polyimide surface that had been given an rf pretreatment prior to metallization. XPS spectra are from either side of the peel. Polymer can be seen on both sides, demonstrating in this case that the failure was cohesive. However, it is noted that some metal can also be seen from the metal peel strip. This

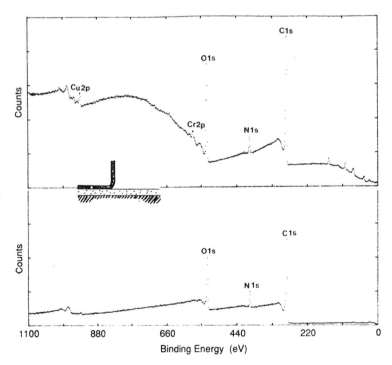

Figure 8.5   Locus of failure analysis by XPS: (*upper*) from the metal peel strip and (*lower*) from the polyimide substrate. The surface had been rf-sputtered prior to metal deposition. (Unpublished results from S. P. Kowalczyk and T. S. Oh.)

indicates that the failure, while in the polyimide, was near the metal–polyimide interface, where the polyimide had been modified. Since the XPS sampling depth is ~100 Å, this indicates where the failure region was.[34] This is corroborated by cross-sectional TEM (XTEM) analysis.[35] The untreated surface exhibited interfacial failure. These tests are ex situ, which limits the usefulness of the information. More chemical information can be obtained by doing these locus of failure analyses in situ; however, setups to do this are currently quite rare.

8.7 Key Adhesion Issues

Once adhesion is measured, the next step is to improve it. There are several techniques to enhance adhesion. Often, the first step is simply surface cleaning prior to the formation of an interface or joint. Surface contamination often can act as an unintentional release layer. Surface-cleaning techniques range from degreasing or degassing to sputter-cleaning. A useful technique is UV ozone treatment, effective against organic contaminants.[36] Surface roughening is another effective approach to enhanced adhesion.[37] Modification of surface chemistry either by dry techniques such as plasma or ion beam treatments[38–40] or wet-chemical treatments such as base hydrolysis reactions on polyimide surfaces are very powerful.[41] These methods are treated in detail in Chapter 7. Seeding by electrochemical or photochemical means has also been successfully utilized.[42–44] Finally, using a monolayer of a coupling agent such as aminosilanes or adhesive layers and primers is an important technique for adhesion enhancement.[45–50] Figure 8.6*a* shows an example of adhesion enhancement by surface modification with rf plasma pretreatment as a function of power; Figure 8.6*b* shows the improvement as a function of duration of treatment.[35]

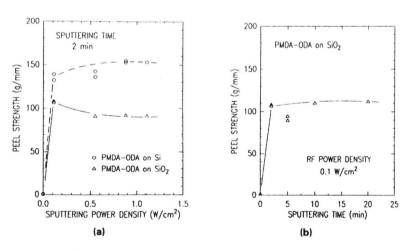

Figure 8.6 (a) Peel strength as a function of rf power density and (b) peel strength as a function of sputter time. (After Reference 35.)

	Au	**Cu**	**Ni**	**Cr**
Metal on PI	0	5	16	42
PI on metal	31	73	86	117

Table 8.3 **Peel strength comparison for several metal-on-polyimide versus polyimide-on-metal interfaces.**

The final major concern is how well adhesion is maintained as a function of time under actual operating ambients. This is usually investigated by using accelerated failure testing, which involves stressing the interface with a desired stressing agent, prior to adhesion testing. Such stresses can include humidity, temperature, solvent exposure, etc. As an example, in microelectronics the interface is often cycled in a controlled temperature/humidity chamber with conditions of 81 °C and 85% humidity.[51] Since some structures inadvertently act as electrochemical cells, resulting in blistering, etc.,[52] it is often desirable to test for the effects of electrical bias. Another more extreme test is the pressure-cooker test.

8.8 Illustrative Examples

In this section we give a few more examples of adhesion from microelectronic-specific applications. In the beginning, we classified adhesion between metal and polymers, between ceramics and polymers, and between polymers and polymers. Metal–polymer adhesion probably has been given the most attention. One important aspect of metal–polymer interfaces not often appreciated is that one must distinguish metal on polymer from polymer on metal, that is, these two interfaces are not always equivalent.[53, 54] Table 8.3 gives the measured peel strengths for several metal–polyimide combinations. In all cases, the polyimide on metal has higher adhesion than the metal on polyimide. The differences between these two types of interfaces include, in the case of the metal-on-polymer example, that the interaction is with metal atoms or metal clusters as the interface grows (which can be dependent on rate or method of deposition). For the obverse case, the metal is already fully formed before the interaction with the polymer. Also, except in a few ultra-high vacuum (UHV) surface science studies, the metal surface is not really a metal surface—it is the native metal oxide and is not always well-defined, at that. In the polyimide-on-metal example, the polyimide is not applied to the "metal." Polyamic acid is applied, and the interface is subjected to heating during the cure to produce the polyimide. This is done in the presence of a solvent. All these factors make the interfaces nonequivalent. There are strong interfacial chemical interactions for the Cr with polyimide, as can be clearly seen in the XPS spectra (Figure 8.7), whereas Cu and Au interact much more weakly.[55, 56] Another factor to be aware of is that the interdiffusion of metals can produce an interface other than

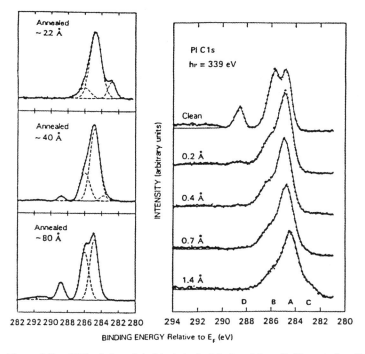

Figure 8.7 XPS of Cr-polyimide interfacial chemistry: (*left*) as a function of increasing polyimide thickness on Cr and (*right*) as a function of increasing Cr thickness on polyimide. (After References 55 and 56.)

what one believes one has. This is a problem, for example, with gold[54, 57]; thus a diffusion barrier is often needed.

8.9 Summary

Although adhesion usually is not the main property of a multilayer system, it is often the limiting property for an application. Adhesion is very complicated, and the actual determination of *interfacial* adhesion is rather tricky, but useful engineering values can almost always be obtained. However, our basic understanding has advanced, driven by the importance of adhesion in multilayer structures in microelectronics, where the layers can be very thin. Measurements of adhesion are improving with the aid of microelectronic fabrication techniques for new types of test structures; the "old standby" peel test has remained quite useful, as our understanding of all contributions to the test has improved. A main proviso is to be precise about what is measured, that is, what the real interface is—polymer on native oxide not metal, thickness of that oxide, condition of preparation, etc. In the near future, truly atomic-level measurements will be possible with scanning probe microscopies, such as atomic force microscopy (AFM).

Complete evaluation and understanding of adhesion and decohesion requires a multidisciplinary approach covering chemistry (interfacial chemistry, thermal hydrolysis, surface chemical modification), physics (thermal), mechanics (crack propagation, viscoelastic behavior), applied physics (stress analysis), and engineering (processing), of which we give only a glimpse here. It is only with such a diverse approach that we can answer affirmatively the "simple" question posed at the beginning: Can we determine adhesion?

Additional Reading

Journals

Adhesion is a multidisciplinary endeavor; thus research is published in a wide range of journals. The following is a list of journals in which research articles on adhesion-related topics are frequently published:

Journal of Adhesion Science and Technology
Journal of Adhesion (UK)
Journal of Materials Science
Journal of Vacuum Science and Technology
Journal of Applied Physics
International Journal of Fracture
Journal of Applied Polymer Science
Thin Solid Films
Surface and Interface Analysis
Materials Science and Engineering
Materials Letter
Journal of Physics D
Journal of Materials Research
International Journal of Solids and Structures
Applied Surface Science
Journal of Polymer Science
Polymer Degradation and Stability
Polymer Engineering and Science
Surface Science

Conference Proceedings

A number or the *Proceedings of the Materials Research Society Symposium Series* are relevant; see volumes 10, 25, 40, 54, 72, 79, 108, 119, 125, 130, 147, 153, 154, 167, 188, 203, 223, 225, 226, 227, and 239.

See also *Proceedings of the Annual Meetings of the Adhesion Society* and *Proceedings of the Electronic Component and Technology Conference.*

Other Societies periodically have special topical conferences or symposia, such as the American Physical Society and the American Vacuum Society.

Books

Adhesion Measurement of Thin Films, Thick Films, and Bulk Coating. (K. L. Mittal, Ed.) ASTM, Philadelphia, 1976.

Adhesion Science and Technology, Vol. 9. (L.-H. Lee, Ed.) Plenum, New York, 1975.

Buckley, D. H. *Surface Effects in Adhesion, Friction, Wear, and Lubrication.* Elsevier, Amsterdam, 1981.

Industrial Adhesion Problems. (D. W. Brewis and D. Briggs, Eds.) Orbital Press, Oxford, 1985.

Israelachvili, J. N. *Intermolecular and Surface Forces.* Academic, London, 1985.

Kinloch, A. J. *Adhesion and Adhesives.* Chapman and Hall, London, 1987.

Metallization of Polymers. (E. Sacher, J.-J. Pireaux, and S. P. Kowalczyk, Eds.) American Chemical Society, Washington, DC, 1990.

Microscopic Aspects of Adhesion and Lubrication. (J. M. Georges, Ed.) Elsevier, Amsterdam, 1982.

Wake, W. C. *Adhesion and Formulation of Adhesives.* Applied Science Publishers, London, 1982.

Wu, S. *Polymer Interface and Adhesion,* Marcel Dekker, New York, 1982.

Articles

Allara, D. L., F. W. Fowkes, J. Noolandi, G. W. Rubloff, and M. V. Tirrell. *Mater. Sci. Eng.* **83**, 213, 1986.

Baglin, J. E. E. E. *Nucl. Instrum. Meth.* **B65**, 119, 1992.

Brown, H. R. *Ann. Rev. Mater. Sci.* **21**, 463, 1991.

Buchwalter, L. P. *J. Adhesion Sci. Technol.* **4**, 697, 1990.

de Gennes, P. G. *Cand. J. Physics.* **68**, 1049, 1990.

de Gennes, P. G. *J. Phys.* (France). **50**, 2551, 1989.

Ingermarsson, P. A. *Nucl. Instrum. Meth.* **B44**, 437, 1990.

Jacobsson, R. *Thin Solid Films.* **34**, 191, 1976.

Kim, J., S. P. Kowalczyk, Y.-H. Kim, N. J. Chou, and T. S. Oh. *Mat. Res. Soc. Symp. Proc.* **167**, 137, 1990.

Kim, K. S., and J.-I. Kim. *J. Eng. Mater. Technol.* **110**, 266, 1988.

Marletta, G. *Nucl. Instrum. Meth.* **B46**, 295, 1990.

Mittal, K. L. *Electrocomp. Sci. Technol.* **3**, 21, 1976.

Mittal, K. L. *J. Adhesion Sci. Technol.* **1**, 247, 1987.

Mittal, K. L. *J. Vac. Sci. Technol.* **13**, 19, 1976.

Oh, T. S., L. P. Buchwalter, and J. Kim. *J. Adhesion Sci. Technol.* **4**, 303, 1990.

Pawel, J. E. and C. J. McHargue. *J. Adhesion Sci. Technol.* **2**, 369, 1988.

References

1   B. J. Feder. *The New York Times.* 31 May 1992, p. 10.

2   K. L. Mittal. *J. Adhesion Sci. Technol.* **1**, 247, 1987.

3   J. Israelachvili. *J. Vac. Sci. Technol.* **A10**, 2961, 1992.

4   K. S. Kim and J. Kim. *ASME J. Eng. Mater. Technol.* **110**, 266, 1988.

5   J. Kim, K. S. Kim, and Y.-H. Kim. *J. Adhesion Sci. Technol.* **3**, 175, 1989.

6   K. S. Kim and N. Aravas. *Int. J. Solids Structure.* **24**, 417, 1988.

7   N. Aravas, K. S. Kim, and M. J. Loukis. *Mater. Sci. Eng.* **A107**, 159, 1989.

8   J. S. Shaffer, A. K. Chakraborty, M. Tirrell, H. T. Davis, and J. L. Martins. *J. Chem. Phys.* **95**, 8616, 1991.

9   A. K. Chakraborty, H. T. Davis, and M. Tirrell. *J. Poly. Sci.* **28**, 3185, 1990.

10   L. J. Gerenser, K. E. Goppert-Berarducci, R. C. Baetzold, and J. M. Pochan. *J. Chem. Phys.* **95**, 4641, 1991.

11   A. R. Rossi, P. N. Sanda, B. D. Silverman, and P. S. Ho. *Organomet.* **6**, 580, 1987.

12   S. A. Kafafi, J. P. LaFemina, and J. L. Nauss. *J. Am. Chem. Soc.* **112**, 8742, 1990.

13   M. M. D. Ramos, A. M. Stoneham, and A. P. Sutton. Harwell Laboratory Report AEA-InTec-1044, 1992.

14   S. R. Cain. *J. Adhesion Sci. Technol.* **4**, 333, 1990.

15   M. Seel, A. B. Kunz, and D. T. Wadiak. *Phys. Rev.* **B37**, 8915, 1988.

16   K. Nath and A. B. Anderson. *Phys. Rev.* **B37**, 7916, 1989.

17   S. Stafström, P. Bodö, W. R. Salaneck, and J.-L. Brédas. In *Metallization of Polymers.* (E. Sacher, J.-J. Pireaux, and S. P. Kowalczyk, Eds.) American Chemical Society, Washington, DC, 1990, p. 312.

18   W. R. Salaneck, S. Stafström, J.-L. Brédas, S. Andersson, P. Bodö, S. P. Kowalczyk, and J. J. Ritsko. *J. Vac. Sci. Technol.* **A6**, 3134, 1988.

19   B. Bhushan. In *Testing of Metallic and Inorganic Coatings.* (W. B. Harding and G. A. DiBari, Eds.) ASTM, Philadelphia, 1987, p. 310.

20   J. K. Strozier, K. J. Ninow, K. L. DeVries, and G. P. Anderson. *J. Adhesion Sci. Technol.* **1**, 209, 1987.

21   R. Farris, M. A. Madden, J. Goldfarb, and K. Tung. *Mat. Res. Soc. Symp. Proc.* **227**, 177, 1991.

22   A. Roche, F. Gaillard, M. Romand, and M. von Fahnestock. *J. Adhesion Sci. Technol.* **1**, 145, 1987.

23   J. E. Ritter, T. J. Lardner, L. Rosenfeld, and M. R. Lin. *J. Appl. Phys.* **66**, 3626, 1989.

24   M. G. Allen, P. Nagarkar, and S. D. Senturia. In *Polyimides: Materials, Chemistry, and Characterization.* (C. Feger, M. M. Khojasteh, and J. E. McGrath, Eds.) Elsevier, Amsterdam, 1989, p. 705.

25   M. G. Allen and S. D. Senturia. *J. Adhesion.* **25**, 303, 1988.

26   H.-S. Jeong, Y. Z. Chu, C. J. Durning, and R. C. White. *Surf. Int. Anal* **18**, 289, 1992.

27   D. Narducci, J. J. Cuomo, D. L. Pappas, and K. Sachdev. To be published in *J. Electrochem. Soc.*

28   J. Kim, P. Buchwalter, H. Clearfield, P. Lauro, K.-W. Lee, S. Nunes, J. Paraszcak, S. Purushothman, A. Viehbeck, and D.-Y. Shih. *Adhesion Test Standardization.* IBM Report RC 17203, 1991.

29   A. J. Kinloch and M. L. Yuen. *J. Mater. Sci.* **24**, 2183, 1989.

30   R. C. Addison, M. W. Kendig, and S. J. Jeanjacquet. *Acoustical Imaging.* **17**, 143, 1989.

31   M. Kendig, R. Addison, and S. Jeanjaquet. *J. Electrochem. Soc.* **137**, 2690, 1990.

32   M. Kendig, M. Abdel-Gawad, and R. Addison. *Corrosion.* **48**, 368, 1992.

33   A. Nieminen and J. L. Koenig. *J. Adhesion Sci. Technol.* **2**, 407, 1988.

34   S. P. Kowalczyk and T. S. Oh. Unpublished results.

35   T. S. Oh, S. P. Kowalczyk, D.J. Hunt, and J. Kim. *J. Adhesion Sci. Technol.* **4**, 119, 1990.

36   R. P. Padmanabhan and N. Saha. *Proceedings.* ISHM, 1989, p. 197.

37   N. Somasiri, R.LD. Zenner, and J. C. Houge. *IEEE Trans.* **CHMT14**, 798, 1991.

38   P. A. Ingemarsson, M. P. Keane, and U. Gelius. *J. Appl. Phys.* **66**, 3548, 1989.

39   N. J. Chou, A. D. Marwick, R. D. Goldblatt, L. Li, G. Coleman, J. E. Heidenreich, and J. R. Parazszak. *J. Vac. Sci. Technol.* **A10**, 248, 1992.

40   A. A. Galuska. *J. Vac. Sci. Technol.* **A10**, 381, 1992.

41   K.-W. Lee and S. P. Kowalczyk. In *Metallization of Polymers.* (E. Sacher, J.-J. Pireaux, and S. P. Kowalczyk, Eds.) American Chemical Society, Washington, DC, 1990, p. 179.

42   R. D. Goldblatt, J. M. Park, R. C. White, L. J. Matienzo, S. J. Huang, and J. F. Johnson. *J. Appl. Poly Sci.* **37**, 335, 1989.

43   T. H. Baum and D. C Miller. *Chem. Mater.* **3**, 714, 1991.

44 A. Viehbeck, C. A. Kovac, S. L. Buchwalter, M. J. Goldberg, and S. L. Tisdale. In *Metallization of Polymers*. (E. Sacher, J.-J. Pireaux, and S. P. Kowalczyk, Eds.) American Chemical Society, Washington, D.C., 1990, p. 394.

45 E. Plueddemann. *Silane Coupling Agents*. Plenum, New York, 1982.

46 H. Ishida. Polym. *Compos.* **5**, 101, 1984.

47 E. Plueddemann. In *Surface and Colloid Science in Computer Technobgy*. (K. L. Mittal, Ed.) Plenum, New York, 1985.

48 H. Ishida and K. Kelley. *J. Adhesion*. **36**, 177, 1991.

49 G. Tesoro, G. P. Rajendran, C. Park, and D. R. Uhlmann. *J. Adhesion Sci. Technol.* **1**, 39, 1987.

50 E. Pennisi. *Science News*. **142**, 171, 1992.

51 T. S. Oh, D. G. Kim, S. P. Kowalczyk, S. Molis, and J. Kim. *J. Mater. Sci. Letters*. **10**, 374, 1991.

52 J. Steinberg and B. Kistler. *Proceedings*. ISHM, 1989, p. 237.

53 S. P. Kowalczyk, Y.-H. Kim, G. F. Walker, and J. Kim. *Applied Physics Letters*. **52**, 375, 1988.

54 Y.-H. Kim, J. Kim, G. F. Walker, C. Feger, and S. P. Kowalczyk. *J. Adhesion Sci. Technol.* **2**, 95, 1988.

55 S. P. Kowalczyk and J. L. Jordan-Sweet. *Chem. Mater.* **1**, 592, 1989.

56 J. L. Jordan, C. A. Kovac, J. F. Morar, and R. A. Pollak. *Phys. Rev.* **B36**, 1369, 1987.

57 G. D. DiGiacomo and S. Purushothaman. *Proceedings*. ISHM, 1989, p. 227.

Chemistry, Reactivity, and Fracture of Polymer Interfaces

STEVEN G. H. ANDERSON and PAUL S. HO

Contents

9.1 Introduction

Excellent adhesion between dissimilar materials is required in many technological applications, such as the bonding of paint to automobiles, chrome plating for corrosion protection, fabrication of high-strength composites, reflective and nonreflective coatings on optical components, magnetic thin films for data storage, and the manufacturing and packaging of microelectronic devices. In each case, the adhesive bonding of the overlayer to the substrate must be strong and stable enough to withstand stresses experienced during both further processing steps and typical use conditions. Clearly, the performance of a particular thin film/substrate combination will depend on its usage. For example, the films may be subject to multiple thermal treatments where differences in the thermal expansion coefficient induce large stresses, aggressive solutions which can erode or swell the thin-film structure to cause delamination, or abrasives during mechanical cleaning processes. Common to most of the failure modes is the initiation and propagation of a crack at or near the interface. The principle factors which affect these two undesirable phenomena at an overlayer-substrate junction include the interfacial chemistry and morphology, interfacial toughness, the use of adhesive layers, contamination, film

215

thickness, elastic moduli, thermal expansion coefficients, and the stress between the layers. In this review, we emphasize some challenges for adhesion in microelectronic packaging applications.

Progress in device technology has continuously improved the device density and speed in integrated circuits (ICs). The leverage of these advances into system performance demands has increased the number of wiring levels in the packaging structure to be defined with greater precision, density, and yield. This has generated a great deal of interest recently in developing polymeric materials as dielectrics in multilayered structures for interconnects and in packaging modules. (Reference 1 provides an excellent overview of the electronic packaging field.) Polyimides are a family of high-temperature polymers with properties well-suited for such applications, including low dielectric constant, high-temperature stability, and extended processability.

Polyimides have developed in two directions. The first is in the development of new polyimides with specific materials characteristics, such as low dielectric constant and photosensitivity.[2] This is motivated by the need for performance enhancement and cost reduction in multilevel packaging structures. The second direction is in the characterization of new polymeric materials and the interaction between polymers and other materials. Although such efforts have been prompted by the practical needs of reliability and processing development, a number of basic issues have emerged regarding the chemistry, structure, and properties of polymer films and interfaces. This chapter reviews the current understanding of this aspect of polyimide properties for packaging applications.

To illustrate the breadth of the materials issues that must be addressed when using polyimides in microelectronics, consider the typical structures that result after fabricating the multilayer structure shown in Figure 9.1. The polymer must first be deposited onto a substrate in the form of a precursor and then stepwise-heated to cure the polymer and develop the desired mechanical and dielectric properties. This is followed by metal deposition and the formation of a metal–polymer interface. After the metal is patterned with polymeric photoresists and lift-off processes, an additional polyimide layer is deposited on the substrate. This process produces two new types of interfaces, one between polyimide and polyimide and the other between polyimide and metal. Each of these three types of interface have been shown to exhibit behavior distinct from the others.

A multilayered structure such as that in Figure 9.1 is subject to large stresses during processing due to differences in the metal and polymer thermal coefficients of expansion (TCEs). The stresses generated in a packaging structure can be sufficient to cause fracture or delamination. This is particularly important at corners and interfaces due to local stress concentrations. To prevent interfacial delamination, we have devoted considerable efforts to understand the chemical interactions and learn how surface modifications prior to or during deposition may affect adhesion. The effect of subsequent thermal processing on interface integrity has also been a major concern, since the presence of water or oxygen has been shown to

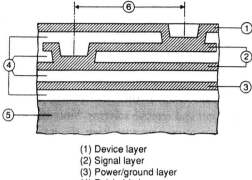

(1) Device layer
(2) Signal layer
(3) Power/ground layer
(4) Polyimide layer
(5) Base plate
(6) Conductor pitch

Figure 9.1   Cross section of a thin-film substrate.

promote further reactions at the metal–polymer and polymer–metal interfaces, lead-ing to degradation of adhesion between the two materials. Evaluation and control of these parameters are essential to maintaining the integrity of the multilayered structures. The combination of surface and interfacial properties with the mechanical properties of thin, anisotropic polymer films presents a number of interesting funda-mental problems involving metal–polymer interfaces. One basic feature distinguish-ing the metal–polymer interface from other solid interfaces comes from the molecu-lar morphology of the polymer. To a significant extent, the molecular morphology can influence the chemical and physical properties of the interface and multilayered structures. Several of these issues have been investigated by our group by examining the contrasting behavior of pyromellitic dianhydride oxydianiline (PMDA-ODA) and bisphenyldianhydride phenyldiamine (BPDA-PDA), two polyimides with simi-lar chemical structures but distinct molecular structure.

In this chapter, the results of studies by us and other groups are summarized in two broad sections. As a preface to these sections, a discussion of the molecular struc-ture, morphology, and mechanical properties of polyimide thin films is presented. The first section then discusses the application of different surface science tools to the problems of polymer–polymer, metal–polymer, and polymer–metal interfaces. Methods for modifying surfaces prior to the deposition of subsequent layers are discussed, as are techniques that have clarified the influence of thermal processing and contaminants on interface stability. The second section examines two aspects of the thermal/mechanical properties of metal–polymer multilayered structures. First, results from measurements of the thermal stress generated in metal–polymer struc-tures during thermal cycling are discussed and related to the polyimide property. Second, the fracture of metal–polymer fine-line structures and its dependence on line geometry are considered.

9.2 Polyimide Structure and Properties

Polymer material properties depend on many parameters, including the chemistry of the repeat unit and its molecular structure, the molecular weight, and the way in which they are processed. Polyimides are generally formed by a two-step process consisting of a low-temperature reaction between various aromatic dianhydrides and diamines to form a polymeric precursor, which is then followed by a high-temperature reaction to promote imidization and remove solvents. Since the thermal stability and mechanical and dielectric properties arise from the imide and aromatic ring structures, variation of diamine or dianhydride moieties provides a method of designing a wide range of properties.[3]

In this regard, it is worthwhile to relate the chemical structure of PMDA-ODA and BPDA-PDA polyimides (shown in Figure 9.2) to their mechanical properties (given in Table 9.1). The PMDA-ODA structure is composed of a relatively rigid PMDA segment and a flexible ODA fragment. Flexiblity arises from the ether oxygen, which joins the two benzene rings of ODA at an angle of about 120° and allows them to twist relative to the PMDA unit. In contrast, BPDA-PDA derives its reduced flexibility primarily from the BPDA unit, which twists in the middle. The differences in molecular structure lead to a stiffer backbone for BPDA-PDA compared to PMDA-ODA and are manifested in the higher Young's modulus, higher tensile strength, and lower strain-to-break values observed for BPDA-PDA. In addition, the rigid-rod structure of BPDA-PDA facilitates ordering of the polymer chains, leading to a higher molecular packing coefficient. Table 9.1 shows that

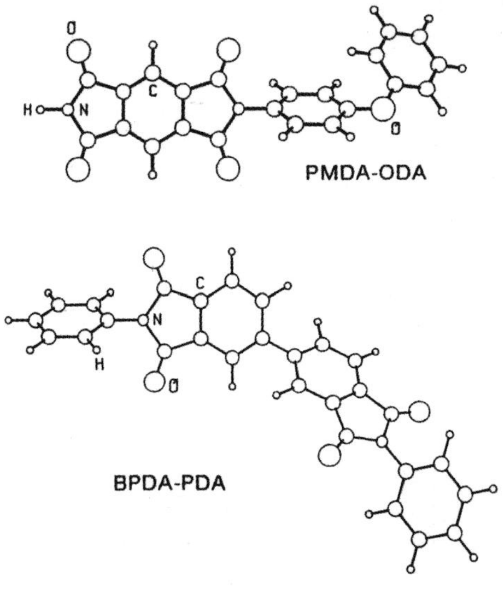

Figure 9.2   The chemical structure of PMDA-ODA and BPDA-PDA polyimides.

CHEMISTRY, REACTIVITY, AND FRACTURE . . .   Chapter 9

the high packing coefficient of BPDA-PDA correlates with a lower water uptake than in the less-packed PMDA-ODA.

Since the molecular packing coefficient depends on the polyimide molecular structure, it is useful to correlate this parameter with water diffusion, elastic modulus, and TCE. To investigate the influence of molecular packing on these material properties, a series of polyimides formed from a variety of diamine and dianhydride groups was studied.[4] A plot of the water diffusion coefficient versus the molecular packing coefficient in Figure 9.3a shows that a more compact structure absorbs water more slowly. The data bracketed by the dashed lines in Figure 9.3b indicate that the TCE decreases as polyimide molecular packing coefficient increases. Molecular packing coefficients greater than those for typical polymers arise from the use of rodlike diamine precursors and suggest that the low thermal expansion is related to the small free volume. The correlation between the elastic modulus and molecular packing shown by the data between the dashed lines in Figure 9.3c supports this conclusion. (Note that these conclusions are consistent with the results of Table 9.1.) However, with the addition of pendant side groups to the rodlike diamines, steric hindrance between the polymer chains is increased and a smaller packing coefficient results without significantly altering the modulus or thermal expansion coefficients. This is indicated by the arrows in Figures 9.3b and 9.3c. These results demonstrate that free volume is not the sole factor in determining bulk polyimide properties, and they illustrate the importance of understanding the chain morphology. A recent simulation study on the packing structure of PMDA-ODA, BPDA-PDA, and PMDA-PDA also confirms the importance of chain linearity in controlling the molecular packing.[5]

Polyimide chain morphology is affected by both the multistep curing processes of polyamic acid (PAA) and the presence of any physical constraints which inhibit film shrinkage during the cure. PAA is synthesized by a condensation reaction

Property	PMDA-ODA	BPDA-PDA
Density (g/cm^3)	1.42–1.44	1.44–1.47
Young's modulus (GPa)	2.5–3.0	9.5–13
Tensile strength (MPa)	180–200	480–600
Strain to break (%)	60–80	30–50
TCE (ppm/°C)	25–40	5–7
Glass transition temperature (°C)	400	320
NMP uptake (wt % at 85 °C)	30–40	1–3
Water uptake (wt % at 22 °C)	3–4	1–1.5
Molecular packing coefficient	0.687	0.719

Table 9.1   Physical properties of PMDA-ODA and BPDA-PDA polyimides.

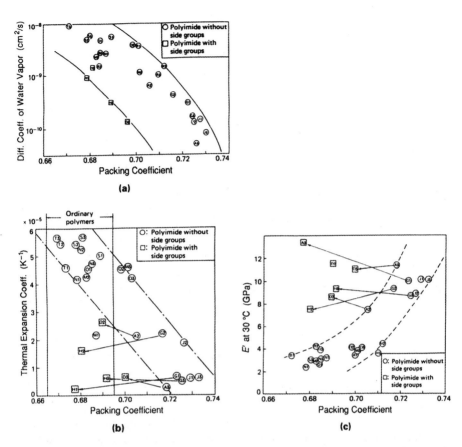

Figure 9.3 (a) The water diffusion coefficient correlates with the molecular packing coefficient for many polyimides. (b) The TCEs for polyimides without side groups scales with the molecular packing coefficient. (c) A general correlation between Young's modulus and the packing coefficient breaks down for polyimides with pendant side groups. (From Reference 4).

between an aromatic diamine and a tetracarboxylic dianhydride in a solution of *N*-methyl 2-pyrrolidone (NMP) and xylene. This PAA solution is spin-coated onto a substrate to form a uniform film and then dried between 75 and 100 °C. Thermogravimetric analysis and laser interferometry studies of the PAA, in conjunction with investigations of the amic acid monomers, showed that the drying process results in the formation of NMP/PAA complexes in about a 2-to-1 ratio.[6, 7] X-ray diffraction (XRD) has shown that the NMP/PAA complexes exhibit an ordered glassy structure with an extended chain conformation along the film plane.[8] The observed repeat distance of 13–14 Å is close to the projected length of the monomer along the chain direction.

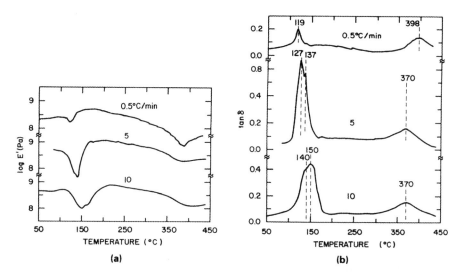

Figure 9.4 DMTA results for freestanding PAA films cured at three heating rates (0.5, 5, and 20 °C/min). (*a*) log *E′* and (*b*) tan δ versus temperature scans exhibit strong dependences on heating rate, reflecting the influence of solvent creation and removal on polyimide matrix mobility. (From Reference 9).

The dynamic mechanical thermal analysis (DMTA) results for the curing of freestanding polyimide films at increasing heating rates, shown in Figure 9.4, reveal that two main processes occur above 110 and 300 °C. In the first process, the NMP/PAA complex decomposes, the amic acid is converted to polyimide to produce water, and both NMP and water leave the film. The release of NMP and water plasticizes the matrix, reduces its initial ordering, and leads to a decrease in the storage modulus (*E′*) and a corresponding increase in tan δ.[9] These solvents are released at different temperatures, as evidenced by the presence of two peaks in Figure 9.4*b* between 120 and 150°C. Note that the heating rate dependence of the DMTA scans reflects the balance between the rates at which NMP and water are produced and the rates at which they diffuse to the surface and leave the film. Signficantly, this balance of plasticizer creation and removal alters the matrix mobility and hence the extent and rate of imidization possible for a given temperature ramp rate and film thickness.

Further annealing of freestanding films between 300 and 400 °C reveals another drop in the storage modulus followed by a slow increase in log *E′* as the polymer passes through the glass transition temperature (Figure 9.4*a*). FTIR studies showed that this modulus increase was not due to further imidization. Instead, XRD experiments showed that the molecular packing in unordered films improves between 350 and 400 °C as the modulus increases.[8] Furthermore, the extent of this high-temperature ordering process depends on the molecular order at the completion of imidization. We note that although the above discussion has focused on freestanding polyimide films, many of the processes hold for films cured on a substrate. There are two significant differences, however. First, transport of the reaction products out of the film cannot take place on the substrate interface, thus modifying the transport and plasticization properties. Second, the substrate inhibits film shrinkage and results in a somewhat ordered film. At the same time, prolonged annealing between 350 and 400 °C promotes further packing and an increase in the storage modulus.

9.3 Polymer–Polymer Interfaces

Adhesion between polymer surfaces is of scientific and technological importance in packaging applications. As sketched in Figure 9.1, the encapsulation of a metal transmission line by a polyimide necessarily results in a polyimide–polyimide interface which must be strong enough to withstand further process steps. In general, the adhesion between the two layers depends on the diffusion and entanglement of polymer chains across the interface.[10] Several methods for improving and controlling polyimide self-adhesion have been demonstrated and are discussed in conjunction with the techniques used to characterize them.

To examine the extent of interpenetration at polyimide–polyimide interfaces, forward recoil spectroscopy (FRES) studies were conducted on bilayer samples composed of a 200–300-Å-thick deuterated PMDA-ODA polyamic acid spun onto protonated PMDA-ODA polyimide.[10, 11] The deuterated polyimide was cured at temperature T_2, and the substrate was cured at various temperatures T_1. The FRES spectra from two different samples cured at T_1/T_2 of 400/150 °C and 150/400 °C are shown in Figure 9.5. To determine the interpenetration distance, the data was fit to the Fickian diffusion equation for thin films on a semiinfinite substrate. The sharp peak near 1.55 MeV for the 400/150 °C sample corresponds to a thin deuterated PMDA-ODA layer confined to the surface. Intermixing is inhibited since the 150 °C cure is below the polyimide glass transition temperature, T_g. In contrast, the spectrum cured at 150/400 °C exhibits a broad peak that tails to lower energies, indicative of significant diffusion. Additionally, the extent of inter-diffusion correlated with the fracture toughness of the joint, which was found to be dependent on the cure schedule.[10] Subsequent experiments demonstrated that the presence of a solvent was necessary to plasticize the films and to promote diffusion across the interface; placing a dried PAA in contact with a dried, partially cured polyimide film results in virtually no interdiffusion because of the poor miscibility between PAA and polyimide.[11] Conversely, the adhesion between polyimide films could be greatly improved by swelling a completely cured polyimide underlayer with solvent prior to the deposition and curing of a second PAA layer. By solvent swelling, the cured polymer, diffusion, and chain entanglement could be enhanced.[12]

Another method that has been developed to facilitate chain entanglement and improve the miscibility between two polyimides involves the chemical conversion

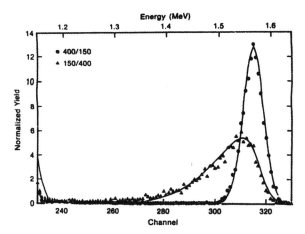

Figure 9.5 FRES spectra for bilayer polyimide films. Samples were cured at temperature T_1, followed by deposition of a second layer and curing at T_2. The solid lines are simulated FRES spectra corresponding to diffusion distances of 22 and 156 nm for the 400/150 and 150/400 °C cured samples. (From Reference 10).

of the top several hundred angstroms of a cured polyimide back into PAA.[13] In this procedure, a polyimide reacts with KOH or NaOH to form a polyamate. Significantly, the reaction does not proceed homogeneously throughout the film. Instead, contact angle and angle-resolved X-ray photoelectron spectroscopy (XPS) measurements revealed that the reaction proceeds along a front that gradually moves into the polymer film. The reaction thus produces a polyamate layer, the thickness of which depends on time, hydroxide concentration, and solution temperature. For reacted layer thicknesses greater than the probe depth of XPS, external reflectance infrared (ERIR) spectroscopy was used to measure the depth of the modified region via a two-step process. First, a linear correlation between the absorbance of imide carbonyl stretching mode with polyimide thickness was found. Next, the difference between absorbance measurements made before and after hydroxide treatment were related to the thickness of polyimide converted to polyamate. The resultant potassium or sodium salt is then protonated with HCl to produce a PAA of controlled thickness on top of unperturbed, cured polyimide.

Although this PAA layer can be reconverted to the imide by heating, it has two properties that make it particularly appealing for improved polyimide–polyimide adhesion. First, since the modified layer is the same material as the PAA subsequently spun onto the surface, immiscibility is not an issue and the adhesion between the two materials is expected to be strong, provided the modified layer is thick enough. Second, the solvent from the spun-on film will first complex with the modified PAA layer and then plasticize both sides of the interface during curing, thus improving interdiffusion and chain entanglement, resulting in excellent self-adhesion.

Polyimide–polyimide adhesion can also be improved by the use of a multicomponent polymer system. Since the required polymer properties often cannot be met by a single polymer, the incorporation of another material into the system either by blending or through the use of a copolymer allows properties such as dielectric constant and self-adhesion to be tailored. As an example, Rojstaczer et al. examined the properties of a blend from a rodlike PMDA-PDA polyimide ($T_g \approx$ 500 °C) and a flexible 6F-BDAF polyimide ($T_g \approx$ 260 °C).[14] These blends exhibited phase separation, forming domains ~1 mm in size as measured by light scattering and phase-contrast microscopy. In addition, wide-angle X-ray diffraction (WAXD) patterns measured in transmission and reflection mode revealed that the blend films exhibit a strong in-plane orientation, analogous to that observed for pure polyimide films cured on a confining substrate.[8] Although bulk material properties like the modulus and CTE scaled with the overall composition, the surface properties did not. In particular, the surface of blends containing only 10% 6F-BDAF were found to be purely 6F-BDAF when examined with XPS, indicating that the fluorine-containing 6F-BDAF component has a strong tendency to segregate to the blend surface. These films provided better self-adhesion properties than films of pure PMDA-PDA because the presence of a low T_g polyimide component on the blend surface enhances interdiffusion and chain entanglement.

9.4 Metal–Polymer Interfaces

Adhesion between polymer dielectrics and the wiring metallurgy is essential in packaging applications. The adhesion of these materials will depend on a variety of factors, including the chemical interactions between them, the way in which they are joined or their surfaces pretreated, and their stability with respect to oxygen and water during different fabrication steps. In this section, the changes in chemical bonding produced in a polyimide substrate by deposition of Cr and Cu are examined. The availability of various methods of surface modification to improve metal–polymer adhesion are then discussed, with particular attention paid to the influence of sputtering the polymer surface prior to metal deposition. Finally, the effects of water and oxygen on metal–polymer interface stability are illustrated.

In some of the first metal–polyimide interface studies, differences in metal reactivities were correlated with differences in metal–polyimide adhesion.[15–17] Consequently, both clean surfaces and metal-covered surfaces have been studied with XPS. This technique has been used because it is relatively nondestructive, surface sensitive, and yet capable of revealing changes in surface chemical bonding configurations. Interpretation of the clean polymer surface requires some care since the polymeric chain consists of strong electron withdrawing groups on an aromatic backbone. The presence of an imide group adjoining a benzene ring results in considerable delocalization of the polymer electronic states, which is revealed under XPS. As shown for the PMDA-ODA and BPDA-PDA polyimides in Figure 9.6, electronic structure differences are reflected in distinct C 1s core level line shapes.

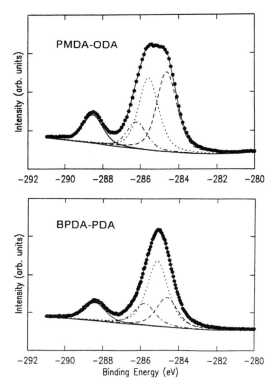

Figure 9.6 XPS spectra of clean PMDA-ODA and BPDA-PDA polyimide films exhibit distinct line shapes, reflecting differences in their electronic structure.

Note that these line shapes are quite different, even though both polymers contain similar chemical groups. At the same time, the peak decompositions shown indicate that contributions from different bonding configurations can be identified.

Each of the C 1s core level line shapes shown in Figure 9.6 have been decomposed into four peaks reflecting the different chemical environments of the carbon atoms. The component ~288.5 eV below the Fermi level corresponds to carbonyl carbon atoms; the peak at −284.5 eV represents carbon atoms in the ODA or PDA segments not bound to nitrogen or oxygen. The peak near −286 eV is assigned to carbon atoms singly bound to nitrogen or oxygen, with the slight peak position differences for PMDA-ODA and BPDA-PDA arising from the lack of an ether oxygen contribution for the latter polymer. Carbon atoms in the BPDA and PMDA portions of their respective polymers appear about 0.5 and 1.0 eV to the left of the emission from C–C bonds in the diamine sections, corresponding to the delocalized influence of 1 and 2 imide groups bound to the central benzene ring.[18]

XPS is particularly powerful when used to examine chemical changes induced by different metals deposited onto similarly prepared polymer surfaces. An example

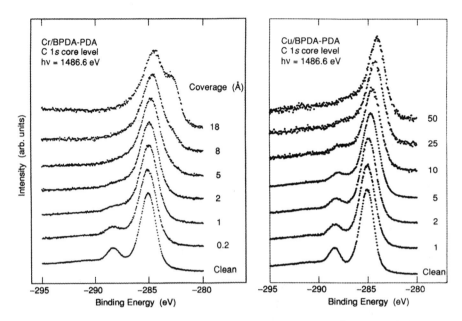

Figure 9.7 The changes in the XPS C 1*s* spectra for BPDA-PDA as a function of increasing (*left*) Cr and (*right*) Cu deposition demonstrate that the interfacial reactivity depends on the overlayer metal. The greater reactivity exhibited by Cr correlates with increased adhesion to polyimides.

of this is shown in Figure 9.7 for Cr and Cu deposited onto a BPDA-PDA polyimide, systems which have been studied extensively for PMDA-ODA.[19–23] A comparison of the line shape changes for these two metals reveals that, for the lowest coverages, Cr induces greater modification of the BPDA portion of the polyimide than does Cu. At the same time, both metals induce a reduction in the carbonyl group intensity and increased emission from the valley between the carbonyl and main C 1*s* lines. This suggests that, during the initial stage of interface formation, metal–polymer bonding may be quite similar. However, differences in the relative rate of change induced in the C 1*s* lines by the metal indicate that the metal morphologies are different. Above 5-Å metal coverage, Cr induces the formation of a sharp feature near –283 eV, which is not observed for Cu for any overlayer thickness. Such line shape differences indicate that, as was observed early on for PMDA-ODA, the metal–polymer interfacial reaction products depend on the overlayer material.

Evidence that similar reactions occur during the initial stages of interface formation for Cu- and Cr-BPDA-PDA interfaces can be drawn from an examination of the N 1*s* core level spectra shown in Figure 9.8. In Figure 9.8*a*, the results for the Cr-BPDA-PDA interface show that a low binding energy shoulder ~1.7 eV from the unreacted N line forms for Cr coverages of 2 Å and below. With subsequent metal deposition, a second component becomes apparent ~3 eV from the

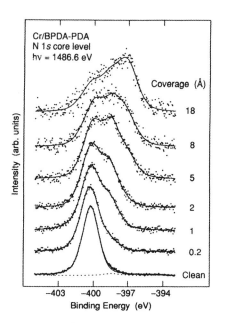
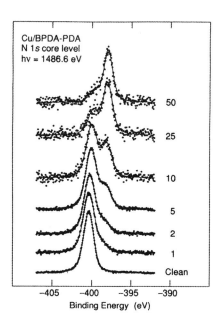

Figure 9.8 Both (*a*) Cr and (*b*) Cu induce distinct chemical modification of BPDA-PDA, as evidenced by the readily resolved changes observed in the N 1*s* core level spectra.

unperturbed substrate peak and becomes dominant by 18 Å. Figure 9.8*b* demonstrates that Cu deposition, like that of Cr, induces the formation of a reaction-induced peak ~1.7 eV below the unperturbed N line. This indicates that the initial metal–polymer interface reaction or reactions produce spectroscopically indistinguishable reaction products. However, unlike Cr deposition, subsequent Cu deposition does not produce another resolvable reaction-induced peak. The continued presence of a small bulk component at coverages as high as 50 Å indicates incomplete overlayer coverage. Indeed, transmission electron microscopy (TEM) has revealed that Cr deposits on PMDA-ODA uniformly, whereas Cu exhibits island formation.[24] We conclude that the interfacial chemistry and morphology are different for these two metal overlayers and that the more reactive metal (Cr) produces the largest metal–polymer adhesion strength.

Metal–polymer adhesion can be modified by combinations of surface pretreatment, evaporation onto substrates held at elevated temperatures, or deposition of the metal in an excited state to promote reaction and intermixing. (Many methods of characterization and modification are discussed in Reference 25.) Sputtering the polymer surface is one of most common surface pretreatment methods. The effect of Ar⁺ sputtering on the adhesion between a series of metals evaporated onto high-density polyethylene (HDPE) is shown in Figure 9.9.[26] The plot of the metal condensation energy versus pull strength shown in Figure 9.9*a* reveals that metal–polymer adhesion is poor for metals whose condensation energies lie below the

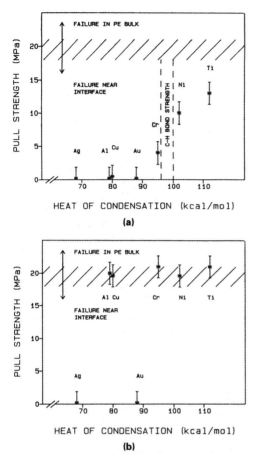

Figure 9.9 A comparison of the adhesion of several metals to HDPE (*a*) before and (*b*) after Ar* ion sputtering. (From Reference 26).

energy necessary to break a C–H bond. Sputtering the surface prior to metal deposition modifies the polymer surface, producing a graphitic or cross-linked structure. This modified surface may be stronger, react more extensively with subsequently deposited metals, or both. One essential point is that the adhesion is improved for most metals once the C-H are modified by the sputtering process. This is illustrated by the pull test results of Figure 9.9*b* for metal–HDPE adhesion, where sputtering has removed the barrier for metal–carbon reactions, and better adhesion results than for the unmodified surfaces. Once these metal–polymer interfaces are strong, failure occurs in the bulk of the polymer, as evidenced by the presence of polymer on both sides of the separated surfaces.

Correlations between the extent of polymer modification and adhesion are often difficult to establish due to differences in the polymer's susceptibility to ion damage

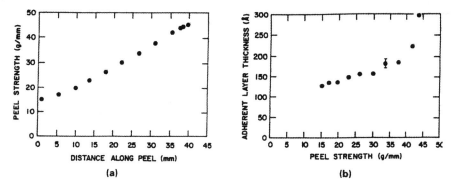

Figure 9.10 (a) The Cu/Cr-BPDA-PDA peel strength increases while moving towards the center of the wafer, where the ion dose is highest. (b) A higher ion dose produces a thicker adherent layer and higher peel strength. (From Reference 27).

and irreproducibilities associated with ion dose and distribution across a sample surface. However, in a recent study of Cu/Cr-BPDA-PDA adhesion, Pappas et al. utilized the known Gaussian distribution of an ion source across a 7.6-cm (3-in.) wafer to examine such a correlation.[27] As shown in Figure 9.10a, the metal–polymer peel strength increases while moving towards the center of the wafer where the ion dose is highest. Using ellipsometry to determine the polymer thickness as a function of distance across the removed peel stripe, one could directly correlate the change in adherent layer thickness with an increase in peel strength (Figure 9.10b).

The presence of an Ar^+ sputtered surface layer has also been shown to have a significant impact on the stability of metal–polyimide interfaces in the presence of isotopically tagged water. This was done by correlating peel test measurements with scanning Auger microscopy (SAM) and secondary ion mass spectroscopy (SIMS) measurements to determine the locus of failure for Cu/Cr–polyimide junctions and the distribution of water throughout the interface.[28] The SIMS depth profiles shown in Figure 9.11 underscore the effects of sputter surface treatments on adhesion, where samples a, b, and c were sputtered for 30 min at 200 W and sample d was sputtered for 10 min at 50 W to reduce the modified layer thickness. Measurements made on sample a represent the naturally occurring abundance of ^{18}O across the as-deposited interface and serve as a reference for subsequent processing. After exposure of the metal–polyimide interface to water[^{18}O] vapor for 48 h, sample b shows a large increase in the ^{18}O concentration near the Cr–polyimide interface. Peel tests made on this structure exhibit no loss of adhesion strength. In contrast, annealing sample c at 350 °C in forming gas for 30 min after water-dosing reduced the peel strength by ~30%. Failure occurs in the polyimide, indicating that water can degrade polyimide. This conclusion was supported by measurements on identically prepared samples that had not been exposed to water and consequently did not exhibit any adhesion degradation after 350 °C annealing.

Examination of the SIMS profiles for samples *c* and *d* in Figure 9.11 demonstrate that the thickness of the sputtered layer affects the transport of water to the interface and in turn affects the adhesion strength. These samples differ only in the amount of Ar^+ pretreatment and hence the altered layer thickness. As shown in the depth profile for sample *c*, the water concentration on the polymer side is greater than that on the Cr side of the interface. However, this concentration profile is reversed when the Ar^+ sputtering pretreatment is reduced. At the same time, failure switches from occurring within the polymer (cohesive mode) to failure occurring along the interface (mixed mode). Clearly, transport of water through the sputter-modified zone alters both the adhesion strength and failure mode. In addition, recent thermal desorption mass spectroscopy (TDS) and gravimetric water uptake measurements showed that the activation energy for water diffusion through a sputtered polyimide surface is about twice the value for the bulk polymer. This suggests that sputter pretreatments may reduce the effects of adsorbed water on interfacial corrosion and the formation of weakly adhering oxides.[29]

The presence of other gases or solvents has also been shown to have significant effects on metal–polymer adhesion. For example, annealing a seed layer of Cu deposited onto PMDA-ODA in air results in a rough, catalytically degraded polymer surface containing copper oxide particles.[30] Although Cu–polyimide interfaces are generally weak, removal of oxides and degradation products followed by redeposition of Cu onto the roughened surface actually exhibits enhanced adhesion. This was attributed to mechanical interlocking and/or greater surface area chemical bonding.

9.5 Polymer–Metal Interfaces

The properties of the polymer–metal interfaces used in packaging and interconnect applications are significantly different from those found for metal–polymer interfaces. In polyimide–metal applications, a polymeric precursor solution is typically

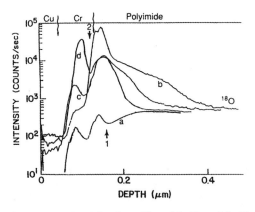

Figure 9.11   SIMS depth profiles of Cu/Cr–polyimide interfaces.[28]

spin-coated onto a metallized surface and then processed at elevated temperatures to produce the desired structure. Factors that affect polyimide–metal junctions include the chemical interactions between the precursor solution and the metal substrate, the influence of the carrier solvent on the redistribution of reaction products throughout the polymer, and the stability of the polymer–metal interfaces under the influence of typical processing conditions. This section first examines the observable differences between metal–polyimide and polyimide–metal interfaces. This is followed by a discussion of the synergistic effects caused by the reaction of a Cu surface with a PAA in the presence of NMP or other carrier solvents. Next, the influence of different metal surfaces and polyimide precursors is examined. Finally, the susceptibility of different polymer–metal interfaces to oxidation at elevated temperatures is discussed.

An examination of the cross-sectional transmission electron microscopy (XTEM) micrographs shown in Figures 9.12a and 9.12b reveals that the Cu–polyimide and

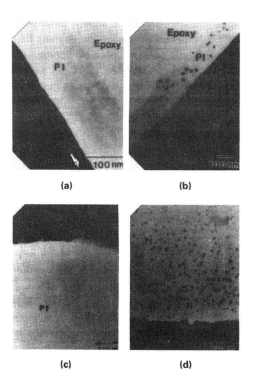

(a) (b)

(c) (d)

Figure 9.12   XTEM micrographs of (a) Cu–polyimide and (b) polyimide–Cu interfaces demonstrate that the process of forming these asymmetric
junctions strongly influences their microstructure. Vapor-deposited
polyimide–Cu interfaces (c) without and (d) with the addition of NMP
prior to curing illustrate the importance of the solvent in the redistribution of reaction products throughout the interfacial region.[33]

polyimide–Cu interfaces have very different morphologies. Although the above discussion of Cu–polyimide interface chemistry showed that reactions do occur, the interface is relatively abrupt when compared to the polyimide–Cu interface. Indeed, the process of curing the PAA precursor on a Cu substrate results in the formation of small particles distributed a micron or more into the polyimide over-layer.[31] Energy dispersive X-ray (EDX) analysis performed in a scanning transmission electron microscope (STEM) showed that the particles were Cu-rich, and a microdiffraction analysis found that they were primarily Cu_2O.[32] The presence of Cu_2O particles significantly increases the dielectric constant measured for the polymer film.[31]

Since the low dielectric constants of polyimides are the primary driving force for their use in microelectronics, many additional tests have been made to further the understanding of the role of the polyimide precursor, solvent, and metallic substrate so that the formation of oxide particles in the polyimide matrix can be minimized. In general, each of these three factors is important. As a first example, after curing a polyamic ethyl ester precursor on Cu, no particles are found in the cured polyimide film. The distinct behavior exhibited by different precursors of the same final polyimide points to the importance of the acid–base reactions possible between the PAA and the basic metallic substrate, which are different from the more neutral polyamic ethyl ester. To examine the influence of the NMP solvent on the eventual formation of Cu_2O particles, PAA was vapor-deposited onto two Cu substrates in the absence of NMP. After the addition of a few drops of NMP to one sample, both samples were cured and examined by XTEM, as shown in Figures 9.12c and 9.12d. These micrographs demonstrate that a carrier solvent is necessary for the transport of material (a Cu-PAA complex) that eventually decomposes to form Cu_2O particles.[33] Finally, the nature of the substrate metal is also important. In practice, the metal substrates have a thin native oxide which may react with the PAA, depending on the resistance of the oxides to dissolution by organic acids.[34] Since Cr oxides exhibit resistance to attack by organic acids (Cu oxides do not), no oxide precipitates are found when PAA is cured on Cr.[32] In conclusion, the nature of the polymer, solvent, and metal substrate all contribute synergistically to the formation of polyimide–metal interfaces.

The stability of polyimide–metal interfaces has also been found to be subject to synergistic interactions between the polymer, substrate metals and their oxides, and a source of oxygen (water, O_2, or air).[35, 36] In particular, interfaces between polyimide and Cu appear to be the most susceptible to degradation. In one study using XPS sputter profiling, Burrell et al. found that polyimide films deposited on Cu (Al) degraded (did not degrade) during aging in air at 200 °C.[35] The enhanced degradation of polyimides by Cu was correlated with the formation of copper carboxylates and oxides which catalyze the degradation process. Since the reactions tend to be extensive, transition metal-catalyzed degradation effects have been examined for many polymers using more bulk-sensitive infrared (IR) studies by several groups.[37, 38]

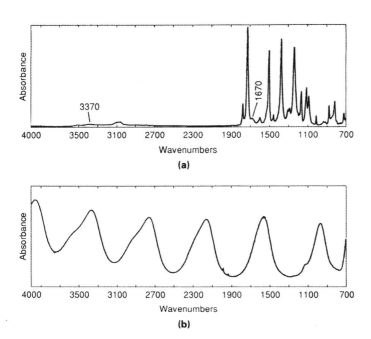

Figure 9.13 FTIR spectra of 1.4-mm-thick PAA films cured on Cu surfaces in (a) forming gas and (b) in N_2 containing 1% O_2. The forming gas cure resulted in a well-imidized film; the presence of oxygen in the curing environment resulted in a completely degraded film. (From Reference 36).

The importance of controlling the environment during annealing can be seen by comparing the two IR spectra in Figure 9.13. After a PAA is cured on Cu to 350 °C in the presence of forming gas, the topmost spectrum exhibits structure characteristic of cured polyimide.[36] Features near 1670 and 3300 cm^{-1} indicate the presence of a small amount of uncured amic acid either in the bulk of the film or at the polyimide–Cu interface. In contrast, the spectrum of a PAA film cured in N_2–1% O_2 exhibits only interference fringes and indicates that complete degradation of the film has occurred. The extent to which degradation takes place has been shown to depend both on the amount of O_2 present in the curing ambient and on the film thickness, where the thickness affects the transport of O_2 to the polyimide–Cu interfaces where most of the degradation reactions occur.

The thermal stability of polymer–metal interfaces in air also depends on the polymer, the substrate metal, and the diffusion layer metal. Reductions in polymer thickness due to catalytic reaction with metal as measured by profilometry for BPDA-PDA and PMDA-ODA polyimides on several different substrate metallurgies are shown in Figure 9.14 after annealing in air at 350 °C. These results show that PMDA-ODA is more susceptible to catalytic degradation by Cu than BPDA-PDA.[39] Except for Co, which exhibits somewhat less enhanced degradation than

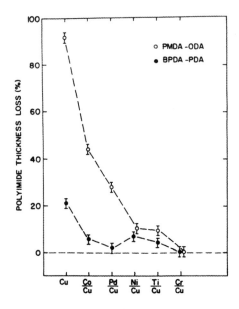

Figure 9.14 Polyimide film thickness reductions for PMDA-ODA and BPDA-PDA films cured on different combinations of substrate metals reveal that BPDA-PDA is subject to less degradation than PMDA-ODA.

Cu, metals such as Pd, Ni, Ti and Cr did not aid film decomposition. However, when thin films of these metals were used as adhesion layers/diffusion barriers on top of Cu, polyimide decomposition proceeds to different extents for each of the substrate metallurgies except for Cr/Cu. The tendency for degradation to occur for Pd, Ni, and Ti overlayers was related to the diffusion of Cu through these thin layers to the polyimide–metal interface, where catalytic degradation then occurs.

9.6 Thermal Stress and Interfacial Fracture

The interfacial chemistry and molecular structure discussed so far play an important role in determining the thermal/mechanical properties of polymer interfaces and multilayered structures. Consider a structure of a metal–polyimide bilayer on a ceramic substrate. Under thermal processing, a stress will be generated and its magnitude determined by the mismatch in thermal expansion and elastic modulus of the materials. For a thick substrate, as encountered in many applications, the deformation of the metal and the polymer is constrained by the substrate. This constraint gives rise to a nonuniform distribution of strain normal and parallel to the plane of the film, with the maximum nonuniformity existing near the interface. The extent of this strain converting to a stress depends on the deformation mechanism and the interfacial integrity. For an elastic film with an ideally strong interface, there is no plastic yield, so all the strain will convert linearly into stress. Generally, this is not the case for the polymer interlayer, which will yield to reduce the stress. The stress buffering effect depends on the thickness of the polymer layer and its viscoelastic properties. This effect is particularly effective in reducing the local stress concentration and lessening the concern for interfacial delamination of a metal–polymer multilayered structure.

A number of techniques have been developed for stress measurements in polymer multilayered structures, and the results have been reviewed recently.[40, 41] A comparative study has been carried out for PMDA-ODA and BPDA-PDA utilizing a bending beam technique.[42] With Cu as the metal layer to minimize the effect of interfacial chemistry, this study focused on the effect of the polymer properties. Results of a Cu(0.53 μm)–PMDA-ODA(2.9 μm)–quartz structure are shown in Figures 9.15a and 9.15b. The stress and strain for the Cu and the PMDA-ODA layers are deduced from the observed beam bending during thermal cycling, where plastic yield of the materials already occurred. In the trilayered structure, the overall deformation is constrained by the thermal expansion of the quartz substrate. Under this constraint condition, the thermal strain suffered by PMDA-ODA is higher than that of Cu by about an order of magnitude, whereas the reverse holds for the thermal stresses. It is worth noting that the stress levels in the trilayer cannot be obtained simply by a superposition of the elastic stresses from the equivalent bilayer structures. Instead, the observed stress in Cu is considerably less, indicating that the polyimide has sustained plastic deformation in order to reduce the overall stress of the structure. The stress reduction occurs, however, at the expense of a relatively large deformation of the polyimide. This stress buffering effect was investigated as a function of PMDA-ODA thickness and was found to saturate with about 3 μm of the polymer for 0.5 μm of Cu.[43] The saturation thickness provides an indirect measure of the range of the nonuniform deformation in the polyimide near the interface, which has been confirmed by an independent observation using transmission electron microscopy (TEM).

The results obtained for a Cu-BPDA-PDA-quartz structure under similar thermal cycling are shown in Figures 9.16a–c. The value of the thermal expansion coefficient of BPDA-PDA is between that of Cu and quartz, in contrast to that of PMDA-ODA, which is about five times higher (see Table 9.1). In addition, the Young's modulus of BPDA-PDA is about four times higher, so its properties are better matched to Cu and quartz than PMDA-ODA. This factor enables BPDA-PDA to reduce the thermal stresses in itself and Cu by about one-third of those in the PMDA-ODA structure, and there are comparable reductions in the thermal strains. In comparison to PMDA-ODA, BPDA-PDA is more effective for buffering the stress in a metal–polymer structure and the saturation thickness was found to be less, about 1 μm.[42]

Results from this comparative study demonstrate the effect on thermal stress behavior due to changes in the polyimide properties originated from the molecular structure. At this time, the mechanism and distribution of the deformation in the multilayered structure are not well understood. Recently, the inelastic deformation

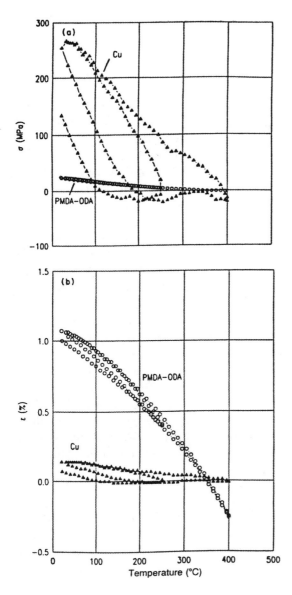

Figure 9.15 Thermal stresses and thermal strains of (*a*) Cu and (*b*) PMDA-ODA layers measured during thermal cycling in a Cu(0.53 μm)–PMDA-ODA(2.9 μm)–quartz structure.[42]

behavior of a Cu-PMDA-ODA bilayer has been studied in tension up to fracture.[44] The relaxation of the polymer was found to be controlled by different mechanisms, depending on the amount of deformation. In the lower range, with strain comparable to that occurring during thermal cycling, the deformation can be characterized

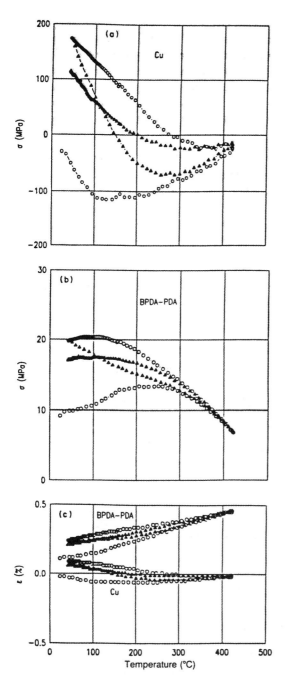

Figure 9.16 (*a*) Thermal stresses of Cu and BPDA-PDA layers measured during thermal cycling in a Cu(0 5 μm)-BPDA-PDA(1.57 μm)-quartz structure and (*b*) the corresponding thermal strains.[42]

by an activated volume corresponding to a shear/slip process in the polyimide. A shear/slip deformation mode has been observed in the Cu–PMDA-ODA–quartz structure by TEM near the metal–polyimide interface after thermal cycling,[43] but the result was not quantitative enough to account for the overall inelastic strain. Similar studies for other metal–polyimide structures would be useful for understanding the deformation mechanism, although such studies have not been reported.

The stress generated due to thermal mismatch, if sufficiently large, can cause interfacial fracture. For layered structures formed with blanket thin films, the magnitude of the strain, as observed for the Cu–polyimide structures, is about 1%, probably insufficient to cause delamination. In packaging structures, however, the local stress near the fine wiring geometries can be substantially higher. In a case study, Lacombe[45] computed the stress distribution using a finite element analysis (FEA) for an isolated Cu via formed in a PMDA-ODA layer on a silicon substrate. For a 2 μm via, the normal and shear stresses can increase 250% at the via corners during thermal cycling between 25 and 400 °C. This raises the stress level to about 4% of the Young's modulus for the polyimide, a magnitude probably sufficient to cause local fracture of the polyimide, particularly when processing defects are present to initiate the fracture process. In this analysis, the deformation of the polyimide was assumed to be linear and elastic; the viscoelasticity of the polyimide which should reduce the magnitude of the thermal stress was not included. Nevertheless, this study shows that interfacial fracture is of concern, particularly for a structure with small geometries where local stress concentration can be significant.

The toughness of a metal–polymer interface has often been evaluated by its adhesion strength. The adhesion testing methods for various materials combinations have been reviewed recently.[46, 47] For compliant and ductile polymeric films, several conventional methods are not well-suited for measuring interfacial fracture toughness due to the large deformation sustained by the polymers in the fracture process. For example, the peel test would require a correction of more than 50% to the peeling force in order to account for the polymer deformation.[48, 49] For this purpose, a stretch deformation method has been developed to measure the fracture energy of the metal–polymer interface.[50, 51] Unlike peel tests, which measure the peeling force to separate the interface, this method measures directly the energy required to fracture the interface. This is carried out by measuring the energy required to stretch a metal–polymer layered structure until complete fracture of the metal film, and subtracting off the energy required to stretch the individual polymer and metal layers. Since this method applies a mode II fracture, it is possible to measure the shear strength of the interface in addition to its fracture energy. The fracture energy measured by this method yields a value of 33 J/m^2 for a Cu/Cr–PMDA-ODA layered structure, as compared with 9.6 J/m^2 for the Cu–PMDA-ODA structure.[50, 51] The difference reflects the stronger chemical bonding between Cr and PMDA-ODA, as discussed previously.

This method has been applied to investigate the fracture behavior of PMDA-ODA and BPDA-PDA interfaces with Au/Cr metal layers as a function of metal line geometry.[52, 53] The metal lines were formed by standard lithographic techniques with widths ranging from 2 to 16 μm. The thickness of the Au layer varied from 0.25 to 1.3 μm and the Cr layer from 50 to 500 Å. The fracture behavior for Au/Cr-BPDA-PDA layered structures is shown in Figure 9.17 as a function of the line width, together with the deformation curve for the BPDA-PDA substrate. A systematic trend of geometrical dependence was observed; with a width increase, the strain at failure decreases while the progression of fracture becomes more abrupt. Compared with the 2-μm lines, the 16-μm line structure fails rather abruptly at about 30% strain with somewhat more stress overshoot due to sudden release of the fracture energy. This reflects a larger deformation zone near the interface due to increasing interfacial confinement. A similar trend was observed when the metal layer thickness decreases. The fracture morphology and energy dissipation rate also show an apparent dependence on the metal dimension. With increasing metal thickness, the crack morphology changes from circular to flat and the energy dissipation rate is higher (Figure 9.18). The values of the fracture energy determined from the energy dissipation rate are in reasonable agreement with those obtained

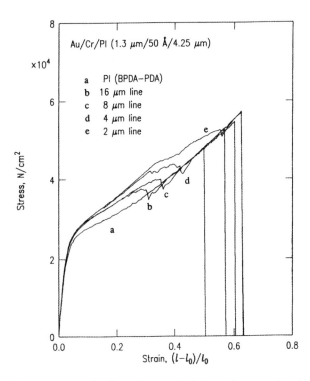

Figure 9.17 Results from the stretch deformation test showing the fracture behavior of Au/Cr-BPDA-PDA line structures as a function of the line width. A deformation curve of the BPDA-PDA substrate is provided for comparison. (From Reference 52.)

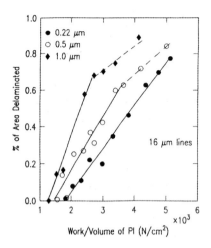

Figure 9.18 Results of delaminated areas as a function of work done per unit polyimide volume for different metal line thicknesses. The slopes of these curves give the energy dissipation rates for crack propagation. (From Reference 50.)

from an energy balance approach. Within the error limits of the experiment (about 50%), the fracture energy seems to be independent of the metal dimension, suggesting that the effect of the metal geometry is primarily on the crack morphology and propagation rate for interfacial fractures. The observed results can be explained by considering the change in stress gradient near the interface due to metal confinement of polymer. This was confirmed by FEA, where plastic deformation of the polyimide was found to be important and must be taken into account.

The results of Au/Cr-PMDA-ODA layered structures show a fracture behavior and metal dimension dependence qualitatively similar to that observed for BPDA-PDA. The fracture energy, however, is about three to four times higher for the BPDA-PDA. This result cannot be attributed to the different chemical bonding between metal and the two polyimides. The bonding contribution to adhesion has been estimated for a Cr-PMDA-ODA interface by calculating its cohesive energy. Assuming a sharp interface with an energy of 3–4 eV per Cr bond to the six-member or the five-member ring in the PMDA or ODA fragment, the interfacial bonding energy is about 3 J/m^2.[54] This corresponds to about 10% of the measured fracture energy. The estimated value can be increased using different interfacial configurations or including some Cr penetration below the top Cr layer; nevertheless, it still can account for only a small percentage of the fracture energy. This shows that most of the fracture energy is accumulated to deform the polyimide near the interface until a threshold is reached; then it is dissipated for fracture propagation. The different fracture energy for the BPDA-PDA and PMDA-ODA interfaces suggests an important role for the molecular structure in controlling fracture characteristics and adhesion strength of the metal–polymer interface.

Acknowledgment

The authors would like to thank many co-workers and colleagues for their contributions and valuable discussions in our study of the metal–polymer interfaces. In particular, we acknowledge M. Moske, C. Feger, F. Faupel, D. Shinozaki, S. T. Chen, J. Leu, B. D. Silverman, R. Saraf, and T. W. Poon. The support of and discussions with M. Haley, K. Srikrishnan, B. Agarwala, K. Sachdev, and H. M. Tong of the IBM General Technology Division are also gratefully appreciated. The material is based in part upon work supported by the Texas Advanced Research Program under Grant No. 470.

References

1 *Electronic Packaging Handbook.* (R. R Tummala and E. J. Rymaszewski, Eds.) Van Nostrand Reinhold, New York, 1989.

2 *Polyimides: Materials, Chemistry, and Characterization.* (C. Feger, M. M. Khojasteh, and J. E. McGrath, Eds.) Elsevier, Amsterdam, 1989.

3 M. I. Bessonov, M. M. Kroton, C. C. Kurdryavtsev, and L A. Laius. *Polyimides: Thermally Stable Polymers.* Consultants Bureau, New York, 1987.

4 S. Numata, K. Fujisaki, and N. Kinjo. *Polymer.* **28**, 2282, 1987.

5 T. W. Poon, B. D. Silverman, R. F. Saraf, A. R. Rossi, and P. S. Ho. *Phys. Rev. B.* **46**, 11456, 1992.

6 M.-J. Brekner and C. Feger. *J. Poly. Sci. Poly. Chem.* **25**, 2005, 1987.

7 M.-J. Brekner and C. Feger. *J. Poly. Sci. Poly. Chem.* **25**, 2497, 1987.

8 N. Takahashi, D. Y. Yoon, and W. Parrish. *Macromolecules.* **17**, 2583, 1984.

9 C. Feger. *Poly. Eng. Sci.* **29**, 347, 1989.

10 H. R. Brown, A. C. M. Yang, T. P. Russell, W. Volksen, and E. J. Kramer. *Polymer.* **29**, 1807, 1988.

11 S. F. Tead, E. J. Kramer, T. P. Russell, and W. Volksen. *Polymer.* **33**, 3382, 1992.

12 K. L. Saenger, H. M. Tong, and R. D. Haynes. *J. Poly. Sci. Poly. Lett.* **27**, 235, 1989.

13 K.-W. Lee, S. P. Kowalczyk, and J. M. Shaw. *Macromolecules.* **23**, 2097, 1990.

14 S. Rojstaczer, M. Ree, D. Y. Yoon, and W. Volksen. *J. Poly. Sci. Poly. Phys.* **30**, 133, 1992.

15 N. J. Chou and C H. Tang. *J. Vac. Sci. Technol.* **A2**, 751, 1984.

16 P. S. Ho, P. O. Hahn, J. W. Bartha, G. W. Rubloff, F. K. LeGoues, and B. D. Silverman. *J. Vac. Sci. Technol.* **A3**, 739, 1985.

17   P. O. Hahn, G. W. Rubloff, J. W. Bartha, F. LeGoues, and P. S. Ho. *Mat. Res. Soc. Symp. Proc.* **40**, 251, 1985.

18   B. D. Silverman, P. N. Sanda, P. S. Ho, and A. R. Rossi. *J. Poly. Sci. Poly. Chem.* **23**, 2857, 1985.

19   R. Haight, R. C. White, B. D. Silverman, and P. S. Ho. *J. Vac. Sci. Technol.* **A6**, 2188, 1988.

20   J. L. Jordan, C. A. Kovac, J. F. Morar, and R. A. Pollack. *Phys. Rev. B.* **36**, 1369, 1987.

21   D. S. Dunn and J. L. Grant. *J. Vac. Sci. Technol.* **A7**, 253, 1989.

22   J. G. Clabes, M. J. Goldberg, A. Viehbeck, and C. A. Kovac. *J. Vac. Sci. Technol.* **A6**, 985, 1988.

23   M. J. Goldberg, A. Viehbeck, and C. A. Kovac. *J. Vac. Sci. Technol.* **A6**, 991, 1988.

24   F. K. LeGoues, B. D. Silverman, and P. S. Ho. *J. Vac. Sci. Technol.* **A6**, 2200, 1988.

25   *Metallization of Polymers.* (E. Sacher, J.-J. Pireaux, and S. P. Kowalczyk, Eds.) American Chemical Society, Washington, DC, 1990.

26   P. Bodö and J.-E. Sundgren. *Surf. Interface Anal.* **9**, 437, 1986.

27   D. L. Pappas, J. J. Cuomo, and K. G. Sachdev. *J. Vac. Sci. Technol.* **A9**, 2704, 1991.

28   B. K. Furman, S. Purushothaman, E. Castellani, S. Renick, and D. Neugroshl. In *Metallization of Polymers.* (E. Sacher, J.-J. Pireaux, and S. P. Kowalczyk, Eds.) American Chemical Society, Washington, DC, 1990.

29   H. M. Clearfield, B. K. Furman, N. Sheth, F. Bailey, and S. Purushothaman. Presentation at the Spring MRS Symposium, San Francisco, 1992.

30   N. L. D. Samasiri, R. L. D. Zenner, and J. C. Houge. *IEEE Trans. Compon. Hybrid Manufact. Technol.* **14**, 798, 1991.

31   J. Kim, S. P. Kowalczyk, Y.-H. Kim. N. J. Chou, and T. S. Oh. *Mat. Res. Soc. Symp. Proc.* **167**, 137, 1990.

32   Y. H. Kim, J. Kim, G. F. Walker, C. Feger, and S. P. Kowalczyk. *J. Adhesion Sci. Technol.* **2**, 95, 1988.

33   S. P. Kowalczyk, Y.-H Kim, G. F. Walker, and J. Kim. *Appl. Phys. Lett.* **52**, 375, 1988.

34   F. Iacona, M. Garilli, G. Marietta, O. Puglisi, and S. Pignataro. *J. Mater. Res.* **6**, 861, 1991.

35   M. C. Burrell, J. Fontana, and J. J. Chera. *J. Vac. Sci. Technol.* **A6**, 2893, 1988.

36 D.-Y. Shih, J. Paraszczak, N. Klymko, R. Flitsch, S. Nunes, J. Lewis, C. Yang, J. Cataldo, R. McGouey, W. Graham, R. Serine and E. Galligan. *J. Vac. Sci. Technol.* **A7**, 1402, 1989.

37 D. L. Allara. In *Characterization of Metal and Polymer Surfaces.* (H. Lee, Ed.) Academic Press, New York, 1977, p. 193.

38 J. D. Webb, A. W. Czanderna, and J. R. Pitts. *J. Vac. Sci. Technol.* **A6**, 997, 1988.

39 D.-Y. Shih, N. Klymko, R. Flitsch, J. Paraszczak, and S. Nunes. *J. Vac. Sci. Technol.* **A9**, 2963, 1991.

40 H. M. Tong and K. L. Saenger. In *New Characterization Techniques for Thin Polymer Films.* (H. M. Tong and L. T. Nguyen, Eds.) Wiley-Interscience, New York, 1990, Chapt. 2.

41 C. L. Bauer and R. J. Farris. In *Polyimides: Materials, Chemistry and Characterization.* (C. Feger, M. M. Khojasteh, and J. E. McGrath, Eds.) Elsevier, Amsterdam, 1989, p. 549.

42 M. Moske, J. E. Lewis, and P. S. Ho. In *Proceedings.* Soc. Plastic Eng., Montreal, 1991.

43 S. T. Chen, C. H. Yang, F. Faupel, and P. S. Ho. *J. Appl. Phys.* **64**, 6690, 1988.

44 D. M. Shinozaki and A. Klauzner. *Mat. Res. Soc. Symp. Proc.* **203**, 15, 1991.

45 R. H. Lacombe. In *Surface and Colloid Science in Computer Technology.* (K. L. Mittal, Ed.) Plenum Press, New York, 1987, p. 178.

46 P. Buchwalter. *J. Adhes. Sci. Technol.* **4**, 697, 1990.

47 K. S. Kim. *Mat. Res. Soc. Symp.* **119**, 31, 1988.

48 K. S. Kim. *Mat. Res. Soc. Symp.* 203, 1991.

49 K. S. Kim and J. Kim. *Trans. ASME.* **110**, 266, 1988.

50 F. Faupel, C. H. Yang, S. T. Chen, and P. S. Ho. *J. Appl. Phys.* **65**, 1911, 1989.

51 P. S. Ho and F. Faupel. *Appl. Phys. Lett.* **53**, 1602, 1988.

52 S. L. Chiu and P. S. Ho. *Mat. Res. Soc. Symp. Proc.* **203**, 1991.

53 J. Leu, S. L. Chiu, and P. S. Ho. *Mat. Res. Soc. Symp.* Anaheim, CA, April 28 to May 2, 1991.

54 P. S. Ho. *Appl. Surf Science.* **41/42**, 559, 1989.

The Polymer–Polymer Interface

TODD MANSFIELD, RUSSELL J. COMPOSTO, and
RICHARD S. STEIN

Contents

10.1 Introduction

The study of polymer–polymer interfaces is being pursued for both theoretical and practical reasons. As in other areas of polymer science, the theory of polymer interfaces preceded the development of experimental techniques with sufficient sensitivity and resolution to test theoretical predictions. For example, although Helfand and Tagami calculated the interfacial profile between two immiscible polymers in 1971,[1] rigorous experimental support did not appear until the recent application of neutron reflection to polymer interface problems. The characterization of the polymer–polymer interface also has technological interest in areas such as impact modification, welding, crack healing, injection molding, and other polymer adhesion problems. As a particular example, accurate data on the diffusion coefficients help the technologist calculate the optimum process conditions—temperature, time, etc.—for the friction welding of automobile parts.

This chapter presents several interface problems of current interest, followed by examples of appropriate characterization techniques. A dramatic example of the importance of interfaces on the mechanical properties of polymer blends is shown in Figure 10.1. This photomicrograph shows the fracture surface of a high-density

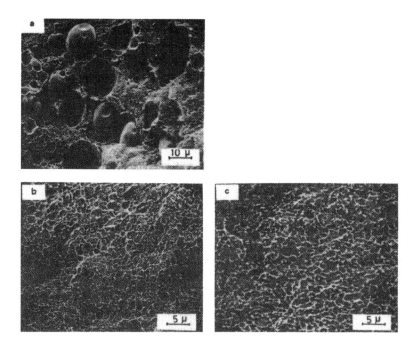

Figure 10.1 SEM micrographs of cryofracture surfaces of (0.8)HDPE:(0.2)PS blends with: (*a*) no copolymer, (*b*) 9 wt % poly(styrene *block*-(1,4)butadiene), and (*c*) 9 wt % S-SB-B triblock copolymer, with the middle block a random copolymer of styrene and butadiene. (From Reference 2.)

polyethylene (HDPE)–polystyrene (PS) blend. Because the interface between the HDPE-rich and PS-rich phases is weak, the sample fails mechanically at low stresses, and fracture is largely confined to the interfacial regions. Thus, a fundamental understanding of the fracture energy of this blend demands a precise characterization of the molecular interpenetration across the HDPE-PS interface. The lower figures show that the addition of a compatibilizing copolymer dramatically changes the texture of the fracture surface and increases the fracture strength. Characterization of the amount and location of the copolymer at the interface is necessary if we are to understand how the copolymer anchors the HDPE and PS phases together.

To limit the scope of this article, we confine polymer interfaces to two geometries: A–B interfaces in which A and B are either miscible or immiscible, and A–C–B interfaces in which C segregates to the interface between the immiscible A–B pair. Figure 10.2 shows the planar sample geometry required by many of the techniques discussed in this chapter. The thickness of A and B depends on the spatial resolution and penetration depth of the technique. The lateral sample dimension depends on the probe and its footprint on the sample. Although consistent with

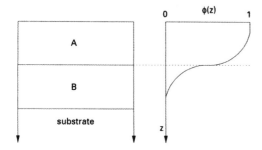

Figure 10.2 Schematic showing sample geometry and concentration profile at the interface between polymers A and B.

many coating and microelectronics applications, the geometries of Figure 10.2 are not of obvious relevance to many three-dimensional problems (e.g. Figure 10.1). Although postmortem analysis of the fracture surface is possible, premortem characterization of the interfacial profile is needed to provide a molecular interpretation of the measured properties such as fracture energy. In most cases, the three-dimensional problem, such as in the HDPE–PS example, can be modeled by the planar geometry.

In choosing the appropriate technique, one should consider foremost the spatial resolution required for the particular interface problem and the contrast between polymers. In general, the higher resolution techniques are more complex (and costly) than lower resolution ones. For example, because neutron reflection (NR) experiments require high neutron fluxes, experiments must be performed at national laboratories such as Brookhaven, NIST, ANL, Oak Ridge, or LANSE. In this chapter, the depth resolution of the selected techniques range from 1 to 1000 nm. Because most polymers contain primarily carbon and hydrogen, contrast between polymers may require labeling of chains by deuteration, staining, or photochromic tagging. Although this is not always possible, labels should be chosen to maximize contrast while minimizing the perturbation of the system. In the best situations, contrast is inherent. For example, in blends of PS with deuterated PS (DPS), the deuteration of (some) chains provides not only contrast but slight chemical differences which can drive such phenomena as partial demixing near surfaces, or phase separation in the bulk—phenomena of considerable interest.

The chapter is divided into problems where the interfacial thickness, as defined in Figure 10.2, is greater than or less than the typical chain size (~10 nm). First, a brief review of polymer–polymer interdiffusion is presented, followed by a discussion of infrared (IR) microdensitometry, electron microprobe, and the ion beam techniques forward recoil spectrometry (FRES), nuclear reaction analysis (NRA), and dynamic secondary ion mass spectrometry (SIMS). Next, sample studies of block copolymer morphology, the interface between immiscible homopolymers, and interfacial segregation are used to introduce neutron reflectivity and fluorescence spectroscopy. The chapter concludes with a technique summary and an explanation of the need for a multitechnique approach.

10.2 Techniques for Measuring Diffusion Distances Much Greater Than Molecular Size

For a pair or semiinfinite solids separated by an interface, the volume fraction profile $\phi(x,t)$ of component A is given[3] by

$$\phi_a(x,t) = \frac{\phi'}{2}[1 + \text{erf}(x/w)] \tag{10.1}$$

where $w = 2(Dt)^2$ is the characteristic diffusion distance, D is the mutual diffusion coefficient, and t is the annealing time. The mutual diffusion coefficient for polymer blends is given[3] by

$$D = (\phi_b D_a^* N_a + \phi_b D_b^* N_b)\Omega \tag{10.2}$$

where D_i^* and N_i are the tracer diffusion coefficient and degree of polymerization of component i, respectively, and Ω represents a thermodynamic driving force for diffusion. For systems with a negative excess free energy of mixing ($\Omega > 1$), as is the case for blends of PS with poly(xylenyl ether),[4] D is increased relative to systems where the driving force is purely entropic. For systems such as PS and DPS, the excess free energy is small but positive and, therefore, diffusion is retarded.[5, 6]

The theory of polymer dynamics is an evolving and expansive topic. Although dated, Graessley's review of the dynamics of entangled systems is excellent.[7] More recently, Doi and Edwards[8] published a detailed account of the dynamical behavior of solids and solutions, and Kausch and Tirrell[9] published a review of diffusion in polymer systems. For long polymer chains, molecules diffuse by the reptation mechanism and thus the D^* can be related to microscopic properties,[10]

$$D^* = D_{\text{rep}} = \left(\frac{4}{15} M_0 M_e \frac{k_b T}{\zeta_0}\right) M^{-2} \tag{10.3}$$

where M, M_0, and M_e are the polymer, monomer, and entanglement molecular weights, respectively, ζ_0 is the monomeric friction coefficient, and k_b and T are the Boltzmann constant and temperature, respectively. It is noteworthy that the parameters that dictate D^*, and also D, are *temperature, molecular weight*, and *chain structure* (through the monomeric friction coefficient). For high-molecular-weight polymers annealed ~70° above the glass transition temperature, diffusion coefficients typically range from 10^{-10} to 10^{-15} cm^{2}s^{-1}. For annealing times of 60 min, the expression for w (the characteristic diffusion distance) predicts that the interfacial thickness will range from 38 µm to 38 nm, respectively. This simple calculation shows that techniques with high spatial resolution (10 nm) are necessary if experiments are to be conducted over a convenient time period. Nevertheless,

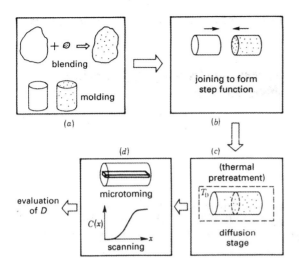

Figure 10.3   Schematic of the sample preparation procedure for an IR microprobe experiment. (From Reference 11.)

many lower resolution yet simple techniques can be used to study polymer–polymer diffusion.

IR microdensitometry is a simple and direct technique for measuring diffusion distances on size scales of millimeters. Figure 10.3 describes the entire experimental procedure.[11] Because of the finite slit width of ~90 μm, diffusion broadened profiles must be at least five times this resolution. Figure 10.4[12] shows the concentration profile for a diffusion couple of 100% polybutadiene (PB) and 95% PB:5% deuterated PB. The concentration profile is measured by scanning the sample using an IR beam tuned to the stretching frequency of the C–D bond. Contrast is provided by the different stretching frequencies of the C–D (2200 cm^{-1}) and C–H (3000 cm^{-1}) bonds. Similar to IR, the modified optical schlieren technique (MOST) relies on a refractive index difference between polymers. In MOST, visible light is deflected as the beam passes through a refractive index gradient. The diffusion coefficient is determined by measuring the angular deflection of light, which in turn is inversely proportional to the concentration gradient. Using MOST, Ye et al. measured the D in a series of PS and poly(vinyl methyl ether) blends.[13, 14] IR microdensitometry and MOST are attractive because they can be set up in-house. Moreover, IR provides the concentration profile directly, without any inverting of spectra. However, because polymers diffuse slowly, long annealing times are required to achieve significant interfacial broadening. Therefore, sample degradation and temperature fluctuations during annealing can be significant problems.

Energy dispersive spectroscopy (EDS) has better spatial resolution than IR and is a direct profiling technique.[15] Also called the electron microprobe because of its 1-μm spot size, this technique makes use of the characteristic X rays from atoms

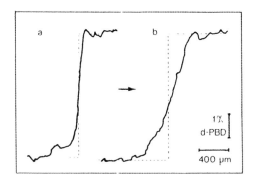

Figure 10.4 The concentration profile measured using IR microdensitometry for a diffusion couple of 100% polybutadiene (PB)/95%PB:5% deuterated PB. (From Reference 15.)

excited by energetic electrons. A scanning electron microscope (SEM) directs a ~1-μm electron beam onto a microtomed or fractured surface and scans the beam across the surface. As shown[16] in Figure 10.5, the electrons deviate strongly from their initial direction due to large angle and multiple elastic scattering. This scattering limits the resolution to ~3 μm. For comparison, the range of 20 keV electrons in polymers is ~10 μm.[15] The characteristic X-ray counts are detected by a Si(Li) solid state detector and converted to concentration by normalizing with respect to the X-ray count far from the interface. Figure 10.6 shows[17] a concentration profile for a poly(vinyl chloride)/poly(caprolactone) couple before and after annealing for 114 h at 91.5 °C. Note that after annealing the interfacial thickness is ~100 μm, about 30 times the resolution. Because the fluorescence signal is weak, the scanning speed of the microprobe is slow, resulting in radiation damage to the sample. Degradation can be reduced by scanning parallel to the interface to increase the signal, allowing for faster scan rates and therefore less radiation damage.

In recent years, the application of ion beam techniques to investigate the interfacial behavior of polymers has become increasingly popular. Because Rutherford

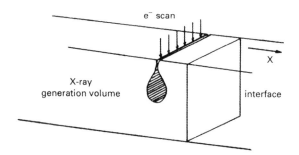

Figure 10.5 Schematic of the "spreading" process of X rays encountering a polymer sample (From Reference 16.)

10.2 TECHNIQUES FOR MEASURING DIFFUSION DISTANCES . . .   249

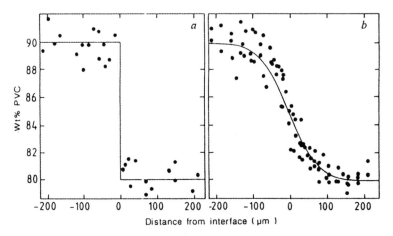

Figure 10.6   A concentration profile for a poly(vinyl chloride)/poly (caprolactone) couple before and after annealing for 114 h at 91.5 °C. (From Reference 17.)

backscattering (RBS) has been developed for semiconductor analysis, several excellent text books are available on this subject.[18] Although RBS has been extremely useful for several polymer interface studies,[19, 20] we discuss the analogous techniques for probing the depth distribution of light elements, namely FRES and nuclear reaction analysis (NRA). A recent review by Shull[21] presents a detailed account of FRES.

In an FRES experiment, a beam of helium ions having an energy E_0 = 3.0 MeV strikes the sample at a glancing angle, usually 15°, as shown in the schematic diagram of FRES geometry in Figure 10.7. Deuterium and hydrogen nuclei from the sample are recoiled out of the sample due to elastic collisions with the incident He. If the incoming He ion collides with a target nucleus at the surface of the sample, the target receives a constant fraction of the energy of the incident ion: 0.67

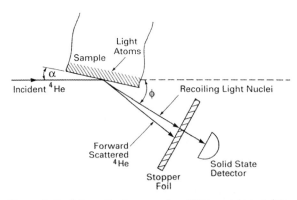

Figure 10.7   Schematic diagram of an FRES experiment. (From Reference 21.)

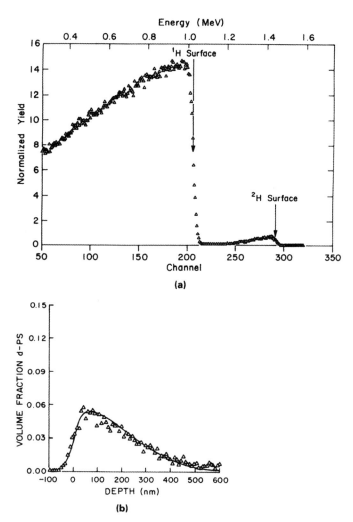

Figure 10.8   FRES spectrum and concentration profile for a thin film of DPS on a
PS/poly(xylenyl ether) matrix after the DPS layer has diffused about
300 nm into the matrix. (From Reference 22.)

for deuterons and 0.48 for protons. Thus, deuterons and protons recoiled from the
surface are well-separated in energy. In practice, a stopper foil (typically a 10-μm
mylar film) is placed in front of the detector to filter the forward-scattered He. Deuterons and protons recoiling from beneath the surface are detected at lower energies because the incoming incident and outgoing target particles lose energy due to
inelastic collisions with electrons in the sample.

Figures 10.8a and 10.8b show an FRES spectrum and concentration profile[22]
for a thin film of DPS on a PS:poly(xylenyl ether) matrix after the DPS layer has

diffused about 300 nm into the matrix. The DPS can diffuse about 700 nm beneath the surface before its signal (deuterium) begins to overlap the signal from hydrogen at the surface. The depth resolution (discussed below) is about 80 nm and is most apparent at the front edge ($z = 0$ nm) of the sample. Note that the energy of the recoiled particles is directly related to their initial position in the sample. Hence, a properly designed FRES experiment can yield the concentration profile of a D- or H-containing component directly and unambiguously. Another advantage of FRES is its sensitivity—volume fractions as low as 0.001 can usually be detected.

We can best understand the depth resolution of FRES[23] in terms of the energy spread of the detected particles. If z is a spatial coordinate normal to the surface, the depth resolution Δz is given as

$$\Delta z = \frac{\delta E_t}{dE/dz} \tag{10.4}$$

where dE/dz is the stopping power of protons (or deuterons) in the sample and the total spread in energy δE_t is given by

$$\delta E_t^2 = \delta E_{\text{detector}}^2 + \delta E_{\text{geometry}}^2 + \delta E_{\text{stopper foil}}^2 + \delta E_{\text{straggling}}^2 \tag{10.5}$$

Here, $\delta E_{\text{detector}}$ is the inherent energy resolution of the detector, and $\delta E_{\text{geometry}}$ results from finite beam divergence and detector acceptance angles. As the recoiled particles traverse the stopper foil, their energies are spread in such a way that, if they are monoenergetic upon entry, they have a (nearly) Gaussian distribution of energy upon exit—they straggle. $\delta E_{\text{stopper foil}}$ is proportional to the thickness of the stopper foil and is usually the dominant term in the expression for δE_t. Straggling can also occur in the sample itself, which gives rise to the last term, $\delta E_{\text{straggling}}$, though this term is usually small. As previously mentioned, the typical depth resolution for FRES is about 80 nm. Recently, the resolution of FRES has been improved[24] to about 30 nm by replacing the stopper foil with a time-of-flight (TOF) filter, which electronically prevents the detection of forward-scattered He particles (TOF FRES). The main disadvantages accompanying TOF analysis are decreased detector sensitivity and the complexity of incorporating TOF instrumentation into the FRES sample chamber.

10.3 Techniques for Measuring Concentration Profiles on the Order of Molecular Size

Consider a partially miscible binary blend of two homopolymers of the idealized geometry shown in Figure 10.2. For $\chi > \chi_c$, the two coexisting phases are separated by an interface of finite width. The volume fraction profile is given by

$$\phi(z) = \tfrac{1}{2}[(\phi_1 + \phi_2) + (\phi_2 - \phi_1)\tanh(z/w)] \tag{10.6}$$

where w is the interfacial width and

$$w \approx \chi^{-1/2} \tag{10.7}$$

Because w depends on χ, the interfacial widths exhibited by high polymers range from a few nanometers (e.g., the PS–polybutadiene interface) to many tens of nanometers (for the PS–DPS immiscible interface). Correspondingly, the resolution required to measure concentration profiles at these interfaces is FWHM = 20 nm and less. In this section, we present several examples of immiscible polymer interfaces to introduce the ion beam techniques of NRA and SIMS, as well as neutron reflectometry.

Though miscible at low and moderate molecular weights, the PS-DPS system exhibits partial miscibility[25, 26] at high molecular weights ($N \gtrsim 10^4$, where N is the degree of polymerization). From Equation 10.6, the small χ value for this system leads us to expect rather large interfacial widths between the PS-rich and DPS-rich coexisting phases. Indeed, this system has proven amenable to investigation using higher resolution ion beam techniques, NRA, and dynamic SIMS.

The experimental geometry for an NRA experiment is similar to the FRES experimental geometry (shown in Figure 10.7) with the exception that the stopper foil is replaced by a magnetic field. The incident particles, typically a collimated beam of ~700 keV ^{3}He ions, enter the deuterium-labeled polymer sample and lose energy (due to inelastic electronic collisions) as they traverse the sample. Deuterons in the sample can react with these incident ^{3}He particles according to the reaction

$$^3\text{He} + {}^2\text{H} \longrightarrow {}^4\text{He} + {}^1\text{H}, \quad Q = 18.352 \text{ MeV} \tag{10.8}$$

The ^{4}He particles produced in the nuclear reaction traverse the sample (again losing energy in the process) and are separated from the incident ^{3}He particles by the magnetic field and then detected using an energy-sensitive detector. The energy spectrum of the detected ^{4}He particles is then analyzed in terms of the calibrated energy loss and reaction cross section to yield the deuterium depth profile. Figure 10.9 shows NRA spectra and their corresponding depth profiles for a high-molecular-weight PS-DPS bilayer sample before and after annealing at 160 °C (in the two-phase region of the phase diagram).[28, 29] A significant advantage of NRA is its depth resolution, typically FWHM = 14 to 20 nm, which can be satisfactorily described using Equations 10.4 and 10.5, setting $\delta E_{\text{stopper foil}} = 0$. (Recall that $\delta E_{\text{stopper foil}}$ was the primary contributor to the energy spread for FRES.) The primary disadvantage of NRA is that the cross section for nuclear reaction is lower than the recoiling cross section for FRES, leading to longer data collection times and a greater potential for sample damage during data collection.

So far we have focused our attention on light ions in the MeV range traversing polymer samples. In these cases, the stopping power of most polymers arises predominantly through electronic interactions between the ions and the material. However, for ions of higher mass (Ar^+ or Cs^+, for example) at lower energies (in the 1 to 10 keV range) the nature of the stopping power of most polymers changes. Stopping occurs via nuclear processes,[11] and its cross section is orders of magnitude

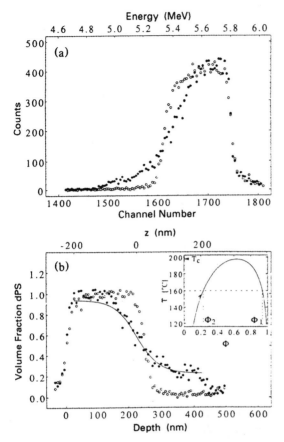

Figure 10.9 **(a) NRA spectra and (b) their corresponding depth profiles for a high-molecular-weight PS-OPS bilayer sample (immiscible interface) before and after annealing at 160 °C for 72.5 h. The coexistence curve for this system is shown at the inset of (b). (From Reference 28.)**

higher than the cross sections for the FRES and NRA processes mentioned in the preceding paragraphs. Under these conditions, the polymer sample is sputtered, creating a crater and a collection of secondary ions (and neutral particles). The ion beam can be rastered over a small area ($\sim$0.2 mm^2) of the sample and, as the crater deepens, the ions of interest can be extracted and detected with a mass spectrometer.[30] This process, dynamic SIMS, has proven useful for depth profiling in polymers and their blends.

The material at or near the base of the crater undergoes mixing during the sputtering process via two mechanisms: reimplantation of sputtered particles by collisions with primary ions and cascade mixing, caused by secondary recoil events set off by the primary ions entering the sample. For polymers, cascade mixing is usually the more important of these two factors and limits the depth resolution of dynamic SIMS to FWHM ≈ 13 nm. The advantages of dynamic SIMS are its high depth resolution and its ability to profile a wide variety of chemical species (^{1}H, ^{2}H, O, N, etc.). Its primary disadvantage is that the sputtering rates of most samples are very sensitive to the sputtering conditions and also depend on local composition, making dynamic SIMS data difficult to analyze quantitatively.

Consider the case of a diblock copolymer composed of two incompatible monomers, PS and poly(methyl methacrylate) (PMMA). When $N_{PS} \gg N_{PMMA} \gg 500$ ($N \gg 1000$), this block copolymer forms microphase-separated lamella of long period $L \approx 40$ nm, separated by interfaces ~5 nm in width. Although dynamic SIMS does not have the resolution required (~1 nm) to measure interfacial width, it is capable of revealing the orientation of 40-nm lamella with respect to the surface. Figure 10.10 shows[31] the SIMS profile of a 100-nm film of PS–PMMA diblock copolymer (with the PMMA block deuterated) on a silicon substrate and annealed at 170 °C for 24 h. These data show that the PS and the d-PMMA segments enrich the air and substrate surfaces, respectively. The D and H signals show pronounced oscillations, indicating that the orientation of the lamella persists over the entire 100-nm film.

In the example discussed in the previous paragraphs, we saw that dynamic SIMS was useful for investigating lamellae (~40 nm in size), but not the interfaces between

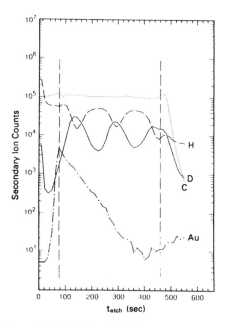

Figure 10.10 SIMS profile of a 100-nm film of PS–d-PMMA diblock copolymer (with the PMMA block deuterated) on a silicon substrate and annealed at 170 °C for 24 h. (From Reference 31.)

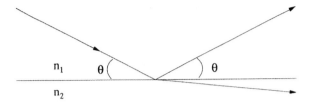

Figure 10.11 Schematic of a simple reflection experiment.

lamella (~5 nm in size, the range of many polymer interfaces). To investigate features such as these immiscible interfaces, we turn to neutron reflection. We begin by outlining basic physical principles of reflection, and then return to the example of PS-PMMA diblock copolymers to demonstrate how experimenters can use these principles to investigate polymer interfaces.

Consider two semiinfinite media with refractive indices n_1 and n_2 separated by a sharp, flat boundary, as shown in Figure 10.11. In general, if a beam of monochromatic light, incident in medium 1, encounters the boundary, part of its intensity will be reflected and part will enter medium 2. If $n_1 > n_2$, there is a critical angle θ_c below which total reflection occurs (i.e., the reflectivity is unity) and above which the reflected intensity falls off rapidly (though remains finite, strictly speaking). The relation between θ_c, n_1, and n_2 is simple and comes directly from Snell's law:

$$\theta_c = \mathrm{Cos}^{-1}(n_2/n_1) \tag{10.10}$$

If we add a third medium so that medium 2 is bounded between media 1 and 3, light that enters medium 2 encounters the second boundary, where it can be reflected or enter the third medium. The light reflected at this boundary can then interfere with the particles reflected at the first boundary. This type of interference is the origin of the colors present in thin oil films on water—the color we observe on the oil film possesses the wavelength satisfying the condition of constructive interference. The reflected light contains information about the thickness and refractive index of the oil film.

These optical principles apply to many types of panicles, including neutrons. Indeed, the example above is very much similar to the case of neutrons as they encounter a film of DPS on a silicon substrate. The reflectance (reflection amplitude) of the air–polymer interface can be precisely calculated using the Fresnel equation[32, 33]:

$$r = \frac{k_0 - k_1}{k_0 + k_1} \tag{10.11}$$

where

$$k_0 = \frac{2\pi}{\lambda}\sin\theta \tag{10.12}$$

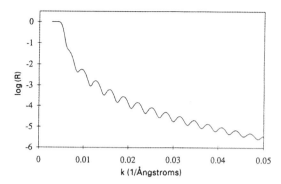

Figure 10.12 Calculated neutron reflectivity as a function of k of a 100-nm DPS film on a smooth, flat silicon substrate.

and

$$k_1 = \left(k_0^2 - 4\pi \frac{b}{V} \right)^{1/2} \tag{10.13}$$

where λ is the wavelength of the neutron and b/V is the density of neutron scattering length of DPS. The Fresnel equation can be extended in straightforward fashion[32] to the case or $n + 1$ media separated by n sharp boundaries. Figure 10.12 shows the calculated neutron reflectivity of a 100-nm DPS film on a silicon substrate using just such an extension ($n = 2$). (Similar results can be obtained with X rays, but we limit our discussion to NR.) Like the optical reflection experiment described in the preceding paragraph, there is a critical value, k_c, below which the reflectivity is unity and above which it falls off rapidly. The oscillations in the neutron reflectivity have the same physical origin as the colors on the thin oil film—interference between the particles reflected from the top and bottom surfaces of the film. In this case, many maxima and minima (corresponding to many orders of constructive and destructive interference) are observed.

Though the physical principles governing reflection apply to a wide variety of particles, neutrons are particularly well-suited to polymer studies. One can change significantly the scattering length density of a polymer (and hence the way it interacts with neutrons) by substituting deuterium for hydrogen. For example, $(b/V)_{PS} = 1.35 \times 10^{-6}$ Å^{-2} whereas $(b/V)_{DPS} = 6.1 \times 10^{-6}$ Å^{-2}. An additional advantage of neutrons is their probe-"buried" interfaces up to 200 nm (or more) deep.

A dramatic example of the sensitivity of NR to a buried interface is shown[34] in Figure 10.13a and Figure 10.13b where the neutron reflectivity is plotted as a function of k for a ~97-nm polymer film on a silicon substrate. The points are the experimentally measured data, and the solid line is calculated from the scattering length density profile shown at the inset. The calculations were performed using the generalization of the Fresnel equation mentioned previously with sufficiently fine spacing to incorporate the features of the scattering length density profile shown in

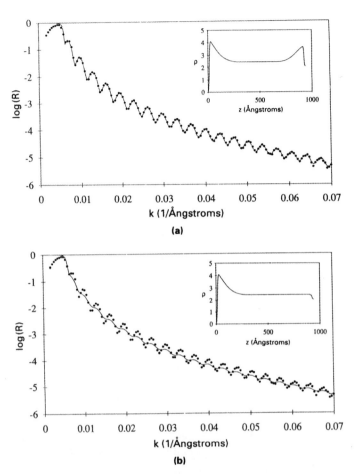

Figure 10.13 Comparisons of measured (*dots*) and calculated (*solid lines*) neutron reflectivity for a 97-nm film of (0.3)*d*-SAN38:(0.7)SAN42 annealed for 72 h at 167 °C. The calculated reflectivities are from the scattering length density profile shown at the insets (*a*) with segregation of *d*-SAN38 to both the polymer–air and polymer–substrate interfaces and (*b*) with the segregation at the polymer–substrate interface removed.

the inset. The film is a miscible blend of two poly(styrene-co-acrylonitrile) (SAN) copolymers: *d*-SAN38 and SAN42, where the numbers signify the weight fraction of AN in the random copolymer (the low-AN content component is deuterated). The overall composition of the blend is $\phi_{d\text{-SAN38}} = 0.30$. The same set of experimentally observed NR data (*dots*) is compared with two different calculated (*solid line*) data sets, based on the scattering length density profiles shown at each inset. The only difference between the calculated data sets is the segregation of *d*-SAN38 at the polymer–substrate interface. The amplitude of the oscillations for the calculated

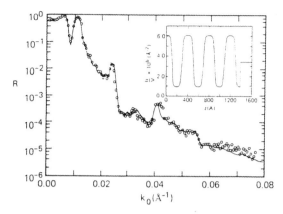

Figure 10.14 NR spectrum for a 140-nm film of poly(*d*-S-*b*-MMA) copolymer on a fused silica substrate annealed for 37 h at 170 °C (From Reference 35.)

reflectivity set decreases dramatically when the segregated layer at the polymer–substrate interface is removed. The amplitude decreases because the refractive index of the polymer blend near the substrate nearly matches that of the substrate. If the refractive indices matched, the oscillations would vanish completely.

Figure 10.14 shows[35] the NR spectrum for a 140-nm film of poly(*d*-S-*b*-MMA) copolymer on a fused silica substrate annealed for 37 h at 170 °C. The reflection data are consistent with a lamellar sample of long period 398 Å (oriented normal to the surface) with interfaces of width 50 ± 3 Å. The data do not, however, distinguish the finest points of the concentration profile. For example, an error function (of width 50 Å) can explain the observed data just as well as a hyperbolic tangent. Nevertheless, the depth resolution of NR is extremely high—no other technique has provided a more precise measurement of polymer interfacial width.

An important limitation of NR is that it does not provide a direct measure of the scattering length density profile. Just as a color on a thin oil film may correspond to more than one order of reflection (and therefore more than one thickness/refractive index pair), an NR spectrum does not necessarily correspond uniquely to a single scattering length density profile. In the analysis of NR data, one must (carefully) proceed with as much independently obtained information as possible. For example, the analysis of the data presented in Figures 10.13 and 10.14 proceeded with the benefits of previous dynamic SIMS results, which provided enough constraints on the profiles (shown at the insets) to lend confidence to the interpretation of the reflectivity results.

10.4 Summary

We have presented polymer interface problems and, in that context, outlined a set of characterization techniques for investigating "buried" interfaces. We have

Technique	Depth Resolution	Probe Radiation	Contrast
IR microdensitometry	90 μm	IR	IR absorption bands
MOST	0.5 mm	visible radiation	refractive index
Electron microprobe	3 um	electron beam	electron density
FRES	80 nm	~3 MeV He$^+$ or He^{2+}	hydrogen/deuterium
TOF FRES	30 nm	~3 MeV He$^+$ or He^{2+}	hydrogen/deuterium
NRA	15–20 nm	~700 keV ^{3}He ions	hydrogen/deuterium
Dynamic SIMS	13 nm	~10 keV ions (Ar$^+$, Cs$^+$, O$^+$, etc.)	H, D, C, O, others
NR	1 nm	neutrons, λ = 1.5 to 15 Å	hydrogen/deuterium

Table 10.1   Characterization techniques for investigating "buried" interfaces.

summarized these techniques in Table 10.1 in terms of size scale, contrast, and probe particles, and hope to impress upon the reader the variety or problems that one can approach with this set of methodologies.

It is worthwhile to bear in mind that the best route to characterizing polymer interfaces usually involves capitalizing on the synergism between two or more different experimental techniques. We have shown how a direct technique (of moderate depth resolution) can be used in conjunction with an indirect one (of high spatial resolution) to provide as complete a picture of the interface or interfaces as possible. Techniques not directly applicable to buried interfaces (contact angle, XPS, ATIR, etc.) can also play roles in multitechnique approaches to polymer interfaces.

References

1   E. Helfand and Y. Tagami. *J. Chem. Phys.* **56** (7), 3592, 1972.

2   R. Fayt, P. Hadjiandreou, and P. H. Teyssie. *J. Polym. Sci. Polym. Chem.* **23**, 337–342, 1985.

3   P. G. Shewmon. *Diffusion in Solids.* McGraw-Hill, New York, 1963.

4   R. J. Composto, E. J. Kramer, and D. M. White. *Macromol.* **21**, 2580, 1988.

5   P. F. Green and B. L. Doyle. *Macromol* **20**, 2471–2474, 1987.

6   P. F. Green and B. L. Doyle. *Phys. Rev. Lett.* **57**, 2407–2410, 1986.

7   W. W. Graessley. *Adv. Polym. Sci.* **16**, 1, 1974.

8   M. Doi and S. F. Edwards. *The Theory of Polymer Dynamics.* Clarendon Press, Oxford, 1986.

9   H. H. Kausch and M. Tirrell. *Annu. Rev. Mater. Sci.* **19**, 341–377, 1989.

10   P. G. de Gennes. *J. Chem. Phys.* **55**, 572, 1971.

11   J. Klein and B. J. Briscoe. *Proc. R. Soc. Lond.* **365**, 53, 1979.

12   E. A. Jordan et al. *Macromol.* **21**, 235–239, 1988.

13   M. Ye, R. S. Stein, and R. J. Composto. *Macromol.* **23**, 1990.

14   M. Ye. Ph.D. dissertation. University of Massachusetts, Amherst, MA, 1989.

15   L. C. Feldman and J. W. Mayer. *Fundamentals of Surface and Thin Film Analysis.* Elsevier Scientific, New York, 1986.

16   F. P. Price, P. T. Gilmore, E. L. Thomas, and R. L. Laurence. *J. Polym. Sci. Polym. Symp.* **63**, 33–44, 1978.

17   R. A. L. Jones, J. Klein, and A. M. Donald. *Nature.* **321**, 161–162, 1986.

18   W. K. Chu, J. W. Mayer, and M. A. Nicolet. *Backscattering Spectrometry.* Academic Press, New York, 1978.

19   R. J. Composto and E. J. Kramer. *J. of Mat. Sci.* **26**, 2815, 1991.

20   Green et al. *Macromol.* **18**, 501, 1985.

21   K. Shull. *Physics of Polymer Surfaces and Interfaces.* (I. Sanchez, Ed.) Butterworth-Heinemann and Manning, Boston, 1992.

22   R. J. Composto, E. J. Kramer, and D. M. White. *Macromol.* **21**, 2580–2588, 1988.

23   P. F. Turos et al. *A. Nucl. Inst. Meth.* **B4**, 92, 1984.

24   M. H. Rafailovich, J. Sokolov, X. Zhao, R. A. L. Jones, and E. J. Kramer. *Hyperfine Interactions.* **62**, 45–53, 1990.

25   F. S. Bates, G. D. Wignall, and W. C. Kohler. *Phys. Rev. Lett.* **55**, 2425, 1985.

26   F. S. Bates and G. D. Wignall. *Phys. Rev. Lett.* **57**, 1429, 1986.

27   S. J. Whitlow and R. P. Wool. *Macromol.* **22**, 2648–2652, 1989.

28   U. K. Chaturvedi et al. *Phys. Rev. Lett.* **63** (6), 616–619, 1989.

29   U. Steiner et al. *Makromol. Chem. MacromoL Symp.* **45**, 283–288, 1991.

30   A. Benninghoven, F. G. Rudenaur, and H. W. Werner. *Secondary Ion Mass Spectroscopy.* Chemical Analysis Series, Vol. 86, John Wiley, New York, 1987.

31   G. Coulon, T. P. Russell, V. R. Deline, and P. F. Green. *Macromol.* **22** (6), 2581–2589, 1989.

32   T. P. Russell. *Materials Science Reports.* **5** (4), 171–271.

33   M. Born and E. Wolf. *Principles of Optics.* 6th ed., Pergamon, Oxford, 1980.

34   T. L. Mansfield et al. *Physica B.* **173**, 207–210, 1991.

35   S. H. Anastasiadis, T. P. Russell, S. K. Satija, and C. F. Majkrzak. *J. Chem. Phys.* **92** (9), 5677–5691, 1990.

Friction and Wear (Tribology)

NORMAN S. EISS, JR.

Contents

11.1 Introduction

In the preceding two chapters, the emphasis has been on the prevention of motion between materials in contact by providing strong adhesive bonding forces. There are, however, applications for polymers in which motion between contacting materials is required. Some examples are gears, bearings, guides, and seals. In these applications it is important to decrease the bonding between the materials so that the relative motion causes low friction and wear. When polymers are selected for these applications, friction and wear (tribological properties) are given a high priority.

The purpose of this chapter is to describe the mechanisms of these tribological properties, indicate some of the correlations which have been found between tribological properties and chemical, physical, and mechanical properties, and discuss the measurement of the tribological properties.

11.2 Basic Concepts

First, some basic definitions.[1] *Friction* is the resisting force tangential to the common boundary between two bodies when, under the action of an external force, one body moves or tends to move relative to the surface of the other. Since friction is a force, the term *friction force* is redundant. *Coefficient of friction* is the ratio obtained by dividing the friction by the normal force pressing these bodies together. *Wear* is the progressive loss of a substance from the operating surface of a body occurring as a result of relative motion at the surface. *Wear rate* is the quantity of material removed in unit distance of sliding, unit time, or one revolution or oscillation. The alteration of the surface topography by plastic deformation is not considered wear by the above definition, but the deformation can render a component unserviceable. Consequently, the terms *wear* and *surface damage* are often used together to indicate conditions which may cause a component to not function properly. The definition of wear by the American Society for Testing and Materials (ASTM) includes both damage and mass loss.[2]

Friction is always present when contacting surfaces are in relative motion. Wear, on the other hand, may or may not occur during sliding. It is clear from the above definitions that friction and wear are different phenomena. Thus, it does not follow that a low value of friction indicates a low value for wear. For example, polytetrafluoroethylene (PTFE) has one of the lowest values of friction and highest values of wear of all polymers. Since wear is usually accompanied by the fracture of the wearing material and fracture occurs when the applied stresses exceed the material strength, friction can affect wear if the friction causes stresses which exceed the strength.

Friction and wear are not properties in the same sense that yield strength and glass transition temperature are. Friction and wear are properties of the system, which consists of the two materials which are in contact, the geometry of the contact, the operating conditions (the force transmitted normal to the surface, the relative sliding speed, and the ambient temperature), and the environment (the solids, liquids, or gases which may be present in the contact). Thus, values of friction and wear are meaningless unless the system in which they were measured is completely described. Consequently, values of friction and wear for two materials measured in a test system may have no relationship to the values of friction and wear for the same two materials in an application.

Because of their system dependence, it is meaningless to ask for the values of the coefficient of friction and wear of a material. For example, the coefficient of friction of PTFE can be as low as 0.05 when it is sliding on polished glass at low sliding speeds. On rougher surfaces the coefficient of friction can be as high as 0.2–0.3. Thus, when looking for a value of coefficient of friction which would be representative of that in an application, one must be quite specific about the system and hope that some previous investigator had measured values in an identical system.

Unfortunately, at this time there are no models which can be used to predict the coefficient of friction or wear given the system characteristics. Thus, experimental techniques for measuring friction and wear are essential.

11.3 Tribological Applications

Tribological applications for polymers can be grouped into two major categories. The first includes applications in which the primary function is tribological in nature. For example, a brake is a device which uses the friction generated at a sliding interface to provide the force necessary to reduce the speed of a mechanism. One of the essential criteria in the design of a material for a brake is the value of friction and its consistency over a wide range of conditions. A second criteria is the wear rate of the material. From the consumers' point of view, the wear rate should be as low as possible so that the brake linings will last a long time. From the suppliers' point of view, the brake lining should have a wear rate which will satisfy a large fraction of consumers and still generate a sufficient market for replacements to make it profitable. Another example is the elastomeric eraser. The eraser removes pencil marks by a controlled wear of the elastomer. This is another case where consumers would like the wear rate to be as low as possible while the supplier would like it to be high enough to ensure continued sales.

The second, and by far the larger, category is the applications in which the primary function of the system is not tribological, but the values of friction and wear have a major influence on the successful performance of the system. Several examples in this category are described next. The primary function of gears is to transmit motion from one shaft to another and to transmit torque from a power source to a load. The gear teeth slide on each other as the gears rotate. The friction energy which is dissipated as heat in the gear materials is added to the output energy to determine the total input energy needed. The sliding action also causes the gear teeth to wear. Gear tooth wear results in a change in the motion transfer characteristics from the design conditions. Thus, one criterion in the system design is to choose materials and environments to reduce friction and wear. The cam and follower is another system in which motion transfer is the primary function and friction and wear must be minimized for good system performance.

A bearing is a device which locates a rotating shaft with respect to a nonrotating support. Hence, its primary function is to maintain the shaft in a desired position. A variant of the bearing is a guide that locates one member which is translating with respect to a stationary member. A seal is a device that prevents a gas or a liquid from escaping from a desired location through an interface between materials in relative motion. Since seals are always used in conjunction with bearings and the bearings are designed to support any forces which are transmitted normal to the interface of the shaft and bearing, the normal load on the seal needs to be high enough to limit the leakage rate to acceptable levels. The bearing, guide, and seal are examples in which the friction and wear must be as low as possible.

In the above examples, polymers were used to make gears, bearings, seals, and guides. Polymers are also used as coatings. In some cases the primary function of the coating is tribological. The coatings for plastic lenses are primarily to provide resistance to wear from abrasives. Polymer coatings on metals can reduce the damage caused by a small amplitude reciprocating motion. In other cases the primary function of the coating is to protect against corrosion or to improve appearance, but the coating must have abrasion resistance to accomplish its primary functions.

In the computer industry the primary function of the magnetic head and the magnetic media (magnetic particles held in a polymeric matrix) is the storage and retrieval of information. In the floppy disk system the head is in continuous contact with the magnetic media, whereas in the rigid disk system the head only contacts the media when the system is shut down. In both systems the friction between the head and media must be kept low enough to prevent damage to the head support. The media must not wear because wear results in the loss of the magnetic particles which store the information.

All of the above applications have one common characteristic: there is relative motion between the components. However, there is a class of systems in which friction between the contacting materials is used to prevent motion between components. Mechanical fasteners such as screws, bolts, and rivets are examples. The friction between the threads of the bolt and nut must be low enough to prevent the shearing of the threads or the bolt during tightening but high enough to prevent loosening during service. Polymer inserts have been used in nuts to provide high locking friction. The bolt threads plastically deform the polymer and create a large contact area over which the friction force acts.

As a final example, there are many systems in which friction-induced noise causes annoyance to consumers. Although the noise does not affect system performance, designers nonetheless work to alter the system to eliminate the noise. In the automotive industry, much effort is expended in eliminating noise caused by instrument panel components rubbing against one another, stabilizer bars rotating in elastomeric bushings when the vehicle is stopped suddenly, and the noise of weather stripping rubbing on the doors when the car is in motion.

In many of the above examples it should be noted that either friction or wear but not both was the tribological concern. This again emphasizes the fact that friction and wear are different phenomena which must be addressed separately. In the next section, the mechanisms of friction and wear are described.

11.4 Mechanisms of Friction and Wear

In applications in which the tribological properties of the system are of interest, it should be noted that one function of the interface is to transmit a load normal to the interface. The load causes the materials to deform either elastically or plastically to create a contact area. At static equilibrium, the sum of the normal stresses over the contact area will equal the normal load. This contact area is called the real area of contact, to distinguish it from the apparent or nominal area of contact. The following example clarifies the difference between the real and apparent contact areas. Consider two cubes, 1 cm^2, which have atomically flat faces. If two faces are placed in contact with no normal load, the real area and the apparent areas will be equal to 1 cm^2. As the normal load is increased, the normal stress on the contact area will increase until the stress equals the yield strength of the material. Further increases in normal load will cause the cubes to deform plastically and the real and apparent areas will increase in order to support the load. Now consider two cubes with rough faces. When they are put together under no load, they will make contact at only a few spots in the interface. As the load is increased these contact spots are stressed beyond the yield point and begin to grow in size as a result of plastic deformation. The area of these contact spots is the real area of contact, which is less than the apparent area of contact of 1 cm^2.

The real area of contact is the area over which the friction occurs and from which the wear particles are formed. Hence, the real area of contact will figure prominently in the description of the mechanisms of friction and wear.

Friction

Two mechanisms of friction are discussed here: adhesion and plowing. In addition, some transient friction effects such as the transition from static to kinetic friction and friction-induced vibrations are presented.

The adhesion theory of friction is based on the forces which are developed between atoms and molecules on the materials in contact. The theory states that the friction is a result of the force required to break these adhesive bonds or to cause a cohesive failure in one of the materials. A simple model of this process is that the magnitude of friction is given by the product of the real area of contact and the shear strength of the weaker of the two materials in contact. This model is for the cohesive failure of the material and not for a failure of the adhesive bonds. It is difficult to use this model to predict the adhesion component of friction because of the difficulty in estimating the real area of contact. An upper bound estimate of the real area is obtained by assuming that the real area is determined by the hardness of the softer of the two materials in contact. The premise is that the harder asperities penetrate the softer material until enough area develops to support the normal load. Because the hardness is measured by the penetration of a hard indenter into a softer material, it is assumed that the stress at the contact points is equal to the hardness. The hardness (which has units of pressure or stress) is divided into the normal load to give the real area.

The plowing theory of friction is based on the roughness of the surfaces in contact. It is assumed that the asperities on the two surfaces interpenetrate. As relative motion occurs, the asperities must either ride up over each other or deform each other to allow the motion to occur. The tangential force required to deform the material is the plowing component of friction. It must be noted that the adhesion friction model assumes that a cohesive failure occurs in one of the materials.

Thus, both theories encompass the idea that the resistance to deformation in the material is the cause of the friction force. It is sometimes stated that the cause of friction is a combination of the energy losses due to hysteresis in elastic deformation, plastic deformation, and fracture.

It is also clear that the adhesive forces on the real contact area affect the deformation of the materials. If the adhesive forces are Van der Waals bonds which are relatively weak, then it is likely that these bonds will be broken before the materials plastically deform or fracture. Stronger adhesive bonds are more likely to cause plastic deformation and thus higher friction.

It is often observed that the force required to initiate sliding (static friction) is higher than the force required to maintain sliding (kinetic friction). Two explanations for this difference have been offered, one based on creep during static contact and the other based on the inherent roughness of the surfaces. When two materials are placed in contact, the normal load and the material properties determine the real area of contact. Since the surfaces remain in contact without moving, creep causes the real area of contact to increase. Thus, when motion is initiated the force required is based on the real area of contact caused by the creep. During sliding, there is no time for creep to increase the real area; thus the force to maintain sliding is smaller.

The roughness explanation is based on observations of small microslips which occur during the increase of a tangential load prior to sliding. It is postulated that as the tangential load is increased, the asperities in contact deform plastically, causing a microslip to occur. As new asperities come into contact, they stop the micro-slip. This process continues until the resistance of the asperities is insufficient to prevent sliding from starting. Once sliding is initiated, friction is the time average of the resistance of the asperities which form the contacts and hence is lower than the friction when sliding was initiated.

Friction-induced oscillations can occur when sliding is initiated (usually called stick-slip friction—see Figure 11.1) or during sliding (harmonic oscillations). The force required to overcome the friction causes the system components to deform elastically. Thus, potential energy is stored in these components. If there is a drop in friction in the transition from static contact to sliding, the stored potential energy is transformed into kinetic energy of the components. This interchange of energy from potential to kinetic and back causes vibrations. Under certain conditions, the vibration only occurs for less than one cycle and the two materials revert to static contact (stick). The external forces build up to a level to initiate sliding (slip) and the process repeats (stick-slip motion). Methods used to minimize stick-slip motion include reducing the magnitude of friction and the difference between static and kinetic friction through the choice of materials or environment and designing a stiff system so that the potential energy storage is small.

The harmonic oscillations which can occur during sliding can be traced to the system dynamics and the relation between the kinetic friction and the relative velocity of sliding. The coupling of the various degrees of freedom of the system,

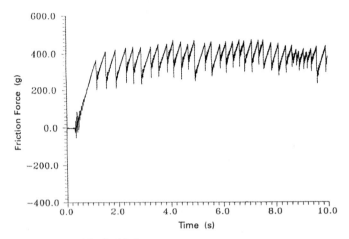

Figure 11.1   Stick-slip friction.

that is, tangential, vertical, and rotational motions, can lead to harmonic oscillations in the presence of a constant value of coefficient of friction. If the kinetic friction decreases as the relative sliding velocity increases, then any disturbance which causes the applied force to not be in equilibrium with the friction can trigger an oscillation. Such disturbances include momentary changes in the normal load, the friction force, or the relative velocity of sliding. Since harmonic oscillations can cause noise to be radiated from the system, control of this source of noise is important. Control can be achieved by redesign to decouple system degrees of freedom and by operating in a relative velocity range where the friction increases as the velocity increases.

Wear

Because wear is the removal of material from a surface as a result of sliding, a cohesive failure of the material must occur. The cohesive failure occurs whenever the stress on the material exceeds its strength. Thus, wear involves the initiation and propagation of cracks that cause the detachment of material as wear debris. The magnitude of the stress relative to the strength determines which wear mechanism will be predominant.

If the maximum stress in the material is less than the yield strength, cracks will not form immediately. However, repeated applications of these stresses will eventually cause cracks which will propagate to form wear particles. This process is called fatigue wear, and this mechanism has been observed in the wear of epoxy and polyimide polymers, which are generally considered to exhibit brittle behavior.[3, 4]

When the polymer is ductile and the maximum stress is above the yield strength but below the ultimate strength, the polymer will plastically deform. Thus, hard asperities sliding on the polymer will plow grooves without causing any removal of the material. Called plowing, this process is a form of surface damage. Repeated plowing of the polymer surface can result in the formation of wear debris through a process similar to low cycle fatigue. In low cycle fatigue, failure occurs as a result of repeated stress cycles which plastically deform the material.

When the maximum stress exceeds the ultimate strength of the material, cracks are initiated and propagated until wear particles are produced. This mechanism is called abrasive wear. Because abrasive wear occurs on the first application of the stress, it has sometimes been considered a limiting case of fatigue, that is, failure after one cycle of stress. If the adhesive bonds between two materials are strong enough, they can cause a cohesive failure in one of the materials. The detached fragment remains attached to the other material. This type of wear is called adhesive or transfer wear. When polymers slide on metals, transfer wear causes a polymer film to transfer to the metal so that the system becomes one of polymer sliding on polymer. Generally, the transferred material results in lower wear than in a system where transfer does not occur.

If the frictional energy dissipated in the material causes the temperature to exceed the glass transition or the melting temperature, the property changes which occur usually result in rapid wear of the material. This softening or melting wear is catastrophic, and components would not be designed to operate at friction levels that cause this type of wear.

Although investigators often try to isolate these mechanisms in the laboratory so that they can be studied, in applications they often occur simultaneously or sequentially. For example, a polymer sliding on a rough metal surface may initially be abraded and plowed by the metal asperities. As the abraded material accumulates in the valleys on the metal surface, the wear caused by abrasion decreases and that caused by fatigue increases. Because of the complexity of the wear process, it has been difficult to develop models to predict wear from basic polymer properties and system parameters. Some possible building blocks for wear modeling have been found in a number of correlations of wear results with polymer properties. Some of these correlations are presented in the next section.

11.5 Correlation of Properties with Friction and Wear

The correlations presented here have been determined in laboratory tests in which the experiment was designed to favor one mechanism. Many of the correlations are for abrasive wear since this mechanism usually causes the greatest wear of any of the other mechanisms except softening or melting. Abrasive wear is likely to be present in a system because of the roughness of surfaces and the possible contamination by abrasive particles.

Energy-to-Rupture and Abrasive Wear

Because abrasive wear involves the initiation and propagation of cracks to form wear particles, the energy required to fracture a polymer should have some influence on the wear rate. Experiments have shown that the wear resistance (reciprocal

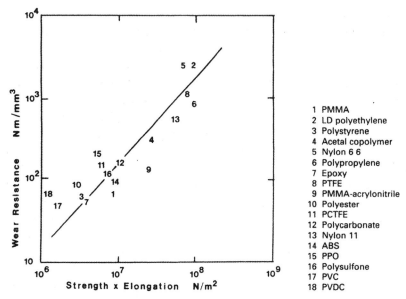

Figure 11.2   Correlation between wear resistance and the product of strength and elongation at fracture during a single traversal of a polymer sliding on mild steel (Ra = 1.2 μm).[5]

of wear rate) for 18 different thermoplastics and thermosets is proportional to the product of the stress and elongation at fracture, as shown in Figure 11.2.[5] These results were measured in a test where polymers were rubbed against a rough mild steel surface in a single-pass test. In a single-pass test the polymer never traverses over

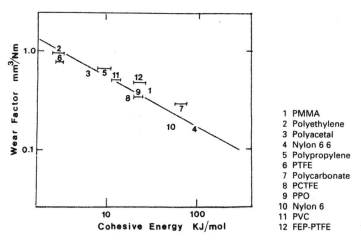

Figure 11.3   Wear factor versus cohesive energy for polymers sliding on 100-grit abrasive paper, single pass.[6]

any part of the steel more than once. This procedure prevents the transferred material from influencing the test result.

Cohesive Energy and Abrasive Wear

Since the fracture of a polymer during sliding should be related to the cohesion between the polymer chains or the cohesive energy, the abrasive wear should correlate with the inverse of the cohesive energy. Twelve different thermoplastic polymers pins were rubbed on silicon carbide abrasive paper in a spiral track to create a single-pass test. The pins were also rotated to prevent debris from accumulating on the trailing edge. The cohesive energies were calculated based on the chemical structures. It is shown in Figure 11.3 that the wear factor (wear divided by normal load and sliding distance) varied inversely with the square root of the cohesive energy.[6]

Ratio of Applied Stress to Failure Strength and Wear

In recent years several investigators have concluded that the ratio of the applied stress to the failure strength should have a direct influence on the wear of polymers. This conclusion was confirmed in several investigations. For example, Figure 11.4 shows that the log wear rates of pins of six different polymers sliding on a steel disk were linear with the log of the ratio. The applied interfacial shear stress was calculated from the normal pressure and the measured coefficient of friction and then divided by the rupture stress.[7]

Siloxane-modified epoxies were worn by a sliding steel ball. Since the epoxy appeared to fail in a brittle manner by cracks that developed perpendicular to the sliding direction and to the direction of the maximum tensile stresses in the surface, the ratio of the calculated maximum tensile stress to the fracture strength was used. The wear rate was linearly related to this ratio. The wear mechanism which appeared to dominate this experiment was fatigue. Several hundred cycles of sliding occurred while the cracks were initiated and propagated before a groove was formed in the epoxy surface[8].

PTFE which was irradiated to give it a range of mechanical properties was slid against wire gauze. The gauze had a known geometry, which permitted the stresses to be calculated from the normal load and the measured friction force. The gauze also permitted any wear debris to fall through the openings and thus not modify the gauze geometry. The log ratio of the mean tensile stress to the failure stress was linear with the log wear rate[9].

It should be noted in the preceding examples that the coefficient of friction was required for the stress calculation. This value could only be obtained from measurements made during the wear tests. Thus, it is not possible to calculate the stress before one has completed the wear experiments.

Molecular Weight and Wear

An increase in molecular weight should decrease the wear because of the increased strength of the polymer caused by the chain entanglement. Two studies on the wear

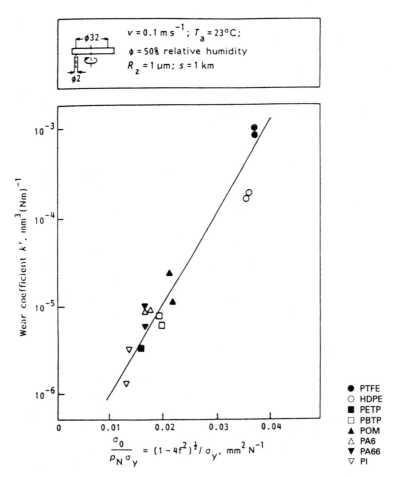

Figure 11.4 Wear behavior of thermoplastics (mechanism of abrasive wear).[7]

of polyethylene have confirmed this. In one test, the molecular weight was varied from 27 000 to 1 000 000 and the wear rate dropped by a factor of 270, as shown in Table 11.1. In this test, polyethylene flats were rubbed against a steel shaft.[10] In the other test, polyethylene was worn by an abrasive wheel and crystallinity was also varied. The molecular weight was inferred from the melt index (i.e., the higher the molecular weight, the lower the melt index). For low-density (0.919–0.925) (i.e., low-crystallinity) polyethylene, the wear doubled as the melt index was raised from 1 to 22. For medium and high densities, there was no significant effect of the melt index changes from 0 to 30 on the wear.[11]

When poly(vinyl chloride) (PVC) pins were worn against rotating steel disks of varying roughness and varying sliding speed, the samples with 70 000 M_n had less wear than those with 40 000 M_n. The reduction on wear rate varied from a factor

M_n	Wear, in.^{3}
32 000	293×10^{-4}
40 000	21
49 000	11
74 000	7.6
250 000	3.0
1 000 000	1.0

Table 11.1   Effect of molecular weight on the wear of polyethylene.[10]

of 7 to 2.5 on the smoothest surface to a factor of 2 on the roughness surface. One test condition showed no difference in wear for the two molecular weights. At the highest sliding speed and on the roughest surface there was massive flow of the PVC out of the interface, indicating that the surface temperatures had probably exceeded the glass transition temperature. A calculated temperature for this condition was above the T_g of 74 °C for these materials.[12]

A study of the wear of polystyrene coatings on steel substrates by a reciprocating steel ball as a function of molecular weight indicated that the wear rate decreased as the molecular weight increased up to 200 000 and then decreased at higher molecular weights. These polymers had narrow dispersions with a MW/M_n of 1.11 ± 0.06 for all except the highest MW (1 440 000), which had 1.56. In these tests the time required to wear through the film to the substrate was proportional to the reciprocal of the wear rate.[13]

Crystallinity and Wear

As noted in the preceding section, wear studies have been performed in which both molecular weight and crystallinity were varied and an interaction between them was found.[9] For the abrasive wear of polyethylene, increasing crystallinity (density increase from 0.92 to 0.96) caused the wear to decrease by a factor of five for high and low melt indexes and a factor of two for medium melt indexes, as indicated in Table 11.2.

The wear of PTFE pins on rotating steel disks was measured as a function of molecular weight and crystallinity. Wear increased as both molecular weight and crystallinity decreased. However, the increases were at most 10% for molecular weight changes from 15 to 5 million and crystallinity changes from 60 to 30%.[14, 15]

The wear of polychlorotrifluoroethylene (PCTFE) pins sliding on rotating steel disks was measured at crystallinities of 45 and 65%. On rough disks the wear rate for 65% was twice that for 45%. Smoother disks produced no significant difference.[16]

Loss Tangent and Friction

If the work of friction is to be dissipated in the polymer, then the loss tangent should correlate with the friction. A steel ball was slid on poly(methyl methacrylate) with sodium stearate as a lubricant. The friction was measured as a function of temperature and sliding velocity. The loss tangent was measured as a function of

Polyethylene Structure		Abrasion Loss,
Melt Index	Density, g/cm^3	g/5000 cycles
22	0.925	0.24
7	0.935	0.08
30	0.965	0.05
3	0.919	0.07
3	0.934	0.03
1.8	0.960	0.02
1	0.924	0.11
1	0.931	0.10
0.9	0.938	0.03
0.15	0.960	0.02

Table 11.2   Effect of melt index and density on the abrasive wear of polyethylene.[11]

temperature and frequency. The sliding velocity was used to estimate the strain rates in the polymer. It was found that the peak values of friction and loss tangent occurred at the same temperature and frequency.

In another study in which a steel ball was rolled on PTFE, a correlation was found between rolling resistance and loss tangent. However, no correlation was found for the sliding friction of polyoxymethylene, PCTFE, polypropylene, or nylon.[18]

Closure

The above correlations are useful if one wishes to change the polymer composition to increase or decrease the friction or wear in a system. For example, if a polymer component is experiencing a high wear rate because of abrasive wear, candidate replacements would be those polymers which had higher energies-to-rupture than the existing polymer. In general, if the wear mechanism in the test which established the correlation is the same as that in the application, the correlation can provide some guidance. Unless the correlations were established in the application or in a test which exactly duplicates the application, the correlations cannot be used to predict wear rates a priori. To determine wear rates and friction values, one must perform experiments. The measurement of tribological properties is the subject of the next section.

11.6 Measurement of Friction and Wear

There are several reasons why the values of friction and wear of a system need to be known. In systems where the tribological properties are of primary importance, their values are needed to establish system performance. The coefficient of friction in a brake system will determine the size of the brakes and the pressure required between the sliding elements to meet the desired stopping distance for various vehicle velocities. The wear of the friction material will determine the number of stops that can be made before the material will need replacement. The coefficient of friction is also needed in power transmission systems so that the heat that must be dissipated can be estimated. These estimates are then used to design the system to reject the heat and thus limit the operating temperature. For transmission systems used in spacecraft equipment, the coefficient of friction is used to determine the size of power sources such as motors. Because minimum weight is an important criterion, accurate friction data are required so that the power source is not oversized.

Measurements of wear rates of components in a system are necessary to determine the durability of the system. The level of acceptable wear rates for each specific system is determined by optimizing several criteria. These criteria include the cost of the worn component relative to the cost (time, labor, and inconvenience) of replacing the worn-out component, the total wear that can be tolerated without degrading the system performance, the ability of the system to manage the wear debris, the noise that the system might produce at various stages of wear, and whether or not the tribological components will be the controlling factor for the system durability.

It should be clear to the reader that friction and wear must be measured in the system in its operating environment if reliable values are to be obtained. It should also be obvious that materials must be chosen for the system in order to obtain these data. These requirements result in an iterative design-testing-redesign process which leads to a system with acceptable friction and wear values.

There are some disadvantages to this iterative design process. Since the materials will be selected to give long life for the components, the generation of wear data will be an expensive process because of the time required to get measurable wear. In order to make the wear measurements, the researcher must disassemble the system so that the dimensions or weight of the worn components can be measured. Friction values are generally inferred from measurements of input and output power of the system. If the system contains several pairs of sliding surfaces, it is impossible to determine the contribution of one pair to the overall friction of the system from the power measurements.

The technique commonly used to shorten the time to get measurable wear in a system is accelerated testing. The principle of accelerated testing is to alter the system operation so that the wear rate is increased. Inherent in accelerated testing are assumptions that the increase in wear is linear with the change of an operating parameter and that the same wear mechanisms are operative. The two most common parameters that are changed are the operating speed and the normal load transmitted by the sliding surfaces.

An increase in the operating speed causes the velocity of sliding to increase. Since wear has been experimentally found to vary linearly with the distance of sliding for metals and polymers, the velocity increase results in more wear per unit time. The higher sliding velocities will cause higher strain rates. In addition, the energy dissipated in the deformation and fracture processes will have less time to be conducted away from surfaces and the temperature will increase. Unfortunately, polymer mechanical properties are significantly affected by both strain rate and temperature increases. Thus, the higher sliding velocities may alter the polymer properties enough to cause the accelerated wear to be nonlinearly related to the sliding velocity.

An increase in normal load causes the real area of contact to increase. Since the wear particles will come from the areas in contact, it follows that an increase in area will result in more wear particles at the same stress level. It has been shown that the area is proportional to the load for asperities which are plastically deformed as well as those which are elastically deformed under low loads. If the increase in normal load causes the wear mechanism to change from fatigue to plowing or fracture, then the wear rate will not be linear with the load changes. Thus, accelerated testing can yield valid wear data only if the material property changes are insignificant and the wear mechanism is unaltered. To reiterate, reliable values for friction and wear can only be obtained from the system in its normal operating environment. However, when friction and wear values have been obtained for a system, there are procedures that can be employed to select alternate polymers which will change the friction or wear to more desirable values. Using data from laboratory tests is one of these procedures discussed in the next section.

ASTM Laboratory Tests

A laboratory test may have one of two purposes. In the context of the previous section, one purpose is to rank friction and wear values of candidate materials for an application. The degree of correlation of the rankings in the lab test and the application largely depends on how well the friction or wear mechanism matches that which occurs in the application. The other purpose is to determine how mechanical, physical, thermal, and chemical properties of polymers affect a specific friction or wear mechanism (see Section 11.5 for examples).

There are very few examples in the literature of laboratory test results which have correlated with the tribological performance in the field. See Reference 19 for a correlation between the wear of carbon fiber-filled polymer bearings worn by the shaft (conformal contact) and the wear of the same materials in a cylinder-on-cylinder lab test (counterformal contact). The major reason for this is that the tribological processes in the application involve several mechanisms either acting simultaneously or sequentially, whereas the laboratory test is often designed to emphasize only one mechanism. A second reason is that the operating conditions in the field are often quite variable and might include conditions which act for a small percentage of the service life but cause the major portion of the wear. Since lab tests are usually performed with a fixed set of conditions, they do not adequately simulate the field conditions.

In recent years laboratory tests have been performed using a matrix of sliding velocities and normal loads. As a result several regimes of wear are encountered. The values of velocity and load which define the boundaries of these wear regimes (or mechanisms) are then plotted on a graph of normal load versus velocity. These "wear maps" and the associated values of wear in each regime for a given material can provide the designer some guidance regarding changes in normal load or sliding speed if it is desired that the wear rate be changed in an application. Unfortunately, other system requirements usually dictate the speeds and loads. Thus the designer needs wear maps for several candidate materials so that one material can be selected which gives the most acceptable tribological performance for the speed and load required by the system. Most of the published wear maps are for metals and ceramics. One would expect to find polymer wear maps in the literature in the future.

Laboratory tests range from being specific to one application to being very general. Some of these tests have been standardized through volunteer organizations such as ASTM. Table 11.3 and 11.4 list ASTM tests for wear and friction of polymers. Some of the more widely used tests are described in the following paragraphs; the wear tests are discussed first, followed by the friction tests. The order in which they are discussed is determined by their generality, the most general being

Number	Test Materials	Method
D 3702	Polymer washer	Thrust washer test
D 1242	Polymer plaque	1. Loose abrasives
		2. Bonded abrasives single pass
D 969	Organic coating	Falling abrasive
D 658	Organic coating	Air blast abrasive
D 4060	Organic coating	Taber Abraser
D 4213	Interior paints	Scrubbed with silica powder
F 735	Transparent plastics	Oscillating sand bed, coatings
D 1044	Transparent plastics	Taber Abraser, coatings
D 3885	Textile fabrics	Reciprocating on hardened edge
D 3886	Textile fabrics	Abrasive paper on flat
D 4157	Textile fabrics	Abrasive paper on oscillating cylinder
D 4156	Textile fabrics	Abrasive paper contacts entire fabric surface
D 4966	Textile fabrics	Abrasive paper horizontal-circular-vertical-circular-horizontal motion
D 3884	Textile fabrics	Taber Abraser
D 3389	Coated fabrics	Taber Abraser
D 1630	Polymers for shoe soles, heels	Abrasive paper on rotating drum
D 2228	Soft vulcanized rubber	Knives rub on surface
F 510	Resilient floor covering	Loose abrasive in Taber Abraser
F 732	Polymers joint prostheses	Reciprocating sliding
F 450	Vacuum cleaner hose, plastic wire-reinforced	Oscillating on abrasive paper
F 595	Vacuum cleaner hose, all plastic	Oscillating on abrasive paper

Table 11.3   ASTM wear tests for polymers.

Number	Test Materials	Method
D 4518	Polymers on steel	1. Inclined plane, static friction 2. Horizontal pull, static friction
D 3028	Polymer on polymer	Slider on rotating cylinder, kinetic friction
D 1894	Plastic film or sheet on self or other material	Slider on flat, static and kinetic
D 2535	Wax coating on itself	Slider on flat, static and kinetic
E 303	Rubber on test surface	Slider pendulum against flat, swing amplitude measured
D 2047	Shoe soles and heels on floor	Slider on flat tilting load, static friction
D 3334	Fabric on itself	Inclined plane, static friction
D 3108	Yarn on solid material	Wrapped on pins, kinetic friction
D 3412	Yarn on yarn	1. Twisted yarn, kinetic friction 2. Yarn on yarn wrapped on a capstan, kinetic friction

Table 11.4   ASTM friction tests for polymers.

listed first. In general, ASTM test methods do not include data for polymers. The data that are typically included are the intra- and interlaboratory measures of variability of the method.

Standard Test Method for Wear Rate of Materials in Self Lubricating Rubbing Contact Using a Thrust Washer Testing Machine (D 3702)   In this test the polymer specimen is machined into the shape shown in Figure 11.5 and loaded and rotated against a steel washer until approximately 0.004 in. (100 μm) has been worn

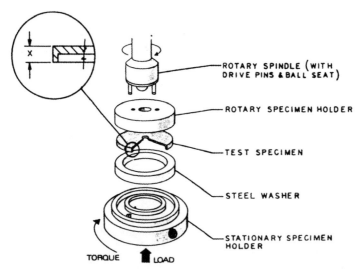

Figure 11.5   Shape of specimen for thrust washer test ASTM D 3702.

Polymer	PV, psi-ft/min	Wear, in./h
Acetal	1250	$5\text{–}10 \times 10^{-6}$
Acetal	2500	10–30
Acetal + 22% PTFE	2500	3–6
Nylon 6,6	2500	10–50
Polyimide + 15% graphite	5000	10–20

Table 11.5   Typical polymer wear rates in thrust washer test (ASTM D 3702).

off the polymer. The specimen requires a 40-h break-in period, and the test can take anywhere from 50 to 4000 h. The wear rate is reported in terms of the linear wear (in inches/hour or meters/hour).

Typical wear rates are given for four polymers in Table 11.5. During the break-in period, transfer of polymer to the steel washer usually occurs. This transfer fills the valleys on the steel surface, making it smoother and thus less abrasive to the polymer. Thus the wear that occurs in this test is usually adhesive or transfer wear and, possibly, fatigue wear. Fatigue wear could be the mechanism of fracture which leads to transfer, as well as the mechanism by which wear particles are generated from the transferred material.

It is recommended that this test be run at a constant value of the product of the apparent pressure P and the sliding velocity V. The apparent pressure is the normal load divided by the apparent area of contact. The value of PV at which the wear rate increases dramatically is called the limiting PV (LPV) of the material. The LPV can be determined by fixing the velocity and incrementally increasing the load or fixing the load and incrementally increasing the velocity until the rapid increase in wear rate occurs. At values of PV less than LPV, the linear wear rate is reported at the specific value of PV used. The implication of the use of the PV parameter is that wear rate is the same for different combinations of P and V as long as PV is constant. This is generally true as long as the wear or failure mechanism is not changed. At high pressures the polymer can yield and cause a catastrophic structural failure. At high velocities the thermal input can cause the polymer to degrade or cause a decrease in mechanical properties which can also lead to a structural failure.

Standard Test Method for Resistance of Plastic Materials to Abrasion (D 1242)
Two types of abrasion tests are specified in this standard—loose abrasive and bonded abrasive. In one test, a rotating plastic specimen is worn by loose abrasives carried into an interface by a rotating plate of metal, ceramic, or other material. In the other test, plastic specimens on a conveyer are sequentially placed in contact with a moving bonded abrasive belt (see Figure 11.6). In this test each portion of the belt only contacts a specimen once (hence it is a single-pass test). A single-pass test has the advantage that any changes in the abrasive surface due to grit breakage or plastic debris transfer will not affect the wear results. The obvious disadvantage is that a new belt

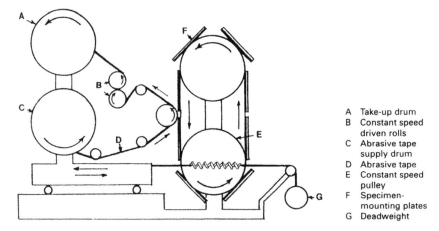

A   Take-up drum
B   Constant speed
     driven rolls
C   Abrasive tape
     supply drum
D   Abrasive tape
E   Constant speed
     pulley
F   Specimen-
     mounting plates
G   Deadweight

Figure 11.6   Bonded abrasive abrading machine ASTM D1242.

must be used for each test. In the loose abrasive test, the grit is not recirculated or reused for the same reason. In both tests the measure of wear is the mass loss. The loose abrasive test is run for 1000 revolutions of the plate; the bonded test is completed when the 50 m of abrasive belt has passed through the machine.

Abrasion resistance of organic coatings and paints   Four ASTM tests present different methods for measuring the abrasion resistance of coatings. These are Standard Test Methods for Abrasion Resistance of Organic Coatings by Falling Abrasive (D 968), by Air Blast Abrasive (D 658), and by the Taber Abraser (D 4060) and the Wet Abrasion Resistance of Interior Paints (D 4213). ASTM D 968 specifies that sand or silicon carbide grit is dropped 36 in on a coated specimen at a 45° angle to the flow. ASTM D 658 requires that silicon carbide be accelerated by an air stream expanding from a pressure of 13 kPa and impinged on a specimen at 45° to the air stream. In both of these tests the measure of wear is the amount of abrasive required to wear through a unit thickness of the coating.

ASTM D 4060 utilizes the Taber Abraser to abrade the coating. The Taber Abraser consists of two rotating abrasive wheels which abrade a rotating specimen. The abrasive wheels are deadweight loaded against the specimen. ASTM D 4213 specifies that a painted surface is scrubbed with a damp cellulose sponge and silica powder. The measure of wear on both tests is the volume or mass lost in a fixed number of cycles or the number of cycles needed to remove the coating.

The measurement of the life of a coating (number of cycles or the amount of grit to wear through the coating) is obviously influenced by the coating thickness. If coatings of different thicknesses are tested, the life is usually normalized by the coating thickness on the assumption that the life is a linear function of thickness. Unfortunately, this assumption is not always true and should be evaluated before reporting normalized results.[20]

Abrasion resistance of transparent plastics   These two tests differ from the tests described previously in that the damage that abrasives cause is measured by a change in light transmission rather than by mass loss. In fact, in these tests there could be no mass loss, only damage due to plowing of the plastic which would decrease the light transmission of the plastic.

The Standard Test Method for Abrasion Resistance of Transparent Plastics and Coatings Using the Oscillating Sand Method (F 735) calls for a reciprocating specimen to be abraded by a bed of sand supported by the specimen. The weight of the sand bed (quartz sized 6/14) provides the contact pressure. The degree of abrasion is measured by the amount of luminous transmission and haze after 100, 200, 300, and 600 strokes.

The Standard Test Method for Resistance of Transparent Plastics to Surface Abrasion (D 1044) utilizes rotating abrasive wheels in contact with a rotating plastic plate, that is, the Taber Abraser, to produce the damage. The surface damage is characterized by the percentage of transmitted light which deviates from the incident beam by forward scattering. Values are given for four different plastics and two coated plastics. In this standard there is a note stating that the use of the Taber Abraser for volumetric or mass loss measurements is inadvisable due to large coefficients of variation of the data and insufficient agreement between participating laboratories. ASTM D 1242 is recommended for mass loss measurement. In view of this advisement ASTM D 4060, which recommends the Taber Abrader to measure the mass loss of coatings, should be viewed with extreme caution.

Abrasion resistance of textile fabrics   Seven ASTM standards address the abrasion resistance of textile fibers using different methods. These are Standard Test Methods for Abrasion Resistance of Textile Fabrics—Flexing and Abrasion Method (D 3885), Inflated Diaphragm Method (D 3886), Oscillating Cylinder Method (D 4157), Uniform Abrasion Method (D 4156), Martindale Abrasion Tester Method (D 4966), and Rotary Platform Double Head Method (D 3884). The Standard Test Method for Coated Fabrics Abrasion Resistance, Rotary Platform Double Head Abrader (D 3389), is similar to ASTM D 3884.

The flexing and abrasion method consists of reciprocal folding and rubbing over a bar with a tool steel or cemented carbide edge. The inflated diaphragm method refers to a means by which the fabric is pressed against a plate covered with abrasive paper. In the oscillating cylinder method, abrasive paper wrapped on a cylinder is oscillated against the fabric. The apparatus used for the uniform abrasion method moves the abradant and the fabric in such a way that the entire surface of fabric is contacted. The Martindale abrasion tester method utilizes the following motion sequence: straight line, elliptical, circular, elliptical, straight line where the second straight line is perpendicular to the first straight line. The rotary platform, double-head method uses the Taber Abrader to abrade the fabric.

Two measures are specified but neither is a mass loss measure. One measure is the degradation of breaking strength caused by the abrasion. The second is the number of cycles to give a specified damage. However, the types of damage which may be used are not described. The standard for the abrasion of coated fabrics uses the mass loss at a given number of cycles or the number of cycles to wear through the coating as its measure of abrasion resistance.

Rubber abrasion resistance Two standards deal with the abrasion resistance of rubber: Standard Test Method for Rubber Property-Abrasion Resistance, NBS Abrader (D 1630) and Rico Abrader (D 2228). ASTM D 1630 is specifically directed toward the testing of materials used for soles and heels of shoes. In this test, a rubber pad is abraded by 40-grit garnet paper on a rotating drum. The number of revolutions to abrade 2.5 mm of the specimen is the measure of wear. ASTM D 2228 is for soft vulcanized rubber compounds. The tester consists of knives that abrade a rotating circular specimen which has been ground to remove the molding skin. The tests are run using a dusting powder of aluminum oxide and diatomaceous earth. The powder helps absorb oils and resins which may otherwise accumulate on the knives.

Standard Test Method for Resistance to Abrasion of Resilient Floor Coverings Using an Abrader with a Grit Feed Mechanism (F 510) The method uses a Taber Abrader modified with leather-clad brass rollers and a grit feed mechanism which deposits 240 alumina on the surface of the floor covering before it encounters the wheels. This test simulates the abrasion caused by grit trapped between a shoe and the floor as the shoe is rotated. The wear is measured by the mass loss for a given number of cycles.

Standard Practice for Reciprocating Pin-on-Flat Evaluation of the Friction and Wear Properties of Polymeric Materials for Use in Total Joint Prostheses (F 732) In this test, a polymer pin is reciprocated on a metal or ceramic flat in the presence of bovine blood serum. The wear measure is the mass loss per unit sliding distance. The friction force is measured continuously by a force transducer. The test can be used to rank polymers relative to ultra-high molecular weight polyethylene (UHMWPE) sliding against stainless steel, to rank polymers relative to each other, or to rank different materials rubbed against UHMWPE.

Vacuum cleaner hose durability Standard Methods for Testing Vacuum Cleaner Hose for Durability and Reliability, Plastic Wire Reinforced (F 450) and All Plastic Hose (F 595), are identical test techniques. The hose is placed in contact with a concave 90° bend coated with 500-grit abrasive paper. The hose is reciprocated and the number of cycles required to wear a hole in the jacket is the measure of wear.

Friction tests Standard Test Method for Measuring Static Friction of Coated Surfaces (D 4518) consists of two techniques. The incline plane method consists of an inclined plane on which the coating has been applied and a stainless steel slider that is placed on the plane. The angle of inclination is increased at a rate of 1.5 ± 0.5 degrees/s until the slider slips. The tangent of the angle at which the block slips is the static value of the coefficient of friction. The second technique is based on the force required to initiate sliding of the steel slider on a horizontal coated

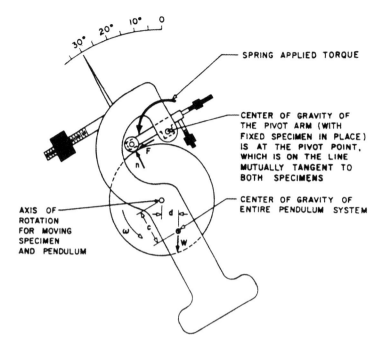

Figure 11.7 Frictionometer components (ASTM D 3028).

surface. A force transducer measures the force which is divided by the weight of the slider to obtain the coefficient of friction.

The Standard Test Method for Kinetic Coefficient of Friction of Plastic Solids (D 3028) utilizes a fixed circular member sliding on a larger rotating circular member, as shown in Figure 11.7. The friction force deflects the pendulum which supports the fixed member. The angular defection of the pendulum is used to calculate the coefficient of friction. The Standard Test Method for Static and Kinetic Coefficients of Friction of Plastic Film and Sheeting (D 1894) measures the friction of a slider on a flat surface. Either plastic film on itself or on another surface can be tested. The Standard Test Method for Coefficient of Kinetic Friction for Wax Coatings (D 2534) also uses the slider-on-flat configuration.

The Standard Method for Measuring Surface Frictional Properties Using the British Pendulum Tester (E 303) measures the travel of a pendulum which is slowed down by the contact of a standard rubber slider with a test surface. The maximum rotation of the pendulum after contact is reported as a reading on a scale of numbers called the BPN (British Pendulum Number). The friction force is not measured directly. The Standard Test Method for Static Coefficient of Friction of Polish Coated Floor Surfaces as Measured by the James Machine (D 2047) is designed for shoe soles and heels on walkway surfaces. The machine consists of a deadweight connected to a slider with a link which is pivoted at the deadweight and the slider.

The surface is moved so that the link gradually moves from the vertical. At the angle at which the slider slips, the friction coefficient is the cotangent of the angle between the link and the vertical.

Three standards address the friction of yarns and fabrics. The Standard Method for Testing Fabrics Woven from Polyolefin Monofilaments (D 3334) describes the measurement of the coefficient of friction of a fabric sliding on itself using the inclined plane test method. The standard test method Coefficient of Friction, Yarn to Solid Material (D 3108) specifies that the yarn is wrapped around chrome pins with either 180 or 360 degrees (θ) of contact. The tension in the yarn is measured before (T_1) and after (T_2) contact with the pins while the yarn is traveling at 100 m/min. The coefficient of friction (μ) is calculated from $\exp(\mu\theta) = T_2/T_1$. The standard test method Coefficient of Friction Yarn to Yarn (D 3412) describes two methods. In one method the yarns are twisted together and the input and output tensions are used to calculate the coefficient of friction. In the other method one yarn is wrapped on a capstan. The other yarn contacts the capstan which is rotated. The tensions before and after the capstan are used to calculate the coefficient of friction as in ASTM D3108.

Designing Laboratory Tests

If standard tests such as those described in the previous sections do not predict the correct ranking of materials in a application or do not predict the correct order of magnitude of wear for a material, then the designer must consider designing a laboratory test which may improve its predictive capability. Before doing this, the designer is advised to review the details and precautions given concerning several of the above standards. These standards have had the benefit of review by members of the working committees of ASTM, and all negative comments must be resolved before the standard is approved. Thus, the standard embodies the considerable experience of reviewers, standards writers, and the personnel performing inter-laboratory tests.

In designing a laboratory test, one should consider the following items:
1 purpose of the test
2 resources available for the test
3 desired discrimination of the test
4 geometry of the test specimens
5 type of motion
6 normal loads sliding speeds and temperature
7 environment surrounding the test specimens
8 methods of measurement.
These items are discussed in the following sections.

Purpose Several purposes have already been presented in Section 11.6. It is highly unlikely that a laboratory test will be used to estimate wear rates or friction forces in an application because of the differences between the two situations. Most probably the test will be used to determine the relative ranking of materials the best of which can then be tried in the application. The test may also be used to determine how other factors such as the mating material, its surface roughness, the environment, and the load and speed change the friction or wear in the test. It is hoped that the magnitude of the changes in the application would reflect those in the lab test, a hope that can only be confirmed in the application.

Resources   Resources include the quantity of materials for the test, the test apparatus, the supporting measuring equipment, the data acquisition and processing system, and the space required for the test apparatus. The time required for technicians to run the tests and for engineers and scientists to interpret the data is also a resource which must be provided. All of the above items can be assigned costs which can be compared to the available allocations. The anticipated value of the data must then be weighed against the estimated cost to obtain the data.

Occasionally the quantity of materials available for testing is extremely limited. This most often occurs when a polymer synthesis group has made a 25-g sample of a new material and this amount must be divided among several groups for characterization. Often the material available for tribological testing is 5 g. The tribological tests must then be designed to obtain the desired data from this small amount of material. One method of testing small quantities is to form coatings and measure the tribological behavior of the coatings. If the coatings show superior performance, then this information could be used to help make the decision to scale up the quantity synthesized.

Discrimination   The size (quantity of materials, number of tests, and time) of an experimental program cannot be estimated until the desired discrimination of the test is specified. In addition, information must be available on the expected variation of repeated test results. In general, as the magnitude of the variation of the repeated results decreases, the number of tests decreases to obtain a desired discrimination.

For example, an economic analysis may indicate that if the wear rate of a component can be decreased to half its present value, an expenditure of a certain amount would be justified for identifying a polymer which would accomplish this reduction. Thus, the test must be able to discriminate between polymers which have wear rates that differ by a factor of two or more. If no experimental data are available on the variability of the test to be used, then preliminary tests would have to be performed to determine the expected variability. With this information, statistical tables could be used to estimate the number of tests which would be required to detect the desired difference.[21] If the estimate of the costs of these tests is less than the allotted expenditure, the test can proceed. If not, then the size of the test must be reduced by better control of the experimental variables to reduce the variability of the results or reduce the number of candidate polymers in the test program.

Most of the ASTM standards give statistical data on the variability to be expected in a test method. This information can be used to estimate the size of a test program to give a desired degree of discrimination. The availability of these statistical data is one reason why the standard tests should be used if they are deemed to simulate adequately the conditions in the application.

Test specimen geometry   The ASTM tests described previously include a variety of specimen geometries. One method of classification for contact geometry is the dimension of the contact under the condition of no normal load. For example, the theoretical contact between convex surfaces (sphere-on-sphere or cylinder) is a point. The contact between a cylinder with its axis parallel to a contacting plane is a line. The contact between a cylinder with its axis perpendicular to a contacting plane is a circle. An advantage which the theoretical point contact has over line and area contacts is that precise alignment of the specimens is not required. For example, a sphere sliding on a plane needs no special alignment, whereas a cylinder sliding on a plane must be aligned exactly with the axis perpendicular to the surface of the disk to achieve an area on contact. Any misalignment would cause the contact to become a point contact.

Of course, as soon as a normal load is applied, all contacts become area contacts since the infinite stresses caused by the point and line contacts cause the materials to deform plastically to create a large enough area so that the stresses are below the flow strength of the material. If the experiment is a wear test, then the wear of the materials creates an area of contact. Thus, one of the reasons a run-in period is required on certain tests is to allow wear to create an apparent area of contact which will remain constant during the wear test. Two reasons for designing specimens which will produce small apparent areas in a wear test is to reduce the magnitude of the normal load and to increase the rate of linear wear (e.g., the reduction in length of a cylindrical pin).

Type of motion   A test apparatus must produce relative motion between two specimens. This is accomplished by moving one or both of the specimens relative to the inertia frame of the apparatus. The motion can be translational (pin moving back and forth on a fixed flat, as in ASTM F 732) or rotational (hollow cylinder rotating on a fixed washer, as in ASTM D 3702). Some tests have both surfaces moving (abrasive wheels rotating against a rotating polymer specimen as in the Taber Abraser).

If friction is to be measured, it is easier to measure the force exerted on a fixed surface than on a moving surface. The electrical connections of the force transducer on the moving surface must pass through a slip ring which can introduce noise on the electrical signal, whereas the transducer can be hard-wired if it is on the fixed surface. Force transducers on moving forces may also be affected by inertial forces if the motion is reciprocating.

The motion can be reciprocating or unidirectional. In reciprocating sliding, the velocity must go to zero at each reversal of motion. Thus, during each cycle of motion the velocity is constantly changing. A velocity control feedback system can be used to maintain constant velocity over most of the cycle, but the reversal requires high accelerations. For materials such as polymers, which have mechanical properties that are sensitive to strain rate, the velocity changes may cause uneven

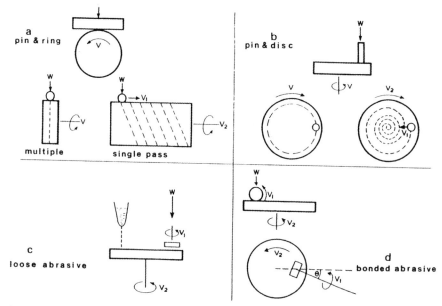

Figure 11.8 Typical laboratory test configurations for single and multiple passes.[17]

wear to occur over the sliding distance. The velocity reversals may also excite stick-slip motion following each reversal. Stick-slip motion introduces an additional component of velocity variation which can affect wear results.

Unidirectional motion can be controlled at a constant velocity provided that the drive motors are not significantly affected by any of the frictional variations which usually occur. In order to cause a measurable amount of wear a sliding distance is required which is too large for a practical translating motion device. Thus, rotary motion is employed to reduce the size of the apparatus.

Two types of relative motion are employed, single pass and multiple pass, as shown in Figure 11.8. In single-pass tests, one of the contacting surfaces is always presenting virgin surface to the other. This can be achieved in a pin-on-rotating disk apparatus by moving the pin radially so that a spiral contact path is formed (Figure 11.8b). The rate of radial motion can be controlled so that the pin contacts any portion of the disk only once during the test. A single-pass test can also be accomplished for a pin with its axis perpendicular to the surface of a cylinder (Figure 11.8a). As the cylinder rotates, the pin is moved parallel to the axis of the cylinder. The single-pass test is used in abrasive wear tests to prevent transferred polymer and fractured abrasive grains from altering the wear process.

Multiple pass tests can have three configurations:

1 The apparent areas of contact for each specimen do not move relative to the specimen during the test—the thrust washer configuration shown in Figure 11.5 is an example of this.

2 The apparent area of contact does not move on one specimen but does move on the other specimen—in the reciprocating pin-on-flat configuration, the end of the pin is always in contact with the flat, but the contact area moves over the surface of the flat as the pin reciprocates.

3 The apparent areas of contact move on both specimens—in the Taber Abraser, since both the polymer disk and the abrasive wheels are rotating, the contact area moves on both as shown in Figure 11.8*d*.

As is discussed in Section 11.7, the interpretation of the wear results depends on whether the apparent area is fixed or moving relative to the wearing specimen. Single-pass tests can have configurations 2 and 3 described above.

Normal load, sliding speed, and ambient temperature   The normal load, the geometry of the contact, and the material properties determine the area of contact. An increase in the normal load causes the real area of contact to increase. The load divided by the area of contact gives the average normal stress on the contact area. As noted in the subsection Wear in Section 11.4, the magnitude of the stresses will determine the mode of wear which will occur. Thus, the load in the laboratory test can be adjusted to give a wear mechanism which appears to simulate that in the application.

The most common method for applying the normal load is by deadweight. The weight can be applied directly to the fixed specimen or indirectly using a lever. If a lever is used, the lever must be designed so that the pivot lies on an extension of the friction force vector. If the pivot does not lie on the friction force vector then the friction force causes a moment about the pivot which changes the normal load at the contact point, as an elementary statics analysis will prove. This type of problem appears most frequently for the block-on-ring configuration, in which the load on the block is applied by a lever.

Because a deadweight has mass, out-of-roundness or out-of-flatness of the rotating surfaces can cause the deadweight to be accelerated perpendicular to the surfaces. The resulting inertial force adds and subtracts from the applied deadweight loading as the surfaces move. For the normal load fluctuation caused by out-of-roundness to be minimized, the normal load can be applied by pneumatic pressure using a piston. The piston mass can be made much less than that required for an equivalent dead-weight loading and thus reduce the inertial load fluctuations. Normal loads are also applied by hydraulic pressure and magnetic attraction.

The relative sliding velocity on the real area of contact must be chosen to simulate that in the application. If higher velocities are used to obtain data in a shorter time, the higher rate of energy dissipated in the polymer may cause temperature increases that will degrade mechanical properties. The product of the friction force and the relative velocity is a measure of the rate of energy dissipation. Temperature increases are most severe in polymer–polymer sliding systems because of the low thermal conductivities of the polymers.

The sliding velocity in systems where one or both specimens are rotating is determined by measuring the rotational speed (usually in rpm) and the radius from the center of rotation to the contact area. For a pin-on-disk configuration the radius is usually taken to be the distance from the center of rotation of the disk to the center of the contact area. Thus, there is actually a velocity distribution across the contact area with the velocity lowest at the closest point to the center of rotation and highest at the farthest point. This distribution is usually neglected since the pin radius is usually small compared to the radius of the wear track.

Ambient temperatures should be the same as those in the application. It must be noted that the temperatures caused by the friction energy will be superimposed on ambient temperature. Thus, the friction and wear will be affected by the total temperature if the mechanical properties are degraded at this temperature. Theories are available which can predict the frictional temperature but they require an estimate of the real area of contact which is often difficult to obtain.

Environment The environment includes any gases, liquids, or solids which can enter the contact area during the test. Since almost all equipment is run in air, this is the most common environment used in tests. For polymers the most important component of air is the water vapor. Since wear takes place at the surface, very small amounts of water vapor can have large effects on the wear of materials. Thus, when one is screening materials for an application, the humidity levels must be known in the application and duplicated in the laboratory test. Other gases which can adsorb on surfaces have the potential for changing intermolecular bonding forces which can affect both friction and wear values.

Some applications are run in the presence of a liquid. A mechanical seal is designed to prevent a process fluid from escaping into the environment or to prevent fluids in the environment from entering the apparatus. Thus, the friction and wear characteristics must be evaluated in the presence of the liquids. The solids in the environment which are of the most concern are abrasive particles. Since polymers are relatively soft materials, a wide variety of harder particles can be abrasive to the polymers. Thus, most of the ASTM tests described in the subsection ASTM Laboratory Tests in Section 11.6 are abrasive wear tests.

Measurement Wear is measured by a change in weight, a change in a linear dimension, or a change in the cross-sectional area of a wear groove. In a pin-on-disk experiment the wear of the pin can be determined by periodically stopping the test and removing the pin so that its weight or change in length can be measured. Since the pin has a constant apparent area as it wears, the change in the length times the area yields the volume worn away.

If the pin is attached to a linear displacement transducer, the change in length can be measured continuously. These wear-time data are important because the wear volume is often not linear with time, particularly during the initial part of a test. Most tests do settle out to a steady-state wear rate after this initial period, but the initial wear rates can be either higher, the same, or lower than the steady-state wear, as shown in Figure 11.9. In a multiple-pass test of a polymer sliding on an abrasive surface, the initial wear rate is high and reduces as the polymer wear debris begins to fill up the voids surrounding the abrasive grits and the sharp points of the

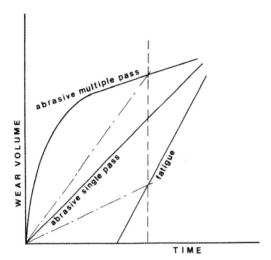

Figure 11.9 **Typical transient wear data.**

grits are broken off or covered with transferred debris. In a single-pass abrasive test, the initial and steady-state wear rates are the same. In a steel ball-on-rotating polymer disk test, the initial wear rate can be zero for several rotations until fatigue begins to cause wear particles to form and a steady-state wear condition is achieved. In this last test, the cross-sectional area of the wear track is measured with a surface profile instrument.

Some polymers can experience surface damage by plastic deformation which destroys their usefulness. For these polymers, measures other than weight loss must be used. The haze or light scattering of damaged transparent plastics (ASTM F 735 and D 1044) are examples of alternate measures. Surface roughness measurements may also indicate damage without significant loss of material.

Friction is measured with a force transducer. Most force transducers are built around an elastic member which is deflected by the force to be measured. The deflection of the elastic member is detected by strain gages, contacting or noncontacting displacement transducers such as linear variable differential transformers (LVDT), eddy current, or capacitance gages. Unfortunately, the elastic component of the force transducer must be put into the test apparatus between the contacting surface and a fixed support, that is, it is in series with the test specimen. The friction force causes elastic energy to be stored in the elastic element. As the friction force changes, this elastic energy is converted into kinetic energy of the specimen and the result is a vibrating motion.

Friction-induced vibrations can cause noise and accelerate wear processes because of the large velocity changes which can occur during the vibrations. For these vibrations to be reduced, the elastic member of the force transduced must be made as stiff as possible. Force transducers which use piezoelectric crystals to detect the forces have very high stiffnesses and thus help minimize vibrations. Piezoelectric crystals generate an electrical charge as a result of the applied forces. This charge decreases as current flows through the electrical components in the measuring circuits. Thus, the transducers are best for measuring rapidly changing forces as compared to measuring forces which are constant.

If the test is designed to measure the wear of materials, then it is better to not include a friction transducer so that the vibrations can be avoided. Most of the ASTM tests for wear do not include a friction measurement for this reason. However, if friction can be measured during a wear test, it can provide valuable information. For example, in the test of a steel ball sliding on a rotating polymer flat, no wear occurs during the initial part of the test. When the wear begins, the friction doubles or triples in value over a few cycles of motion, thus providing an accurate indication of when the wear process actually started.

11.7 Reporting and Interpretation of Measurements

In experiments, the wear (weight, linear dimension, or cross-sectional area change) is measured as a function of time or the number of cycles of rotation of the moving member. The wear rate (weight per unit time) is obtained by dividing the weight lost by the time. However, it is more common to calculate a sliding distance by multiplying the sliding velocity by the time, and then dividing the weight lost by the sliding distance. Since volume lost can often be directly related to dimensional changes of the specimen, the weight loss is often divided by the material density to obtain the volume loss. Then the wear rate is expressed in terms of volume loss per unit sliding distance (e.g., cubic millimeters/meter). Most ASTM wear standards recommend the calculation of volume loss from weight loss data.

In the pin-on-disk configuration the wear rate for the disk is often modified by expressing the volume of wear and sliding distance by πdA and πdN, respectively, where d is the sliding track radius, A is the cross-sectional area of the wear track, and N is the number of cycles of rotation. Thus the wear rate can be expressed as the cross-sectional area per cycle. In some experiments, the wear volume of the disk is calculated from the measured cross-sectional area of the wear track.

Often the wear rate is normalized by the normal load used in the experiment. The rationale for this is the large body of data on the wear of metals which indicates that wear rate is proportional to the normal load. Thus, results from tests performed at different normal loads could be compared if the wear rates were divided by the normal load. Many polymers show a nonlinear dependence of wear rate on load, so the normalization is not strictly applicable. However, many papers in the literature report normalized wear rates for polymers. The normalized wear rate is called the specific wear rate or the wear factor and is expressed in units such as cubic millimeters per meter per Newton.

A nondimensionalized measure of wear can be obtained by multiplying the specific wear rate by the hardness of the material if the hardness is expressed in units of force/area. This measure is called the wear coefficient. The wear coefficient is also the proportionality factor in the Archard wear equation:

$W = KLD/H$

where W is the wear volume, K is the wear coefficient, L is the normal load, D is the sliding distance, and H is the material hardness. Solving the equation for K gives

$K = WH/LD$

which is used to calculate the wear coefficient from experimental data. The use of the wear coefficient to compare the wear of two materials can be misleading if the material hardnesses are quite different. For example, a hard material may have small wear, whereas a soft material may have high wear. However, if the product of wear and hardness is the same for each, the wear coefficients would be the same. Thus, the wear coefficient would not indicate that the soft material did have a higher wear rate.

Many papers in the literature express wear data in terms of wear coefficients. However, the wear factor and the wear rate are preferred because their values directly indicate the amount of wear that has occurred.

As indicated in Figure 11.9, a wear test usually has an initial transient phase followed by a steady-state phase. If only one measurement of wear is made at a fixed time, number of cycles, or sliding distance, it is not known whether this measurement was made in the transient or steady-state phase. The wear rate calculated from this one measurement could be higher than, equal to, or lower than the steady-state wear. Thus, it is important to design a test with a run-in period to establish that steady-state wear has been achieved and then take a wear measurement during the steady-state period.

An even better procedure is to make several wear measurements that span the transient period and extend well into the steady-state period. Then a regression curve can be fit to the data points in the steady period. The slope of the regression curve is the wear rate. The statistical regression calculation can also give the confidence limits on the calculated slope.

As mentioned in the subsection Abrasion Resistance of Textile Fabrics in Section 11.6, the apparent area may either be stationary or moving on a surface. The major differences between these two conditions is that the surface with the moving apparent area experiences cyclic temperature and stress variations. These variations could cause wear by the fatigue mechanism. Thus, if identical materials were used as the pin and disk, for example, the wear rates could be different because of the cyclic temperature and stress variations experienced by the disk.

The friction force during sliding varies because of irregularities in surface topography and composition, variations of normal load, contaminants which change adhesive forces, variations in properties due to localized surface temperatures, and the excitation of system natural frequencies. If the purpose of the friction measurement is to estimate the power required to drive a system, then average values of friction force over time can be used. The most common technique for obtaining

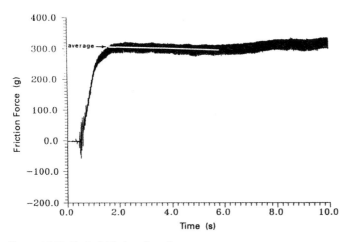

Figure 11.10 Typical friction-time data.

this average is to estimate the centerline value on a friction force versus time graph, as shown in Figure 11.10. The averaging can also be done by filtering out the high-frequency components of the transduced signal electronically before it is recorded. The result is an average value of the friction as a function of time. Because the transducer has limitations on its ability to record frequencies above a certain value, it also acts a filter. Thus, if the researcher chooses a transducer which cannot record high frequencies, the friction force that is recorded will be the average value.

11.8 Summary

This chapter presents a brief introduction to the subject of tribology. It is emphasized that friction and wear are system properties, not material properties. Therefore, they must be measured in the system or in some device which simulates the tribological conditions in the system. Standardized tests such as those of ASTM have the advantage that interlaboratory testing has established the expected variability of the method. If the friction and wear mechanisms are the same in a system and in a standard test, then the standard test may be used for an initial screening of materials. In the absence of a suitable standard test, a specialized test which simulates the friction and wear mechanisms in the system can be designed for screening materials. The screening tests would not be expected to give the values of friction and wear that will occur in the system. These can only be determined by making measurements in the system.

References

1   "Glossary of General Tribological Terms," *Wear Control Handbook*. (M. B. Peterson and W. O. Winer, Eds.) ASME, New York, 1980, p. 1152.

2   "Standard Terminology Relating to Wear and Erosion," ASTM G40-90*a*, ASTM, Philadelphia, 1991.

3   J. W. Jones and N. S. Eiss, Jr. In *Polymer Wear and Its Control* (L.-H. Lee, Ed.) American Chemical Society, Washington, DC, 1985, p. 135.

4   M. R. Chitsaz-Zadeh and N. S. Eiss, Jr. *Wear.* **110**, 359, 1986.

5   J. K. Lancaster. *Wear.* **14**, 233, 1969.

6   J. P. Giltrow. *Wear.* **15**, 71, 1970.

7   E. Santner and H. Czichos. *Tribology International.* **22**, 103, 1989.

8   M. R. Chitsaz-Zadeh and N. S. Eiss, Jr. *Tribobgy Trans.* **33**, 499, 1990.

9   B. J. Briscoe and P. D. Evans. *Wear.* **133**, 47, 1989.

10 G. C. Pratt. In *Lubricants and Lubrication.* (E. R. Braithwaite, Ed.) Elsevier Applied Science, New York, 1967, p. 403.

11 R. D. Deanin and L. B. Patel. In *Advances in Polymer Friction and Wear.* (L.-H. Lee, Ed.) Plenum, New York, 1974, p. 569.

12 N. S. Eiss, Jr., and G. S. Vincent. *Am. Soc. Lubrication Engineers Trans.* **25**, 175, 1982.

13 R. S. Bradley. M.S. thesis. Virginia Polytechnic Institute and State University, Blacksburg, VA, 9 Dec. 1988.

14 T. Y. Hu and N. S. Eiss, Jr. In *Wear of Materials 1983.* (K. C. Ludema, Ed.) Am. Soc. Mechanical Engineers, New York, 1983, p. 636.

15 T. Y. Hu and N. S. Eiss, Jr. *Wear.* **84**, 203, 1983.

16 N. S. Eiss, Jr., J. H. Warren, and T. F. J. Quinn. In *Wear of Non-metallic Materials.* (D. Dowson et al., Eds.) Mechanical Engineering Publishers Ltd., London, 1978, p. 11.

17 A. M. Bueche and D. G. Flom. *Wear.* **2**, 168, 1959.

18 K. C. Ludema and D. Tabor. *Wear.* **9**, 329, 1966.

19 J. K. Lancaster. *Tribobgy.* **6**, 219, 1973.

20 R. Raciti, N. S. Eiss, Jr., H. H. Mabie, and M. J. Furey. *Wear.* **132**, 49, 1989.

21 M. G. Natrella. "Experimental Statistics," *National Bureau of Standards Handbook 91.* U.S. Government Printing Office, Washington DC, 1966, p. 2–9.

References for Future Study

HO-MING TONG and DA-YUAN SHIH

This chapter provides readers from diverse polymer industries with references for their future studies in the various subject areas covered by this volume. Also given are references germane to the investigation of polymers in general, to polymer characterization and test methods, to the advances in the use of computers for polymer research, as well as to environmental effects on polymers. Even though they may not be directly related to the theme of this volume, references that shed light on bulk properties or characteristics of polymers foster well-rounded approaches for tackling surface- and interface-related problems. Details of some common techniques used for the surface and interface characterization of materials in general are the subject of the lead volume in this series, *Encyclopedia of Materials Characterization*. Characterization techniques or test methods specially applicable to polymers were discussed in the various chapters of this volume. Additional references are provided herein to augment the literature despite the nonsurface- or noninterface-specific nature of some references. The advent of faster and more powerful computers has added a new dimension to their use in polymer design and investigations. References in this area are now available to document new developments. The increasing use of polymers in aggressive technological and operational environments is a cause of concern to polymer technologists. Useful references are provided below for the assessment of environmental effects that can be either detrimental or beneficial, depending on the function for which the polymer is intended.

The field covered by this book is so diverse and rapidly evolving that one cannot possibly cover the literature in its entirety. A sizable portion of the references presented below were recommended by the authors of the preceding chapters.[†] Although most references covered by this chapter were published after 1982, some older references are included for the values they retain.

In this chapter, the references are grouped in one of the following areas: general-interest polymers, polymer characterization and test methods, computers in polymer research, environmental effects, and chapter subjects in the order of their appearance. Articles having to do with adhesion and polymer-related interfaces arc grouped according to their relevance. For each area, the references can be found in one of the following categories:

---

† The name of the author recommending a reference is given at the end of the reference, preceded by a dash.

- books and proceedings—although there are exceptions, books written by one or several authors tend to be more organized compared to *Proceedings*, which, if recently published, can provide more up-to-date and specific information on new developments

- handbooks, compendiums, and encyclopedias—these source books typically present a large amount of technical data specially organized for easy referencing

- bulletins or highly specialized reports—in this case, the coverage of a particular subject is often in depth

- journal articles—all journal articles listed were provided by chapter authors

Only a handful of references repeat themselves as a result of their relevance to different subject areas. The information contained in these references was typically extracted from sources such as technical literature, society meetings, materials suppliers, industrial experience, trade associations, academic research, or government programs.

General-Interest Polymers

Dictionary of Rubber Technology. A. Craig, Philosophical Library, New York, 1969. *A dictionary that includes trade names.*

Encyclopedia of Polymer Science and Engineering. 2nd ed., Vols. 1–15 and Suppl. (H. Mark, N. Bikales, C. Overberger, C. Menges, and J. Kroschwitz, Eds.) Wiley, New York, 1985–90. *A comprehensive compilation of a broad range of polymer-related work and practices.*

Engineering Resins: An Industrial Guide. E. Flick, Noyes Publications, Park Ridge, NJ, 1988. *Describes over 2500 engineering resins for industrial usage and their specifications.*

Handbook of Common Polymers. J. Roff, J. Scott, and J. Pacitti, CRC Press, Cleveland, 1971. *Extensive information on fibers, films, plastics, and rubbers. Good sections on polymer fabrication. Properties covered include general properties (e.g., refractive index, birefringence, and permeability), thermal properties, electrical properties, and mechanical properties.*

Handbook of Composites. (G. Lubin, Ed.) Van Nostrand Reinhold, New York, 1982. *Organized into sequential chapters of materials, processes, design and analysis, and typical applications, with an appendix containing basic data. Also covered are novel processes, recently developed materials, environmental suitability processing aids, practical solutions to manufacturing problems, and examples of composite materials.*

Handbook of Fillers and Reinforcements for Plastics. (H. Katz and J. Milewski, Eds.) Van Nostrand Reinhold, New York, 1978. *A unified compilation of information and data to aid in rapid selection of a satisfactory filler or reinforcement for use with plastics. Also included is detailed information on Kevlar filaments, ribbon reinforcements, microfibers, and basalt fibers and flakes.*

Handbook of Thermoplastic Elastomers. (B. Walker, Ed.) Van Nostrand Reinhold, New York, 1979. *A reference source that gives detailed guidance in the selection and use of thermoplastic elastomers.*

Handbook of Thin Film Technology. (L. Maissel and R. Glang, Eds.) McGraw-Hill, New York, 1970. *Excellent, in-depth chapters on various aspects of thin film processing and characterization.*

Handbook of Water-Soluble Gums and Resins. R. Davidson, McGraw-Hill, New York, 1980. *Organization and writing tailored for fast referencing.*

Macromolecules. Vol. 1: Structure and Properties; Vol. 2: Synthesis, Materials and Technology, H. Elias, Plenum, New York, 1984. *An excellent reference that offers an integrated representation of chemistry, physics, and technology based on a broad survey of the whole field of macromolecules. Serves as a bridge between highly simplified introductory textbooks and highly specialized textbooks and monographs.*

Paint and Surface Coatings: Theory and Practice. (R. Lambourne, Ed.) Ellis Horwood Ltd. Publishers (Halsted Press), Chichester, 1987. *A general introduction on various aspects of paint and surface coatings, including film formers, pigments, solvents, additives, and industrial paint-making processes, as well as rheological and mechanical properties.*

Plastics for Electronics: Desk-Top Data Bank. International Plastics Selector, San Diego, 1979. *Aids those who select, specify, and order plastics in the electrical and electronic industries. Includes plastics for extruding and molding, adhesives, casting, encapsulating and potting resins, films, laminates, and prepregs.*

Plastics Polymer Science and Technology. (M. Baijal, Ed.) Wiley-Interscience, New York, 1982. *A general reference. Deals with material science and product technology aspects of polymers.*

Structure–property Relationships in Polymers. R. Seymour and C. Carraher, Jr., Plenum, New York, 1984. *Covers the chemical and physical structure of polymers, their optical, mechanical, electrical, and thermal properties, their solubility and permeability and their chemical resistance. The effects of additives on important classes of polymers are also discussed.*

Technicians Handbook of Plastics. P. A. Grandilli, Van Nostrand Reinhold, New York, 1981. *Concise reference book on methods, materials, equipment, and techniques involving industrial plastics. Presentation is down to layman's level of comprehension.*

Whittington's Dictionary of Plastics. L. Whittington, Technomic Publishing, Stamford, CT, 1968. *Terms are arranged in alphabetical order.*

Polymer Characterization and Test Methods

Characterization of Plastics by Physical Methods: Experimental Techniques and Practical Applications. G. Kampf, Hanser Publishers, New York, 1986. *A survey of physical methods for the characterization of the solid structure of plastics with examples for technical practice. Overview of methods is kept simple and without recourse to theoretical principles.*

New Characterization Techniques for Thin Polymer Films. (H. Tong and L. Nguyen, Eds.) Wiley, New York, 1990. *Compilation of newly developed characterization techniques for the determination of the bulk and surface properties or characteristics of polymer thin films.*

1990 Annual Book of ASTM Standards. ASTM, Philadelphia, 1990. *Documents widely accepted standards and tests in polymer-related areas:*

- *Paints, Related Coatings, and Aromatics (Section 6)*
- *Textiles (Section 7)*
- *Plastics (Section 8)*
- *Rubber (Section 9)*
- *Electrical Insulation and Electronics (Section 10)*
- *General Methods and Instrumentation (Section 14). Includes analytical methods (Volume 14.01), general test methods, nonmetals, laboratory apparatus, statistical methods, and durability of nonmetallic materials (Volume 14.02), and temperature measurement (Volume 14.03)*
- *Adhesives (Volume 15.06).*

Non-Destructive Testing of Fiber-Reinforced Plastics Composites. (J. Summerscales, Ed.) Elsevier Applied Science Publishers, London, 1987. *A comprehensive and in-depth review of the subject art far the complete range of fiber-reinforced plastics. Techniques covered include radiography, acoustic emission, thermal, optical, vibration, and chemical methods.*

Thermal Characterization of Polymeric Materials. (E. Turi, Ed.) Academic Press, New York, 1981. *Excellent, in-depth chapters on the principles of thermal analysis and characterization of polymers, including thermoplastics, block copolymers and polyblends, thermosets, elastomers, fibers, and additives.*

Computers in Polymer Research

Applications of the Monte-Carlo Method in Statistical Physics. K. Binder, Springer Verlag, Berlin, 1984. *Includes computer simulation of complex physical systems in polymer research.*

Computer-Aided Design of Polymers and Composites. D. Kaelble, Marcel Dekker, New York, 1985. *Discussions based on atomic and molecular models. Analyzes the origins of stiffness, strength, extensibility, and fracture toughness in composite materials in terms of chemical composition and molecular structure. Includes effects of temperature, stress-strain, time, and environmental exposure.*

Computers in Polymer Sciences. (J. Mattson, H. Mark, Jr., and H. MacDonald, Jr., Eds.) Marcel Dekker, New York, 1977. *Computer studies of reactions on synthetic polymers, stochastic calculations of polymer structure, kinetic studies in thermal analysis and radiation chemistry of high polymers, digital spectral peak resolution, and more.*

Designing with Plastics and Advanced Plastic Composites. M. Dorgham, Interscience Enterprises, Geneva, 1986. *Selected articles on plastic and thermoplastic developments, thermoplastic composites, design methods—including computer-aided design—and automotive and aircraft applications.*

Environmental Effects

Chemical Resistance of Polymers in Aggressive Media. Y. Moiseev and G. Zaikov (R. Moseley, English trans.), Consultants Bureau, New York, 1987. *Covers chemical effects on polymers. Provides scientific understanding of chemical resistance phenomena and their impact on material technologies as well as their beneficial aspects, far instance, in designing biocompatible materials.*

Corrosion Resistant Materials Handbook. 3rd ed., I. Mellan, Noyes Publications, Park Ridge, NJ, 1976. *A practical and useful reference for the selection of proper commercially available corrosion-resistant materials, including suitable plastics, rubber, glass, ceramics, and stainless steels. Corrosion inhibitors are treated in a separate volume, also published by Noyes Publications.*

Corrosion Resistance Tables. 2nd ed., P. Schweitzer, Marcel Dekker, New York, 1986. *A guide on the corrosion resistance of metals, plastics, nonmetallics, and rubbers.*

Diffusion of Electrolytes in Polymers. G. Zaikov, A. Iordanskii, and V. Markin, VSP BV, Utrecht, The Netherlands, 1988. *Describes transport of chemically active fluids through polymers and prediction of effective lifetimes.*

Durability of Macromolecular Materials. (R. Eby, Ed.) ACS Symposium Series, ACS, Washington, DC, 1979. *Covers aging, fatigue, adhesion, degradation, etc.*

Environmental Effects on Polymeric Materials. 2 volumes, (D. Rosato and R. Schwartz, Eds.) Interscience Publishers, New York, 1968. *Covers different environments and materials, testing methods, service-life determination, fundamentals of degradation, rheology, design criteria, material selection, and technical business decisions.*

Physical Aging in Amorphous Polymers and Other Materials. L. Struik, Elsevier Applied Science, New York, 1978. *Good discussions on aging. Contains four parts: origin and basic aspects of aging in amorphous polymers, generalization, influence of aging on the long-term deformation behavior of amorphous polymers, and evaluation of results.*

Plastics Industry Safety Book. (D. Rosato and J. R. Lawrence, Eds.) Cahners Practical Plastics Series, Cahners Books, Boston, 1973. *Concerned with accident prevention involving the use of plastics.*

Structure, Relaxation and Physical Aging of Glassy Polymers. (R. Roe and J. O'Reilly, Eds.) MRS Proceedings, 1991. *Experimental theoretical, and molecular modeling of the nature of the glassy state and glass transition of polymers, and associated problems such as aging.*

Water in Polymers. (S. Rowland, Ed.) ACS Symposium Series, ACS, Washington, DC, 1980. *Clear illustration of the significance and role of water in synthetic polymers and commercial resins.*

Chain Structures and Polymer Synthesis

Chemical Reactions on Polymers. (J. Benham and J. Kinstle, Eds.) ACS Symposium Series, ACS, Washington, DC, 1988. *Contains work on chemical structures, chemical reactions, and polymer modification.*

Comprehensive Polymer Science. 7 volumes, (G. Allen, S. Aggarwal, J. Bevington, J. Booth, G. Eastmond, A. Ledwith, C. Price, S. Russo and P. Sigwalt, Eds.) Pergamon, Oxford, 1989–90. *An immense compilation of information on all aspects of polymers. Presentation is at the graduate and post-graduate level*—C. Rogers

Encyclopedia of Polymer Science and Engineering. 2nd ed., Vols. 1–15 and Supplement (H. Mark, N. Bikales, C. Overberger, C. Menges and J. Kroschwitz, Eds.) Wiley, New York, 1985–90. *Provides the definitive encyclopedic coverage of polymer science and engineering. Presentation is at the graduate level. A useful place to start a study of the literature on a given specific topic. Another very useful source is* *the* **Kirk-Othmer Encyclopedia of Chemical Technology**, *which contains many articles on polymer-related topics.*—C. Rogers

Modern Plastics Encyclopedia. McGraw-Hill, New York, published annually. *A yearly review of all commercially available plastics and resins and of processing technology. Presentation is on a pragmatic, industrial level without advanced scientific concepts. Invaluable for tables of data for materials selection, properties, design, and processing for a wide range of applications.*—C. Rogers

Modification of Polymers. (C. Carraher, Jr., and M. Tsuda, Eds.) ACS Symposium Series, ACS, Washington, DC, 1980. *Specific aspects of polymer modification are discussed.*

Plastics Materials. 5th ed., J. Brydson, Butterworths, London, 1989. *Very good discussions of the compositions and properties of most commercially available plastics. Also, there are discussions of general property-structure relationships, processing methods and design considerations. The presentation is at the graduate and post-graduate level.*—C. Rogers

Polymer Chemistry. B. Vollmert, Springer Verlag, New York, 1973. *Good reference of polymer chemistry, polymer syntheses, and bulk structures.*

Polymer Handbook. 3rd ed., (J. Brandrup and E. Immergut, eds.) Wiley, New York, 1989. *This is the largest and most useful compilation of data relating to the composition, properties, and other characteristics of polymers.*—C. Rogers

Polymer Materials: An Introduction for Technologists and Scientists. 2nd ed., C. Hall, Halsted Press, Wiley, New York, 1989. *A broad survey of the materials science of polymers on a fairly elementary level but with good coverage of the topics. An excellent introductory textbook.*—C. Rogers

Polymers: Chemistry and Physics of Modern Materials. 2nd ed., J. Cowie, Trans-Atlantic Publ., Philadelphia, 1991. *A good recent textbook giving a nicely balanced coverage of significant aspects of polymer science, including synthesis, reaction mechanisms and kinetics, physical characterization, and polymer structure and properties. Presentation is at the advanced undergraduate level*—C. Rogers

Principles of Polymer Engineering. N. McCrum, C. Buckley, and C. Bucknall, Oxford University Press, New York, 1988. *An up-to-date discussion of polymer structures, viscoelastic and mechanical properties, processing techniques, and design considerations for polymers. Presentation is at the advanced undergraduate level A good introduction to engineering aspects of polymers.*—C. Rogers

Principles of Polymerization. 2nd ed., G. Odian, McGraw-Hill, New York, 1970. *A presentation of the organic chemistry, reaction mechanisms, and kinetics of polymerization on the graduate level. Good coverage at the time of publication with clear and relevant discussions.*—C. Rogers

Properties of Polymers. 3rd ed., D. Van Krevelen, Elsevier, New York, 1991. *An exceptionally useful book which gives methods, correlations, and data for the prediction or estimation of properties of polymers. The discussions give good insight into the basis of property-structure relationships. Presentation is at the graduate and post-graduate level*—C. Rogers

Fabrication Techniques

Chem. Lett. M. Kakimoto, M. Suzuki, T. Konishi, Y. Imai, M. Iwamoto, and T. Hino, 823, 1986. *Good introduction on the formation of Langmuir–Blodgett films.*— L. Matienzo

Composite Materials Handbook. M. Schwartz, McGraw-Hill, New York, 1984. *Chapters include fabrication of composite materials, composite designs, processing, automated fabrication methods, machining, and future potential.*

Encyclopedia of Materials Science and Engineering. 2nd ed., Vol. 12, (H. Mark, N. Bikales, C. Overberger, and G. Menges, Eds.) Wiley-Interscience, New York, 1987. *General review on fabrication methods and polymer properties.*—L. Matienzo

Extruder Principles and Operation. M. Steven, Elsevier Applied Science, New York, 1985. *Describes general principles of extrusion engineering.*—L. Matienzo

Fiberglass-Reinforced Plastics Deskbook. N. Cheremisinoff and P. Cheremisinoff, Ann Arbor Science Publishers, Ann Arbor, MI, 1978. *A practical and useful manual on the processes and products in industrial applications.*

15th Nat. SAMPLE Tech. Conf. Proc, L. Matienzo, T. Shah, and J. Venables, Cincinnati, 4–6 Oct. 1983. *Covers analytical methods to identify formulations of commercial mold release agents on surfaces.*—L. Matienzo

J. Therm. Insul. L. Matienzo, T. Shah, A. Gibbs, and S. Stanley, **9**, 30, 1985. *On the requirements placed on isocyanurate foams for insulation.*—L. Matienzo

Modern Plastics Encyclopedia. McGraw-Hill, New York, 1990. *Excellent and comprehensive source book updated yearly on plastics and their fabrication and processing.*

Molecular Engineering of Ultrathin Polymeric Films. (P. Stroeve and E. Franses, Eds.) Elsevier Applied Science, New York, 1987. *Proceedings of a workshop. Contains various film-forming techniques (including Langmuir–Blodgett), state-of-the-art reviews, and an assessment of current trends and future research directions.*

Plastics Engineering Handbook. 4th ed., (J. Frados, Ed.) Van Nostrand Reinhold, New York, 1976. *Chapters are devoted to topics such as plastic materials and their applications, popular methods of processing plastics, radiation processing, mold-making and materials, interrelationships between the basic material, the process, the design, and the finishing and assembly as well as compounding and materials handling.*

The Plastics Engineer's Handbook. A. Glanvill, Industrial Press, New York, 1971. *Contains basic plastics engineering data for those who design and process plastics. Information tabulated for the processing of large-volume plastics, such as polyolefins and polystyrene, using molding, extrusion, thermoforming, film coating, and other techniques.*

Plastics Technology Handbook M. Chanda and S. Roy, Marcel Dekker, New York, 1987. *Excellent reference book for plastics. Chapter 4 reviews a large number of polymers, highlighting their chemical structures, characteristic properties and uses. Chapter 2 is on molding processes.*

Plastics Tooling and Manufacturing Handbook. ASTME, Prentice Hall, Englewood Cliffs, NJ, 1965. *A reference book on the use of plastics and engineering materials for tool and workpiece fabrication.*

Polyimides. (D. Wilson, H. Stenzenberger, and P. Hergenrother, Eds.) Blackie & Son, Glasgow, 1990. *Excellent reference for high-temperature polyimides.*—L. Matienzo

Polym. Composites. A. Lustiger, F. Uralil, and G. Newaz, **11**, 65, 1990. *Pertains to the development of high performance composites through engineering principles.*—L. Matienzo

Polymers in Microelectronics: Fundamentals and Applications. D. Soane and Z. Martynenko, Elsevier Scientific, Amsterdam, 1989. *Discusses polymers and relevant processes for microelectronic applications.*—L. Matienzo

Polymers, Liquid Crystals and Low-Dimensional Solids. N. March and M. Losi, Plenum, New York, 1984. *Account is given of the structure of polymeric solids, the influence of processing on polymers, and the electronic structure of biopolymers.*

Silane Coupling Agents. 2nd ed., E. Plueddemann, Plenum Press, New York, 1991. *On the role and behavior of silane coupling agents in composites and other polymers.*—L. Matienzo

Vac. Sci. Technol. J. Salem, F. Sequeda, J. Duran, and W. Lee, **A4**, 369, 1986. *On vapor deposition of polymers from starting materials, and routes to dry processes in microelectronics.*—L. Matienzo

Chemical Composition

Metallization of Polymers. (E. Sacher, J. Pireaux, and S. Kowalczyk, Eds.) American Chemical Society, Washington, D.C., 1990.—S. Kowalczyk

Photon, Electron and Ion Probes of Polymer Structure and Properties. (D. Dwight, H. Thomas, and T. Fabish, Eds.) ACS Symposium Proceedings, 1981. *Focuses on the experimental and theoretical techniques used to describe the anion and cation states in polymers.*

Polymer Surfaces and Interfaces. (W. Feast and H. Munro, Eds.) Wiley, New York, 1987.—S. Kowalczyk

Spectroscopy of Polymers. J. Koenig, American Chemical Society, Washington, DC, 1992.—S. Kowalczyk

Surface Science Investigations in Tribology. (Y. Chung, A. Homola, and G. Street, Eds.) American Chemical Society, Washington, DC, **1992**.—S. Kowalczyk

Microstructures and Related Properties

Applied Polymer Light Microscopy. (D. Hemsley, Ed.) Elsevier Applied Science, New York, 1989.—E. Thomas

Electron Microscopy. E. Thomas, Vol. 5 of Encyclopedia of Polymer Science and Engineering, Wiley, New York, 1986.—E. Thomas

Methods of Experimental Physics. Vol. 23: Neutron Scattering, Part C, K. Skold and D. Price, Academic Press, Orlando, FL, 1987. *Chapters on neutron scattering investigations of polymers, proteins, biological structures, magnetic materials, and solids undergoing phase transition.*

Polymer Microscopy. L. Sawyer and D. Grubb, Chapman and Hall, New York, 1987.—E. Thomas

Practical Scanning Electron Microscopy. (J. Goldstein and H. Yakowitz, Eds.) Plenum, New York, 1975.—E. Thomas

Principles of Polymer Morphology. D. Bassett, Cambridge University Press, New York, 1982. *Good general reference on crystalline orders in polymers.*

Rubb. Chem. Tech Reviews. J. White and E. Thomas, **57**, 457–506, 1984.—E. Thomas

Science. R. Pool, **27**, 634–636, 1990.—E. Thomas

Structure Analysis by Small-Angle X-Ray and Neutron Scattering. L. Feigin and D. Svergun, (G. Taylor, Ed.) Plenum, New York, 1987. *Part III deals with the application of the general theory of small-angle diffraction to polymers. Also discussed are the basic requirements for measuring devices and the principles of design and construction.*

Structure of Crystalline Polymers. (I. Hall, Ed.) Elsevier Applied Science, New York, 1984. *Centers on the shape of molecular chains in crystalline polymers, the experimental techniques employed to study polymer structure, and the uncertainties and ambiguities that remain.*

Articles on Reflectivity

Phys. Rev. A. A. Braslau, P. Pershan, C. Swislow, B. Ocko, and J. Als-Nielsen, **38**, 2457, 1988. *Detailed article describing the reflectivity method, the resolution of the method, and its limitation.*—R. Saraf

Phys. Rev. B. S. Sinha, E. Sirota, S. Garoff, and H. Stanley, **38**, 2297, 1988. *This article deals largely with the roughness method.*—R. Saraf

Phys. Rev. B. I. Tidswell, B. Ocko, P. Pershan, S. Wasserman, G. Whitesides, and J. Axe, **41**, 1111, 1990. *An interesting article that demonstrates how reflectivity may be applied in measuring density variations in organic films.*—R. Saraf

Articles and Books on Grazing Incidence X-Ray Scattering (GIXS)

Critical Phenomena at Surfaces and Interfaces: Evanescent X-Ray and Neutron Scattering. H. Dosch, Springer Verlag, 1992. *A recently published book that describes GIXS in great detail, from both the theoretical and experimental viewpoints.*—R. Saraf

Phys. Rev. B. G. Vineyard, **26**, 4146 (1982) *A theoretical paper describing the wave distortion approximation that makes the analysis of GIXS possible.*—R. Saraf

Phys. Rev. B. H. Dosch, **35**, 2137 (1987) *This article describes the actual experimental GIXS method.*—R. Saraf

Articles on Optical Methods

Appl. Phys. Lett. P. Tien, R. Ulrich, and R. Martin, **14**, 291, 1969. *Demonstrates the use of wave guiding and m-line spectroscopy to measure organic thin films.*—R. Saraf

IBM J. Res. Develop. J. Swalen, R. Santo, M. Tacke, and J. Fischer, p. 168, March 1977. *Deals exclusively with 12 different polymers prepared by various methods.*—R. Saraf

The Principles of Nonlinear Optics. Y. Shen, Wiley-Interscience, New York, 1984. *Deals with the use of the nonlinear optical probe to measure organic surfaces and interfaces. Chapter 25 is more relevant in the present context.*—R. Saraf

Thermodynamics

Inverse Gas Chromatography. (D. Lloyd, T. Ward, and H. Schreiber, Eds.) ACS Symposium Series 391, Washington, DC, 1989. *This monograph is a good summary of theory methodology, and results of experiments using inverse gas chromatography (IGC). IGC is becoming a very promising technique for the chemical characterization of polymer, filler, and adhesive surfaces.*—D. Dwight and T. Lloyd

J. Adhesion Sci. Tech. F. Fowkes, **4** (8), 669–691, 1990. *This reference summarizes many concepts and procedures used in the characterization of polymer surfaces and the materials they contact.*—D. Dwight and T. Lloyd

J. Adhesion Sci. Tech. R. Good, **6**, 1992. *This review by a long-time expert in the field brings the thinking up-to-date and makes suggestions for future work.*—D. Dwight and T. Lloyd

J. Am. Chem. Soc. F. Riddle, Jr. and F. Fowkes, **112**, 3259–3264, 1990. *The corrected acceptor numbers of Mayer and Guttmann are given by this reference. Those who are interested in using Guttmann-numbers should pay attention to this paper, since this is a new concept.*—D. Dwight and T. Lloyd

Principles of Colloid and Surface Chemistry. P. Hiemenz, Marcel Dekker, New York, 1977. *This is a good textbook dealing with many aspects of surface science, especially adsorption isotherms.*—D. Dwight and T. Lloyd

Thermodynamics of Polymer Solutions. M. Kurata, Harwood Academic Publishers, New York, 1982. *Devoted to the thermodynamic theories of osmotic equilibrium, phase segregation, concentration fluctuations, and sedimentation equilibrium in polymer solutions.*

Physical Properties

Applications of Time-Resolved Optical Spectroscopy. B. Bruckner, K. Feller, and U. Grummt, Elsevier Scientific, Amsterdam, 1990. *Topics on methods of time-resolved optical spectroscopy and of data analysis and interpretation.*

Dielectric Properties of Polymers. (F. Karasz, Ed.) Plenum, New York, 1972. *Proceedings of a symposium on dielectric behavior.*

Electrical and Electronic Properties of Polymers: A State-of-the-Art Compendium. J. Kroschwitz, Wiley, New York, 1988. *Coverage includes conductive and insulating polymers, their syntheses, properties, and uses—a wealth of physical and mechanical properties as well as standards and specifications for materials. This work is of value to those investigating the potential for combining the properties of metals and semiconductors with those of polymers.*

Electrical Properties of Polymers. (D. Seanor, Ed.) Academic Press, New York, 1982. *Chapters on electrical conduction, photophysical processes, polymer electrets, and contact electrification of polymers and its elimination, as well as dielectric breakdown phenomena.*

Electrical Properties of Polymers: Chemical Principles. C. Ku and R. Liepins, Hanser Publishers, New York, 1987. *Excellent review with extensive tables of data on fundamental parameters of polymer electrical properties and properties directly effecting them. Of predictive value in the control of dielectric constant, tangent of dielectric loss angle, dielectric breakdown, and electrical conduction.*

Electronic Properties of Polymers and Related Compounds. (H. Kuzmany, M. Mehring, and S. Roth, Eds.) Springer Verlag, Berlin, 1985. *Deals with the study of the electronic structure of polymers and with physical and chemical properties directly related to this structure. A majority of the work presented has to do with conjugated systems.*

Electronic Structure of Polymers and Molecular Crystals. (J. Andre and J. Ladik, Eds.) Plenum, New York, 1975. *Basic principles of the field of the electronic structure of polymers.*

Electron Microscopy in Solid State Physics. H. Bethge and J. Heydenreich, Elsevier Scientific, Amsterdam, 1987. *Methods of investigation in electron microscopy are given in Part 1. Part 2 considers applications to polymer morphology.*

Guide to Plastics—Property and Specification Charts. (Editors of Modern Plastics Encyclopedia) McGraw-Hill, New York, 1981. *Charts for quick reference on properties, design data, chemicals and additives, machinery specifiers, and manufacturers/suppliers.*

Introduction to Synthetic Electrical Conductors. J. Ferraro and J. Williams, Academic Press, Orlando, FL, 1987. *Concerned with conductive organic polymers, organic charge transfer metals, and the effects of temperature and pressure on electrical conductors.*

Mechanical Properties of Solid Polymers. 2nd ed., I. Ward, Wiley, New York, 1983. *Good reference on mechanical properties and their measurements and associated theories.*

Nonlinear Optical Properties of Organic and Polymeric Materials. (D. Williams, Ed.) ACS Symposium Series, ACS, Washington, DC, 1983. *Presents findings of researchers on the preparation and characterization of new organic and polymeric nonlinear optical media and reviews progress in this field.*

Nonlinear Optical Properties of Organic Molecules and Crystals. D. Chemla and J. Zyss, Academic Press, Orlando, FL, 1987. *Topics covered include the structure and properties of the organic solid state, quadratic nonlinear optics, molecular hyperpolarizabilities, and electro-optic materials.*

Optical Techniques to Characterize Polymer Systems. (H. Bassler, Ed.) Elsevier, Amsterdam, 1989. *On recent developments in spectroscopic methods due to the advent of modern light source and detection technology. Chapters include absorption, luminescence, and diffraction techniques, such as nonlinear optics for physical structure probing of polymers.*

Physics of Amorphous Solids. R. Zallen, Wiley, New York, 1983. *A tutorial on a broad range of topics concerning the physics of amorphous solids. Structural aspects of organic polymers are discussed along with their physical properties.*

Polymer Handbook. 3rd ed., (J. Brandrup and E. Immergut, Eds.) Wiley, New York, 1989. *Excellent compilation of fundamental constants and parameters for synthetic polymers, poly(saccharides) and derivatives, and oligomers. Some data on biopolymers are also provided.*

Solitons and Polarons in Conducting Polymers. L. Yu, World Scientific, Singapore, 1988. *Introduction to basic concepts on bipolarons, soliton and polaron dynamics, optical, magnetic, and transport properties, metal-insulator transitions, quantum field theory etc.*

Thermal Conductivity—Nonmetallic Solids. Y. Touloukian, R. Powell, C. Ho, and P. Klemens, Vol. 2 of Thermophysical Properties of Matter, IFI/Plenum, New York, 1970. *A comprehensive compilation of data for a wide variety of materials, including polymers. Other volumes in this series involve thermal expansion, radiative properties, specific heat, etc. For instance. Vol 13 is on the thermal expansion of nonmetallic solids.*

Surface Modifications

Ion Bombardment Modification of Surfaces: Fundamentals and Applications. (O. Auciello and R. Kelly, Eds.) Elsevier Applied Science, New York, 1984. *On surface modifications by ion beams.*

Journal of Adhesion Science Technology. *This journal contains many good up-to-date papers on adhesion and related phenomena.*—C. Chang

Laser Controlled Chemical Processing of Surfaces. (A. Johnson, D. Ehrlich, and H. Schlossberg, Eds.) MRS Proceedings, 1983. *Compilation of papers on laser modification of surfaces.*

Nuclear Instruments and Methods in Physics Research. *This journal describes work involving the use of high-energy beams to modify surfaces for enhanced adhesion.*—C. Chang

Photon, Beam and Plasma Stimulated Chemical Processes at Surfaces. (V. Donnelly, I. Herman, and M. Hirose, Eds.) MRS Proceedings, 1987. *Work compiled ranges from very fundamental studies of surface modification to microelectronics applications including VLSI and optoelectronic devices. Proceedings on Metallization of Polymers and Plastics, American Chemical Society, or Electrochemical Society These references contain specific work on metal–polymer reactions and adhesion.*

Surface Contamination: Genesis, Detection and Control. Vols. 1 and 2, (K. Mittal, Ed.) Proceedings of the 4th International Symposium on Contamination Control, 1979.

Surface Modification and Alloying By Laser, Ion and Electron Beams. (J. Poate, G. Foti, and D. Jacobson, Eds.) Plenum, New York, 1983. *Excellent review on microstructures and associated physical properties produced by the different energy deposition techniques.*

Surface Modification Technologies. (T. Sudarshan and D. Bhat, Eds.) Proceedings of the 1st International Conference on Surface Modification Technology, The Metallurgical Society, Warrendale, PA, 1988.

Adhesion and Polymer-Related Interfaces

Adhesion and Adsorption of Polymers. (L. Lee, Ed.) Plenum, New York, 1979. *Broad coverage in areas such as statistical thermodynamics, surface physics, surface analysis, fracture mechanics, viscoelasticity, failure analysis, surface modification, adsorption kinetics, and biopolymer adsorption.*

Adhesion Aspects of Polymeric Coatings. (K. Mittal, Ed.) Plenum, New York, 1983. *A Proceedings with good overviews.*

Adhesion International 1987. (L. Sharpe, Ed.) Gordon and Breach Science Publishers, New York, 1987. *A Proceedings.*

Adhesion 12. (K. Allen, Ed.) Elsevier Applied Science, New York, 1987. *On adhesion of polymers to different materials such as copper and mold materials.*

Adhesives Handbook. J. Shields, CRC Press, Cleveland, 1970. *Includes chapters on joint design, adhesive selection, adhesives and properties, adhesive product directory, surface preparation, bonding processes, physical testing of adhesives, and adhesive trade sources.*

Appl. Phys. Lett. S. Kowalczyk, Y. Kim, G. Walker, and J. Kim, **52**, 375, 1988. *TEM is used to clearly demonstrate the microscopic differences between metal–polymer and polymer–metal interfaces (metal–polymer implies polymer deposited prior to metal). In addition, the influence of solvent on the redistribution of reaction products in the interfacial region is shown for the two types of interface.*—S. Anderson and P. Ho

Atomic Level Properties of Interface Materials. P. Ho, B. Silverman, and S. Chiu, (D. Wolf and S. Yip, Eds.) Chapman and Hall, New York, 1992. *Contains a chapter that provides a concise review of metal–polymer interface chemistry, metal diffusion*

into polyimides, thermal stresses in thin-film structures, and fracture characteristics of fine metal line–polymer structures.—S. Anderson and P. Ho

Characterization of Metal and Polymer Surfaces. (L. Lee, Ed.) Academic Press, New York, 1977. *A Proceeding.*

Electronic Packaging Handbook. (R. Tummala and E. Rymaszewski, Eds.) Van Nostrand Reinhold, New York, 1989. *This book provides an excellent overview of the electronic packaging field.*—S. Anderson and P. Ho

Fundamentals of Adhesion. L. Lee, Plenum, New York, 1991. *The chemistry and physics of adhesion between solids is discussed in great detail. Particular emphasis is placed on the theories of adhesion.*—S. Anderson and P. Ho

Handbook of Adhesives. 2nd ed., (I. Skeist, Ed.) Van Nostrand Reinhold, New York, 1977. *Thirty-five chapters that deal with chemically distinct families of adhesive materials, differing in origin, cost, suitability for each substrate, modes of application and setting, and properties. The remaining 21 chapters are concerned with theories, economics, applications, testing, and key end products.*

Handbook of Pressure-Sensitive Adhesive Technology. (D. Satas, Ed.) Van Nostrand Reinhold, New York, 1982. *Includes most aspects of the pressure-sensitive technology, such as mechanical details of equipment, polymer chemistry, and physics, as well as business and marketing.*

High-Performance Adhesive Bonding. (G. DeFrayne, Ed.) Society of Manufacturing Engineers, Dearborn, MI, 1983. *Emphasizes basic adhesive technology and state-of-the-art adhesive applications in manufacturing instead of the science and chemistry of adhesives. Well-suited for determining what sticks best to what under which conditions.*

Interfaces in Polymer, Ceramic and Metal Matrix Composites. H. Ishida, Elsevier Applied Science, New York, 1988. *A proceedings.*

Interfaces in Polymer Matrix Composites. (E. Plueddemann, Ed.) Academic Press, New York, 1974. *Presents a picture of the interface region as deduced from studies of composite properties, surfaces, and surface modifiers.*

J. Vac. Sci. Technol. D. Shih, N. Klymko, R. Flitsch, J. Paraszczak, and S. Nunes, **A9**, 2963, 1991. *This paper emphasizes the influence of different substrate metallurgies on the stability of polymer–metal interfaces. The role played by oxygen in the catalytic degradation of polyimides is described.*—S. Anderson and P. Ho

Mat. Res. Soc. Symp. Proc. J. Kim, S. Kowalczyk, Y. Kim, N. Chou, and T. Oh, **167**, 137, 1990. *Reviews the properties of metal–polymer and polymer–metal interfaces. Several correlations between chemistry, microstructure, electrical properties, adhesion, and reliability are made.*—S. Anderson and P. Ho

Metallization of Polymers. E. Sacher, J. Pireaux, and S. Kowalczyk, American Chemical Society, Washington, DC, 1990. *This excellent book emphasizes how surface-sensitive spectroscopies can be used to characterize metal–polymer interfaces. These techniques are then utilized to determine surface morphology, characterize the effects of surface modification, and investigate the interactions between metals and polymers. An additional section is devoted to correlations between metal–polymer adhesion and surface properties.*—S. Anderson and P. Ho

Microscopic Aspects of Adhesion and Lubrication. (J. Georges, Ed.) Elsevier Applied Science, New York, 1982. *Covers both adhesion and lubrication.*

Physicochemical Aspects of Polymer Surfaces. 2 volumes, (K. Mittal, Ed.) Plenum, New York, 1983. *Covers areas such as acid–base interactions, adsorption and contact-angle studies, surface characterization techniques, and surface modification.*

Polymer. S. Numata, K. Fujisaki, and N. Kinjo, **28**, 2282, 1987. *Polyimide chemical structure and molecular packing coefficient are correlated with macroscopic properties such as thermal expansion coefficient, modulus, and water absorption.*—S. Anderson and P. Ho

Polymer Surfaces and Interfaces. (W. Feast and H. Munro, Eds.) Wiley, New York, 1987. *Deals with polymers, polymerization, and surface chemistry.*

Polyimides: Materials, Chemistry and Characterization. (C. Feger, M. Khojasteh, and J. McGrath, Eds.) Elsevier Scientific, Amsterdam, 1989. *Proceedings from the Third International Conference on Polymides focusing on the synthesis, curing, and characterization of new polyimide materials.*—S. Anderson and P. Ho

Polyimides: Thermally Stable Polymers. M. Bessonov, M. Kroton, C. Kurdryavtsev, and L. Laius, Consultants Bureau, New York, 1987. *An overview of polyimide chemistry and properties.*—S. Anderson and P. Ho

Friction, Abrasion, and Tribology

Boundary Lubrication, An Appraisal of World Literature. (F. Ling, E. Klaus, and R. Fein, Eds.) American Society of Mechanical Engineers, New York, 1969. *Excellent chapter on test methods that has 536 references.*—N. Eiss

CRC Handbook of Lubrication. Vol. 1: Application and Maintenance; Vol. 2: Theory and Design, (E. Booser, Ed.) CRC Press, Boca Raton, FL, 1984. *Detailed coverage is given to textiles. Provides general discussions on friction, wear, and lubrication, and industrial practices, and a chapter on the wear of nonmetallic materials.*— N. Eiss

Friction and Wear Devices. ASLE SP-4, 2nd ed., Society of Tribologists Lubrication Engineers, Park Ridge, IL, 1976. *A compendium of devices and test conditions and the significance of test results.*—N. Eiss

Friction and Wear of Polymer Composites. (K. Friedrich, Ed.) Elsevier Applied Science, New York, 1986.

Friction and Wear of Polymers. G. Bartenev and V. Lavrentev (D. Payne, English trans.), Elsevier Scientific, Amsterdam, 1981. *Detailed account is given on the nature of friction and wear, the influence of temperature on friction, the effects of sliding velocity and the duration of contact with a special focus on the relations between the friction properties and the structure of polymers.*

Fundamentals of Tribology. (N. Suh and N. Saka, Eds.) MIT Press, Cambridge, MA, 1980. *Proceedings of an international conference.*

Lubricomp Internally Lubricated Reinforced Thermoplastics and Fluoropolymer Composites. Bulletin 254–688, ICI Advanced Materials, Malvern, PA. *Provides friction and wear data from the thrust washer test for a wide variety of filled and unfilled polymers at room and elevated temperatures. Mechanical and thermal properties are also given.*—N. Eiss

New Directions in Lubrications—Materials, Wear and Surface Interactions: Tribology in the 80's. (W. Loomis, Ed.) Noyes Publications, Park Ridge, NJ, 1985. *The science and technology of lubrication, friction, and wear is described.*

Physical Methods of Chemistry. 2nd ed., Vol. 7: Determination of Elastic and Mechanical Properties, (B. Rossiter and R. Baetzold, Eds.) Wiley, New York, 1991. *Chapter on friction and wear.*—N. Eiss

Tribology. J. Lancaster, **6**, 219, 1973. *Polymers, carbon-graphites, solid film lubricants and composites, and ceramic-cermets used in bearings are included in this discussion of performance criteria and the testing of dry bearings.*—N. Eiss

Tribology, A System Approach. H. Czichos, Elsevier Scientific, Amsterdam, 1978. *Excellent chapter on measurement of friction and wear.*—N. Eiss

Tribology in Particulate Technology. (B. Briscoe and M. Adams, Eds.) Adam Hilger, Bristol, U.K., 1987. *This book contains four parts: friction in powder flows, adhesive forces in powder flow, powder compaction and interface wear, and attrition and wear, including agglomeration.*

Wear Control Handbook. M. Peterson and W. Winer, American Society of Mechanical Engineers, New York, 1980. *While primarily oriented towards the wear of metals, the sections on wear theory and mechanisms and design considerations for effective wear control are applied to polymers. Excellent glossary and listing of international voluntary standards.*—N. Eiss

Other Volumes in This Series

The readers are also referred to other volumes of this series for work involving other types of materials:

- **Encyclopedia of Materials Characterization**
- **Characterization of Metals and Alloys**
- **Characterization of Ceramics**
- **Characterization in Silicon Processing**
- **Characterization in Compound Semiconductor Processing**
- **Characterization of Integrated Circuit Packaging Materials**
- **Characterization of Composite Materials**
- **Characterization of Tribological Materials**
- **Characterization of Optical Materials**
- **Characterization of Catalytic Materials**
- **Characterization of Organic Thin Films**

Acknowledgments

The authors of this book are gratefully acknowledged for providing valuable references for inclusion in this chapter.

Index